A COLLECTION OF MINING LAWS OF SPAIN AND MEXICO.

COMPILED AND TRANSLATED

BY H. W. HALLECK, A. M.

COUNSELLOR AT LAW.

SAN FRANCISCO:

O'MEARA & PAINTER, PRINTERS, 132 CLAY STREET.

1859.

PREFACE.

The code of Mining laws established by Spain and adopted by Mexico and other Spanish-American States, is, in itself, well worthy of attention, as a distinct and peculiar branch of jurisprudence. In the United States, from our recent acquisitions of territory containing large Mining interests, which territory has heretofore been governed by Spanish and Mexican law, the study of this code has become a matter of special interest, not only to the legal profession, but also to statesmen and legislators.

It is the object of the present work to furnish a compilation and translation of the Mining laws of Spain, from the earliest period to the independence of Mexico, and those of the general government of Mexico, from the independence of that country to the year 1853. The Mining laws of the several States, Departments, and Territories of that Republic, do not come within the design of this compilation.

The division which has been adopted is somewhat arbitrary, but will be found convenient in consulting the writings of commentators on these several codes. Part I contains the laws, edicts, &c., prior to 1584, the most important of which are the Ordinances of March 18th, 1563, usually denominated "The Old Code." This code has been explained and commented on by Lagunez, Solorzano, Escalona, Cardinal de Luca, Gamboa, and others. Part II contains the Ordinances of August 22d, 1584, commonly called "The New Code," and the subsequent laws and orders, to the publication of the Mining Ordinances of New Spain, in 1783. The laws of the new code furnish the text for the valuable Commentaries of Gamboa, which were published in 1761. Part III contains the Ordinances of May 22d, 1783, formed especially for the Mining of New Spain, but subsequently enforced in other Spanish-American countries. This code has been commented on by Velasquez de Leon. Part IV contains the laws, orders, &c., issued by the King, and by the Viceroys of New Spain, from 1783 to the independence of Mexico, in 1821. The collection for this period is very incomplete, as the compiler has not been able to procure numerous orders and decrees which have been referred to by text-writers. Part V was intended to contain all the laws, orders, &c., of the general government of Mexico, from the independence of that country to 1853. But this part is also very incomplete, as many of the orders and regulations relating to Mining matters, especially those of an administrative character, are not published in the collections of Mexican laws. It was also intended to extend this part to the present time (1859), but the compiler has not been able to obtain the Mexican laws since 1853, in an authentic form, in time for this publication.

The Appendix contains the Spanish Mining Law and Regulation of 1849, and the Mining Ordinances of Peru, published in 1785–6. The former has no application to Mexican territory, but is interesting, as showing the changes which have taken place in the Spanish Mining code since the independence of Mexico. Nor have the Peruvian Mining Ordinances of 1785–6 ever been in force in Mexican territory, but they are of particular interest, as showing the interpretation given to certain articles of the Ordinances of New Spain. The old Mining Ordinances of Peru are frequently referred to by Gamboa, in his commentaries on the Mining of New Spain, for the purpose of explaining the laws and usages in the latter country. The compiler has been unable to obtain a copy of the old code.

The several laws, decrees, &c., are arranged in the order of their dates, and without regard to the subject matter upon which they treat. In order to determine what Mining laws were in force at any particular period it will be necessary to examine only that portion of the work which precedes the date of such period. This arrangement will therefore be found of great convenience, for reference, in studying the writings of Gamboa and others who have, at different times, commented on the Mining laws of Spain and Mexico.

This, it is believed, is the first attempt to compile and translate the whole body of Mining laws in force in Mexico, although parts of these laws have been previously translated. Mr. Heathfield translated, in his Commentaries of Gamboa, the Ordinances of 1584, and Mr. Thompson, another English lawyer, published a translation of the Ordinances of 1783. Both of these translations are embodied in Mr. Rockwell's compilation of Spanish and Mexican laws relating to Mines and Real Estate.

Without undertaking to criticise the labors of others, the compiler will merely remark that the translations now offered differ in some particulars from those of Mr. Heathfield and in many more from those of Mr. Thompson, (at least from those reprinted in Mr. Rockwell's compilation, the compiler not having seen Mr. Thompson's original publication). Where these differences occur the compiler has taken great pains to secure correctness in his own version, and he hopes that these efforts have been generally successful. Nevertheless, it is very possible, that while avoiding some of the errors in former translations, he may have committed others equally glaring. He has endeavored, however, to give the exact meaning of the original, by an almost literal translation, and has not hesitated, for this purpose, to sacrifice style in composition. To express in English the precise meaning of technical words and phrases used in the Spanish Mining laws, is, in many cases, very difficult, and, in some, impossible. In such instances the words of the original are inserted in italics, and are explained in a Glossary which will be found at the end of the book. The explanations and definitions there given are taken principally from the Glossary appended to Gamboa's Commentaries, and from the Law Dictionary of Escriche.

In every case, the source from which the law, order, &c., was derived, is indicated at the beginning of the translation of such law or order. This arrangement will be found of great convenience to those who may wish to consult the original text.

San Francisco, April, 1859.

TABLE OF CONTENTS.

PART I.

Laws and Ordinances from 1263 *to* 1584.

1263.

[PARTIDA III, TITLE XXVIII, LAW XI.]

1383.

[BOOK VI, TITLE XIII, LAW II, RECOPILACION DE CASTILLA.]

1387.

[BOOK VI, TITLE XIII, LAW III, RECOPILACION DE CASTILLA.]

January 10th, 1559.

[BOOK VI, TITLE XIII, LAW IV, RECOPILACION DE CASTILLA.]

March 4th, 1559.

[Comentarios de Gamboa, Cap. II, § 44.]

March 18th, 1563.

[Book VI, Title XIII, Law V, Recopilacion de Castilla.]

ORDINANCES OF MINES;

(*Denominated bg Gamboa, The Old Ordinances, or Old Code of Mines.*)

August 10th, 1564.

[Book IX, Title VIII, Law XIX, Recopilacion de Castilla.]

May 8th, 1572.

[COMENTARIOS DE GAMBOA, CAP. II, § 51.]

PART II.

Laws and Ordinances from 1584 *to* 1783.

August 22d, 1584.

[BOOK VI, TITLE XIII, LAW IX, RECOPILACION DE CASTILLA.]

August 18th, 1607.

[Book VI, Title XIII, Law X, Nueva Recopilacion.]

Modification of some of the provisions of the Ordinances of August 22d, 1584, with respect to the Royal Dues, and to the working and reduction of Mines of gold, silver and other metals.

May 18th, 1680.

[Recopilacion de Leyes de las Indias.]

Book I, Title XII.

Book II, Title I.

Book II, Title XVI.

Book IV, Title III.

Book IV, Title XIX.

February 6th, 1705.

[Pandectas Hispano-Mexicanos, No. 586.]

July 5th, 1718.

[Comentarios de Gamboa, Cap. II, § 40.]

April 3d, 1747.

[Book IX, Title I, Law VII, Novisima Recopilacion.]

December 19th, 1754.

[Book IX, Title XVIII, Law V, Novisima Recopilacion.]

May 30th, 1778.

[Mexican Ordenanzas de Mineria.]

May 20th, 1780.

[Book IX, Title XX, Law I, Novisima Recopilacion.]

PART III.

Mining Ordinances of New Spain.

May 22d, 1783.

Royal Ordinances for the Direction, Regulation and Government of the Important Body of Mining of New Spain, and of its Royal Tribunal General.

Title I.—*Of the Tribunal General of Mining in New Spain.*

Title II.—*Of the Judges and Deputies of the Mining Reales.*

Title X.—*Of Mines which require draining.*

Title XI.—*Of Mines worked in partnership.*

Title XII.—*Of operatives in Mines, Haciendas, and reducing works.*

Title XIII.—*Of the supply of water and provisions in Mining establishments.*

Title XIV.—*Of the Millers and Ore-Traders.*

Title XV.—*Of Suppliers of Mines and Dealers in bullion.*

Title XVI.—*Of the Fund and Bank of supplies of Mines.*

Title XVII.—*Of Experts for working Mines and reducing ores.*

Title XVIII.—*Of the education and instruction of youth intended for the Mines, and the encouragement of industry in them.*

Title XIX.—*Of the privileges of Miners.*

PART IV.

Decrees, Orders, &c., from 1783 *to* 1821.

January 15th, 1784.

[MEXICAN ORDENANZAS DE MINERIA.]

March 4th, 1785.

[MEXICAN ORDENANZAS DE MINERIA.]

September 2d, 1785.

[PANDECTOS HISPANO-MEXICANAS, No. 2314.]

February 7th, 1786.

[MEXICAN ORDENANZAS DE MINERIA.]

May 9th, 1786.

[MONTEMAYOR Y BELENA'S RECOP. SUMMARIA DE A. A.]

June 17th, 1786.

[MEXICAN ORDENANZAS DE MINERIA.]

December 4th, 1786.

[REAL ORDENANZA DE INTENDENTES.]

On the powers and duties of Intendants with respect to Mines and Mining

July 18th, 1789.

[MEXICAN ORDENANZAS DE MINERIA.]

December 26th, 1789.

[BOOK IX, TITLE XX, LAW II, NOVISIMA RECOPILACION.]

Rules for the working of Mines of Stone-coal.

January 28th, 1790.

[MEXICAN ORDENANZAS DE MINERIA.]

August 18th, 1790.

[BOOK IX, TITLE XX, LAW III, NOVISIMA RECOPILACION.]

October 19th, 1790.

[BOOK IX, TITLE XVIII, LAW VI, NOVISIMA RECOPILACION.]

December 25th, 1790.

[PANDECTAS HISPANO-MEXICANAS, No. 221.]

November 12th, 1791.

[MEXICAN ORDENANZAS DE MINERIA.]

August 24th, 1792.

[BOOK IX, TITLE XX, LAW IV, NOVISIMA RECOPILACION.]

Free trade in coal, and rules for working coal Mines.

February 5th, 1793.

[MEXICAN ORDENANZAS DE MINERIA.]

Explanations, additions and modifications of the Mining Ordinances with respect to jurisdictions, &c.

August 5th, 1793.

[BOOK IX, TITLE XX, LAW V, NOVISIMA RECOPILACION.]

September 11th, 1794.

[PANDECTAS HISPANA-MEXICANAS, No. 3751.]

November 17th, 1796.

[MEXICAN ORDENANZAS DE MINERIA.]

December 6th, 1796.

[MEXICAN ORDENANZAS DE MINERIA.]

February 12th, 1797.

[MEXICAN ORDENAZAS DE MINERIA.]

June 10th, 1797.

[MEXICAN ORDENANZAS DE MINERIA.]

February 6th, 1798.

[MEXICAN ORDENANZAS DE MINERIA.]

April 27th, 1801.

[MEXICAN ORDENANZAS DE MINERIA.]

January 19th, 1805.

[MEXICAN ORDENANZAS DE MINERIA.]

August 22d, 1809.

[MEXICAN ORDENANZAS DE MINERIA.]

January 26th, 1811.

[MEXICAN ORDENANZAS DE MINERIA.]

January 13th, 1812.

[GALVAN'S DECRETOS DE LAS CORTES DE ESPANA.]

August 8th, 1814.

[MEXICAN ORDENANZAS DE MINERIA.]

October 13th, 1816.

[MEXICAN ORDENANZAS DE MINERIA.]

January 7th, 1818.

[MEXICAN ORDENANZAS DE MINERIA.]

December 19th, 1818.

[MEXICAN ORDENANZAS DE MINERIA.]

PART V.

[LAWS, DECREES, ORDERS, &c., FROM 1821 TO 1853.]

February 20th, 1822.

[MEXICAN ORDENANZAS DE MINERIA.]

March 24th, 1823.

[MEXICAN ORDENANZAS DE MINERIA.]

October 7th, 1823.

[MEXICAN ORDENANZAS DE MINERIA.]

October 29th, 1823.

[GALVAN'S COLECCION DE ORDENES, &c. TOMO II.]

February 13th, 1824.

[GALVAN'S COLECCION DE ORDENES, &c. TOMO III.]

July 24th, 1824.

[GALVAN'S COLECCION DE ORDENES, &c. TOMO III.]

November 16th, 1824.

[GALVAN'S COLECCION DE ORDENES, &c. TOMO III.]

May 20th, 1826.

[GALVAN'S COLECCION DE ORDENES, &c. TOMO IV.]

September 15th, 1829.

[Galvan's Coleccion de Ordenes, &c. Tomo V.]

April 14th, 1832.

[Galvan's Coleccion de Ordenes, &c. Tomo VI.]

May 22d, 1833.

[Galvan's Coleccion de Ordenes, &c. Tomo VII.]

July 2d, 1833.

[Galvan's Coleccion de Ordenes, &c. Tomo VII.]

July 27th, 1833.

[Galvan's Coleccion de Ordenes, &c. Tomo VII.]

November 18th, 1833.

[Galvan's Coleccion de Ordenes, &c. Tomo VII.]

May 12th, 1838.

[Arrillaga's Recopilacion de Leyes, &c., 1838.]

November 22d, 1838.

[Arrillaga's Recopilacion de Leyes, &c., 1838.]

December 11th, 1838.

[Arrillaga's Recopilacion de Leyes, &c., 1858.]

October 19th, 1841.

[Lara's Colecion de Decretos, &c., No. 56.]

November 27th, 1841.

[Lara's Coleccion de Decretos, &c., No. 183.]

March 11th, 1842.

[LARA'S COLECCION DE DECRETOS, &c., No. 325.]

Rights of foreigners to acquire real estate in Mexico.

July 12th, 1842.

[LARA'S COLECCION DE DECRETOS, &c., No. 414.]

August 31st, 1842.

[LARA'S COLECCION DE DECRETOS, &c., No. 452.]

October 18th, 1842.

[LARA'S COLECCION DE DECRETOS, &c., No. 496.]

October 26th, 1842.

[LARA'S COLECCION DE DECRETOS, &c., No. 520.]

December 2d, 1842.

[LARA'S COLECCION DE DECRETOS, &c., No. 549.]

Regulations of the Junta of Encouragement and Administration of Mining.

January 14th, 1843.

[LARA'S COLECCION DE DECRETOS, &c., No. 574.]

January 31st, 1843.

[LARA'S COLECCION DE DECRETOS, &c., No. 579.]

February 11th, 1843.

[LARA'S COLECCION DE DECRETOS, &c., No. 588.]

February 17th, 1843.

[LARA'S COLECCION DE DECRETOS, &c., No. 592.]

April 7th, 1843.

[LARA'S COLECCION DE DECRETOS, &c., No. 639.]

May 23d, 1843.

[LARA'S COLECCION DE DECRETOS, &c., No. 657.]

May 24th, 1843.

[LARA'S COLECCION DE DECRETOS, &c., No. 659.]

July 5th, 1843.

[LARA'S COLECCION DE DECRETOS, &c., No. 698.]

July 8th, 1843.

[LARA'S COLECCION DE DECRETOS, &c., No. 703.]

July 11th, 1843.

[LARA'S COLECCION DE DECRETOS, &c., No. 708.]

July 14th, 1843.

[LARA'S COLECCION DE DECRETOS, &c., No. 717.]

July 28th, 1843.

[Lara's Coleccion de Decretos, &c., No. 726.]

August 24th, 1843.

[Lara's Coleccion de Decretos, &c., No. 756.]

September 6th, 1843.

[Lara's Coleccion de Decretos, &c., No. 768.]

September 25th, 1843.

[Lara's Coleccion de Decretos, &c., No. 786.]

Decree to encourage the discovery and working of Mines of quicksilver.

October 3d, 1843.

[Lara's Coleccion de Decretos, &c., No. 812.]

December 26th, 1843.

[Lara's Coleccion de Decretos, &c., No. 865.]

December 30th, 1843.

[Lara's Coleccion de Decretos, &c., No. 880.]

December 30th, 1843.

[Lara's Coleccion de Decretos, &c., No. 882.]

April 30th, 1844.

[Mexican Ordenanzas de Mineria.]

June 28th, 1852.

[Navarro's Legislacion Mexicana.]

September 2d, 1852.

[NAVARRO'S LEGISLACION MEXICANA.]

June 1st, 1853.

[NAVARRO'S LEGISLACION MEJICANA.]

June 28th, 1853.

[NAVARRO'S LEGISLACION MEJICANA.]

July 16th, 1853.

[NAVARRO'S LEGISLACION MEJICANA.]

July 30th, 1853.

[NAVARRO'S LEGISLACION MEJICANA.]

APPENDIX No. 1.

[SPANISH LAW AND REGULATION OF MINING, 1849.]

April 11th, 1849.

Law of Mining.

July 31st, 1849.

Regulation of Mining.

APPENDIX No. II.

MINING ORDINANCES OF PERU, 1785—1786.

December 8th, 1785.

August 1st, 1786.

October 7th, 1786.

(NOTE.—The Declarations or Articles of these Ordinances, follow the order of the Titles of the Ordinances of New Spain, several Declarations being made for each Title).

GLOSSARY

PART I.

MINING LAWS OF SPAIN AND MEXICO,

FROM 1263 TO 1584.

PART I.

MINING LAWS OF SPAIN AND MEXICO,

FROM 1263 TO 1584.

1263.

[From Partida III, Title XXVIII, Law XI.]

In what things Emperors and Kings properly have seigniory.

The revenues of ports, and the port duties (*portadgos*) which the merchants pay for the things which they export and import, and the rents of salt-works (*salinas*) or of fisheries, and of iron works (*ferrerias,*) and of other metals, and the taxes and tribute (*pechos y tributos*) which individuals pay, belong to the emperors and kings; and all these things were tacitly granted (*otorgadas*) to them in order that they might have wherewith to maintain themselves with dignity in their expenditures; and wherewith to protect their lands and their kingdoms, and to carry on war against the enemies of the faith; and in order that they might exempt their people from paying to them heavy taxes, and from other grievous burthens.

1383.

Don Alonzo XI, Pragmatica, of the year 1383.

[From Laws 47 and 48, Title 32 of the Ordenamiento de Alcala, y Cortes alli pet 32.

Vide also, Book VI, Title XIII, Law II, Recopilacion de Castilla; Book IX, Title XVIII, Law I, Novisima Recopilacion.]

Right of the Kings in mines of gold, silver, and other metals, salt waters and springs; and prohibition to work them without a royal license.

All Mines of silver, and gold, and lead, and of any other metal whatever, of whatsoever kind it may be, in our Royal Seignory, shall belong to us; therefore, no one shall presume to work them without our special license and command; and also the salt springs, basins and wells, which are for making salt, shall belong to us; wherefore, we command that they revert to us with the produce of the whole thereof; and that no one presume to intermeddle therein except those to whom the former kings, our predecessors, or we ourself may have given them as a privilege, or who may have held them from time immemorial.

1387.

Don Juan I., at Birbiesca, in the year 1387, pet 52.

[From Book VI, Title XIII, Law III, Recopilacion de Castilla.

Vide also, Book IX, Title XVIII, Law II, Novisima Recopilacion.]

Power to search for mines in private inheritances, and in those of others, and to work them, with the reward which is assigned.

Whereas, we are informed that these, our kingdoms, are abundant and rich in Mines; therefore, in order to concede grace and favor to our said kingdoms, and to the inhabitants and residents of the cities, *villas*, and places thereof, and to ecclesiastical persons, notwithstanding that by us and by the kings whom we succeed, in the privileges which have been given as pure gifts (*mercedes*,) there have been reserved for us Mines of gold and of silver, and of other metals whatsoever, it is our gift that in future all said persons, and all others whosoever of these, our said kingdoms, shall be at liberty to search for and examine (*catar*,) and dig in their said lands and inheritances, the said Mines of gold and silver, and quicksilver and tin, and stones and other metals; and that they may also search and excavate in all other places whatsoever, not causing injury one to another in their searching and excavations, and doing so with the permission of the owner; and all that shall be found and taken out from the said Mines shall be divided as follows: First, that there shall be delivered and paid therefrom to the person who has taken out the mineral, all the expenses which may be incurred in excavating and extracting it; and of that which may remain after deducting said expenses, the third part shall belong to the person who has taken it out, and the other two-thirds to ourself.

January 10th, 1559.

Don Phillip II, and in his absence the Princess Doña Juana, at Valadolid, January 10th, 1559.

[From Book VI, Title XIII, Law IV, Recopilacion de Castilla.

Vide also, Book IX, Title XVIII, Law III, Novisima Recopilacion.]

Incorporation of mines of gold, silver, and quicksilver, into the Crown and Royal Patrimony; and the mode of working them.

We command that the members of our council, our Contadores Mayores, the President and Oidores of our Audiencias, the Alcaldes, the Alguaciles of our *Casa*, and Court, and *Chancillerias*, and all the Councils, and *Corregidores*, *Assistentes*, Governors, Alcaldes, and other Judges and Courts of all kinds of all the cities, *villas*, and places of our kingdoms and *señorios*, and all persons of whatever rank and condition they may be, to whom the contents of this our Letter may come or concern, and to each one of you, health and grace. It is a thing known and very notorious, the great benefit and advantage, which, as well to us and to our royal patrimony, as to our subjects and native citizens (*naturales*) and the public welfare of these our kingdoms, result from the discovery, working and reduction of the Mines (*Mineros*) of gold and silver and quicksilver, and other metals, in these our kingdoms, which, as is learned from very ancient times, are very rich and abundant; and as, by the law which His Majesty (*el Señor Rey*) Don Juan I., enacted, all persons are permitted to search for, and dig and work said Mines and metals, and as by the same law the apportionment which is always to be made is designated, nevertheless, experience has shown and shows, that there are few Mines which have been discovered and worked; and yet it is said that there are some persons who have information of Mines rich and advantageous, and keep them concealed and do not wish to discover and make them known; which, we are informed, among other causes, has resulted and results from the fact that a greater part of said Mines have been granted (*hecho merced*) to noblemen and other persons in this kingdom, by granting them with Bishoprics and Archbishoprics and Provinces, so that with regard to said Mines almost the whole kingdom has been divided up and distributed. And considering that, as the Mines are granted to individuals, others are unwilling to interfere with or engage in their discovery and working, particularly, as in many of said grants it is expressly and particularly provided, that without their (the grantees') license no one shall be at liberty to discover or work them; and the noblemen and persons who hold the said grants, either in order to avoid the expense and labor, or because they do not give it their attention, have devoted and devote very little care and diligence to the discovery, working and reduction of said Mines; and therefore from said grants very little advantage has resulted or results to them, while the benefit which we and our subjects and citizens might derive therefrom has been and continues to be impeded; and it is said that others also are unwilling to engage in the discovery, working and reduction of said Mines, because, although by the said law of His Majesty, the King, Don Juan, the apportionment which is to be made is designated; yet as it is so ancient, and has been so little in use and practice, and as neither in that nor any other laws of this kingdom are the doubts and difficulties, which may occur, determined, from which arise occasions for law-suits and differences, they fear and hesitate to expend their capital and devote their labor in such discovery and working; and especially having doubt whether the said law and its provisions apply to and include Mines which might be rich, and from which great and excessive profit might be expected and obtained; and that all the above may be provided for, so that said impediments and difficulties may cease, and the rewards and advantages may be effectually secured, and many persons of wealth and capital may assist such discovery, working and reduction of Mines, by means of whose diligence and labor God may be pleased to make known the riches and wealth which are hid and buried in the earth, and our royal patrimony be increased, and our subjects benefitted, and these our kingdoms enriched: And having commanded that our *contadores mayores* be consulted on the foregoing, jointly with some of the members of our council; and these having discussed and conferred on it as on a matter of the greatest importance, and having consulted with us, it has been resolved, that we should command this our Letter (*Carta*) to issue, and insert in it the following provisions, and we have approved the same, and we will that it have the force and effect of law, the same as if it had been made and enacted in Cortes, on petition of the delegates (*Procuradores*) of the cities and villages of these our kingdoms.

Art. 1. First: We reclaim, resume and incorporate in ourself in and our crown and patrimony, all the Mines of gold and silver and quicksilver of these our kingdoms, in whatsoever parts and places they may be, and are found, whether in royal lands or in those of lordships, or of the clergy (*Abadengo*,) and whether in public, municipal or vacant lands, or in inheritances, places and soils of individuals, notwithstanding the grants which by us, and by the kings, our predecessors, have been made to any persons, of whatsoever condition, rank and dignity they may be, and for whatsoever causes and reasons, as well those for life, and for years, and on condition, as those in perpetuity, and free, and without condition; all which said grants, in view of the facility and generality with which they have been made, and the prejudice which to us and to our crown and royal patrimony has ensued and ensues, and the damage and injury which has resulted and may result to the public good and to the common welfare of our subjects and citizens, and for other just causes which thereunto move us: we revoke, and annul, and make void; and we will that the said Mines shall and may henceforth (without any other act of seizin (*aprehension*) and possession) belong to our said crown and patrimony, in accordance with, and as by the laws of these kingdoms and ancient usage and custom, they do belong to us, the same as if the said grants, or any of them, had never been made or conceded; those only remaining in force and vigor respecting Mines of silver and gold, which, by the persons to whom the said grants were made, or by others in their name and with their consent, have been begun to be worked and are actually worked now at the date of this our Letter (*Carta.*) And also, it is our will to recompense and satisfy the noblemen and persons to whom the said grants, which we thus revoke, have been made, according to what their titles of grant, and the causes and reasons for making them, and the conditions and limitations contained in them, and what, on their part, has been done and complied with, being examined, may seem just and reasonable; and, for this purpose, we command that those who shall hold said grants, and who shall claim said recompense, shall present them within one year, in order that, the above matters being examined, they may receive the recompense which is their due.

Art. 2. Inasmuch as the reclamation and incorporation of the said Mines in us, and in our royal patrimony, as aforesaid, are not made with the object or purpose that we alone, or others in our name alone, shall search for, discover and work such Mines; but as, on the contrary, it is our intention and will that our subjects and citizens shall participate, and have a part in said Mines, and engage in the discovery and working of them; therefore, by these presents, we give permission and authority to our said subjects, and native citizens, that they may freely, without our license, or the license of any one else, examine, search for, and dig the said Mines of gold and silver, in all parts whatever of royal lands, or lands of lordships, or of the church, or of any others, as well in public, municipal and common lands, as in the inheritances and estates of individuals, paying to the owners the damage; and that no person, or persons, shall be at liberty to interpose any impediment or embarrassment, either on account of the said grants which have been made, and which, as aforesaid, we revoke, or for any other cause or reason whatever. And also, we give free authority and permission to all our said subjects, and native citizens, that, in relation to Mines of gold and silver which they shall have discovered, having registered them in the manner hereafter declared, they may dig and take out from them the ores, and work and reduce them; and construct in them all the machinery, works and operations (*diligencias*) which may be necessary, without these being by us, or by any other person in our name, occupied, embarrassed or impeded, and without any person being at liberty to enter, in order to dig, or search, or work, or reduce, within the limits and bounds of the Mine, the said discoverer observing what, in this our provision, shall hereafter be stated and ordained; by which it is to be understood, that they shall be at liberty to make, examine and discover said Mines in said parts and places, except in the Mines of Guadalcanal, and one league around them, and the Mines which are discovered within the limits of Cazalla, and Aracena, and Galaroca, and a quarter of a league around each one of them. All which must have full and complete effect, notwithstanding any lease which we have ordered to be made of any of the Mines of the kingdom.

Art. 3. And with respect to the part which such discoverers and workers of said Mines are to have, conforming ourselves in this to what the said king, Don Juan, in said law ordained and directed, it is our pleasure, and we will that those who shall thus discover and work the said Mines of silver and gold, first deducting the expenses which, in excavating, working and reducing said Mines, shall have been incurred, of the remainder, after said expenses are deducted, they shall have the third part, and the other two-thirds shall be for us, which shall be understood as general, and without regard to the quality or the richness of said mines, even though it be very great and excessive: for, without any exception or distinction as to richness or quality, we will, that they shall have the third part, and that, for no cause or reason, shall it be taken, or be permitted to be taken from them, or withheld or encumbered, and that no other intendment, or interpretation, or declaration can be given to this, our law; but that in every case, and in all ways they be certain of, and secure in the said third part, with this limitation and restriction, that he who may thus work and reduce the said Mine, having enjoyed and received, after deducting the expenses, of his third part, one hundred thousand ducats, while he may enjoy the said Mine, shall thereafter have only the fourth part, in the same way and form as he had the third part; and that having enjoyed and received, after deducting expenses, another hundred thousand ducats, being in all two hundred thousand ducats, thereafter shall have only the fifth part, which in future shall continue the same without diminution or reduction, even though the Mine continue to be of very great profit, in whatsoever amount, or in whatsoever manner.

Art. 4. And with respect to the order and form which are to be observed in the discovery and registry of the said mines, and the time in which it must be made by such discoverers, in order that in this matter there may be no doubt, and it may be understood what they ought and must do, we command that he who shall discover Mines of gold or silver, within twenty days after he shall have discovered them, and have found metal, shall be obliged to register them before a Royal Notary, or before the Court in the jurisdiction of which such Mine may be, carrying and presenting the metal which he shall have found, and in the register shall be stated the person who discovered it, the place where it was and was found, and the ore which was presented; and that, within sixty days after the date of such registry, it be sent to our administrator, or the person or persons who by him shall be appointed for Bishoprics or Provinces, in order that it may be noted and put in the book and general register which he must keep for the said Mines, and be of record, and that he may know all those which shall be discovered; and, that not making the registry in said form and time, and not doing what is prescribed, any other person may make said registry, and have and acquire the right which such discoverer would have had by making such registry; and whereas, heretofore and prior to the date and publication of this our Provision, there have been discovered and registered many Mines, which are thus occupied and encumbered, and of them there is no complete notice, and the registries have been made differently and in various ways: We command that all those who prior to the date and publication of this our Letter, shall have discovered and registered Mines, shall be obliged within two months, to renew and make again the said registries, in the manner and form above prescribed for those who may hereafter discover and register Mines; and that, within three months more, the said registries be sent to our said Administrator, and that if this be not done, there be observed what is said with respect to those who shall discover and register anew.

Art. 5. And whereas, by not designating the limit and space which the Mines that shall thus be discovered are to have, there may result great confusion, differences and lawsuits; and the first discoverer may pretend that his Mine, and the right which by discovery may belong to him, cover and include the whole extent and continuation of the metalic vein, and that in the whole of such extent and continuation no person can interfere to try (*catar*,) search or work, from which may result great embarrassment and impediment to the discovery, and working and reduction of said Mines: We declare that, the discoverer of the Mine or Mines of silver, having made his discovery and registry as prescribed in the preceding article, the said Mine which he shall have so discovered and registered shall be one hundred varas long and fifty varas wide, as he may select it, so that within said limit and space no other person can interfere to try, dig or work, and that within said space and limits the said discoverer alone shall have such right and power, without that he can be impeded or prevented by us, or in our name, or by any other person, as aforesaid; and outside of said space any other person shall be at liberty to enter and discover, and have the same space and privilege in the manner aforesaid.

Art. 6. Also, inasmuch as the discoverers of the said Mines, after having thus discovered and registered them, pretend that, by that act alone, they have acquired such a right to them that no other person can, within the limits and space of such Mines, enter, or try, or work, and that they can thus keep them encumbered, without working them themselves or permitting others to do so, by which they prevent the principal produce and profit which belongs as well to us as to our subjects and to the public welfare, since that principally consists in the working and reduction of Mines and metals, and not merely in their discovery; we declare and command that such discoverer of the Mine or Mines of silver, after having made registry in the manner prescribed, shall be obliged, within six months, to sink and excavate to the depth of three *estados*, and not sinking and excavating his Mine to the depth of three *estados*, it may be denounced before the Judge, and registry made of it as of a vacant and undiscovered Mine; and that, after having sunk the Mines or wells to the said depth of three *estados*, they shall be obliged to have them occupied (*pobladas*) and to work them in the manner and order and time which shall be declared by the ordinances, in which shall be given the order which, in the above, ought to be observed, without that in them shall be annulled or changed anything contained in this our Letter.

Art. 7. And whereas, by this matter being of great importance, both to our service and to the general benefit of our subjects and vassals, it is just that those who attend to the working and discovery of said Mines should be assisted and favored: we command you, our said courts, councils and individuals, to give them, and cause to be given to them, all the favor and aid which may be necessary to carry out the provisions of this our Letter: and that you give them, and cause to be given to them, all the wood and coal, and animals, and tools, and materials, and other things which may be necessary for the working and operation of our Mines, they paying for them their fair value; without that, in the matters aforesaid, there be placed, or permitted to be placed, any obstacle or impediment, but, on the contrary, that in everything they be favored and assisted.

March 4th, 1559.

The Princess Regent, at Valladolid, March 4th, 1559.

[From the "Comentarios de Gamboa," Cap. II, § 44.]

Reasons for sending quicksilver to New Spain, to be sold on account of the Crown, and prohibition to others to send it there.

To our Royal Officers of New Spain:

Having seen what you and the Viceroy of that country have written to us, concerning the great necessity which exists for sending a quantity of quicksilver thither for reducing the silver, which is extracted from the Mines of that New Spain, and concerning the conveniences which would result from our sending it there to be sold and disposed of by the hands of our officers; because, by this means, besides the great benefit which would be conferred upon the inhabitants of that country, we ourself may derive great advantage, and, on the said quicksilver, may make a profit of twice what it costs here: We have provided that all the quicksilver which has been produced, or shall hereafter be produced in the Mine of Almaden, be sent to you, and a further quantity, which our Factor General has now purchased by our command, in order that you may sell and make use of the same; and our officers of the *casa de contratacion* of Seville, will, therefore, send you the said quicksilver by virtue of the order which has been written to them; and we have prohibited the sending of any quicksilver to those parts, except in our name, and by our command. Wherefore, I command you, that you provide some person to receive, at the city of Vera Cruz, all the quicksilver which our said officers of Seville shall send to you there, and the said person shall send it to you at that city; and when, and so often as you shall receive what may be sent to you, you will sell it at the best advantage you can, and making the profit out of it which may be expedient, as it is entrusted to your fidelity and care; and the money which may be derived from it, our treasurer will charge to you, and you will always render us an account of the quantity you receive, and the price at which it is sold; and whereas, the quicksilver costs here from fifty-five to fifty-eight ducats per quintal, having reference to this and to the cost of freight and other expenses, of which the said officers of Seville will advise you, you will dispose of it at the best advantage you can.

Dated at Valladolid, March 4th, 1559.

LA PRINCESSA.

By command of His Majesty, Her Highness acting in his name.

OCHOA DE LUYANDO.

March 18th, 1563.

Don Philip II, at Madrid, March 18th, 1563.

[From Book VI, Title XIII, Law V, Recoplacion de Castilla.]

Ordinances of Mines; (denominated by Gamboa the old ordinances or old code of Mines.)

Art. 1. In the first place, in order to confer a benefit and favor upon our subjects and natives, and upon all other persons whomsoever, even though they be aliens to these our kingdoms, who shall discover and work any Mines of silver, notwithstanding the part which, by said Edict (*Pragmatica*) is designated, we will and command that they shall have and take the following:

Art. 2. If the metals, which are taken out of the said Mines, shall yield at the rate of a mark per quintal of lead-silver (*plomo-plata,*) or under, they shall pay to us the eighth part of the silver which is taken from said Mine, without any deduction for expenses, or anything else; because all these shall be at the charge of the said persons who shall discover, work and reduce the said Mines; and all that remains, after taking out the said eighth part of the said silver, they may have and take for themselves.

Art. 3. In Mines which yield more than a mark per quintal of lead-silver, and up to three marks, there shall be paid to us the fourth part of the silver which is extracted, without deducting expenses; and the said persons shall take the remainder, as aforesaid.

Art. 4. In Mines which yield more than three marks per quintal of lead-silver, and up to six marks, there shall be paid to us the third part of the silver which is extracted, without deducting expenses; and the said persons shall have the remainder, as aforesaid.

Art. 5. In Mines which yield more than six marks per quintal of lead-silver, whatever their goodness, quality or richness may be, or become, whether expected or not expected, there shall be paid to us the half of the silver which is extracted, without deducting expenses; and the said persons shall have the remainder, as aforesaid.

Art. 6. In Mines of gold, whatsoever their ley, quality, quantity or richness may be, or may become, there shall be paid to us the half of the gold which is extracted, without deducting any expenses; and the persons who shall discover and work them shall have the other half.

Art. 7. And inasmuch as there are some old Mines which, before the publication of the said Edict (*Pragmatica*) herein above incorporated, were worked and reduced, but, at present, are not worked or reduced by their owners, and, in fact, were not worked at the time said Edict was issued; and, inasmuch as refuse heaps (*terreros*) and slag have been taken from them, we command that such persons as may desire to work said Mines, and to reduce said refuse heaps and slag, without prejudice to the right which their owners may have in them, agreeably to said Edict, may do so; and that, out of the metals which are taken from them, there shall be paid as follows:

Art. 8. In Mines which, before the publication of the said Edict, were abandoned and not worked, those which were sunk to the depth of twenty *estados* and below that to any depth whatsoever, of the metals taken out which yield a mark and a half per quintal of lead-silver, and under, there shall be paid to us, of the silver which is extracted, the eighth part; and if they yield more than said mark and a half per quintal, there shall be paid to us the same rates as from Mines newly discovered, as is declared above, without deducting any expenses.

Art. 9. And of the silver which is taken out of the said refuse-heaps of old Mines, which before the publication of the said Edict had been worked and were abandoned as aforesaid, there shall be paid to us the fifth part; and of the silver which is taken out of said slag, there shall be paid to us the twentieth part in all, free of expenses.

Art. 10. And the lead, chalk, ashes and sweepings, and all else that may result from the refinings, after extracting the silver, of which there is to be paid to us the proportions declared above, free of all expenses, must and shall remain, with the proportions which shall pertain, to the owners of the said Mines, without that in this matter any restriction can or shall be imposed on them.

Art. 11. And whereas, poor lead, which is not worth refining from its containing little or no silver, and antimony and copper, are required for the reduction of the Mines of silver: we command that Mines of such lead, antimony and copper, which have been or may be found in places where there have been made grants of metals, may be searched for and worked; and that of these, there be paid to us, of the copper the twentieth part; and of the antimony the eighth part; and of the poor-lead, which is to be understood such as will not yield more than four *reales* of silver per quintal, the fifteenth part; all free of expenses; provided that, if the said copper should contain gold, there shall be paid to us of this gold the fourth part, in addition to the duty on the copper; and that if it should contain silver, there shall be paid of it the half of the duty above directed to be paid on the silver, according to the number of marks per quintal, and in addition, the duty on the copper.

Art. 12. It is to be understood that all the above mentioned proportions which we are to receive out of all the said Mines, new and old, and the refuse-heaps and slag, must be paid to us in silver at the refining houses and smelting works, which we are to have for the said refinings, and not in ore nor in lead-silver; and those out of the poor-lead and copper, in sheets; and those out of the antimony, in metal, and clear of all expenses.

Art. 13. And whereas, according to the said Edict above incorporated, those who had grants of Mines were to enjoy all that was not gold, or silver, or quicksilver, according to their privileges; and also were to enjoy Mines of gold and silver which had been begun to be worked, and were actually worked by them, or by other persons in their name, prior to said Edict, which is above incorporated; and whereas, some doubts have arisen respecting these words, it being said that it might happen that they had been discovered and begun to be worked prior to said Edict, and the working had been discontinued sometime before it was issued, and consequently they were excluded by the said Edict, as not being actually worked at the time it was issued; therefore, it is declared that said Mines of gold and silver which are to be enjoyed by the owners of said privileges, shall be those which were actually worked and producing at the time said Edict was issued, or within four months previous, and no others.

Art. 14. And whereas, in the said Edict it is provided that no person shall be at liberty to search for Mines, within one league around the Mine of Guadalcanal, or a quarter of a league around those of Cazalla, Galaroca, and Aracena, respectively; and whereas, it has since been found to be expedient for our service to enlarge the said limits of the said quarter of a league, and to declare from what point they are to run; we therefore command that in said three places, and in that of Guadalcanal, and in each one of them, no person shall take or hold a Mine within the space of one league around each one of the said four places; and that the said leagues are to be computed and measured in the following manner: That of Guadalcanal from the house which is built there for the workshop of the said Mines; and that of Cazalla from the house which is over the Mine of Pedro Candil; and that of Aracena from the house which is built at the Mine of El Cerro de los Azores; and that of Galaroca from the first Mine which was discovered there, and which is near the town (*Lugar;*) and the said leagues shall be of the legal length of fifteen thousand feet, each foot being one-third of a *vara*, land measure; and all the Mines which shall be found within those limits shall be for us; but, if previous to the day of the promulgation of this our Letter (*Carta,*) any Mines should be found outside of the said quarters of a league, and within the spaces of a league, which are now designated, the discoverers thereof shall enjoy them in accordance with said first Edict (*Pragmatica.*)

Art. 15. Also, we ordain and command, that all persons whosoever, even though they be foreigners, may freely search for Mines of gold and silver, and such others as by these our ordinances are declared, and make trial-pits (*catar*) and do all things necessary for the discovery of the said metals in all our said kingdoms and lordships (*señorios*) of the crown of Castile, outside of the places excepted, in the fields, woods, vacant places, outside lands (*egidos,*) and pastures, whether belonging to us, to Pueblos, or to private persons, and in any inheritances whatsoever, without that in it, any of the owners, or any person whosoever can impose any impediment or prohibition, and that if it should be necessary to dig or sink pits in said pastures and inheritances, they may do so; provided, that if any damage be done, the Court of Mines will appoint two persons of trust to appraise the damage, who shall inspect it, and declare on oath what it is; and if they should not agree in their declaration, there shall be appointed a third one or more sworn persons, until there is an agreement, and what the majority shall agree upon declaring, shall be ordered to be paid, and shall be paid for it; and if they shall find ore, which appears worth following out, and they shall make an establishment (*Asiento*) and other things necessary for the working and reduction of said ore, the said two persons shall inspect the damage which by reason thereof such pasture ground or inheritance may have received, or may receive, and, upon a due consideration of the whole matter, under their said oaths, shall appraise such damage, and the Court shall order it to be paid as aforesaid, provided, that besides paying the said damage, from all the silver which is taken out of the Mines, which shall fall in said pasture grounds or inheritances, there be paid to the owner of said pasture ground or inheritance, one per cent. clear of all expenses, and before our duty is deducted and paid to us, for above all, it is our will and we command that the said one per cent. be paid: which we command shall also be observed in all Mines which have heretofore been discovered.

Art. 16. Also we ordain and command that whosoever may discover a Mine of gold or silver, shall within twenty days after he has discovered or found the ore, be obliged to register it before the Court within whose jurisdiction such Mine may be situate, and in the presence of a Notary, producing the ore which he shall have found; and in the register shall be declared the person who discovered and registered it, and the place where it is situate and where was found the ore presented; and that within sixty more days from the date of such registration, he who made it shall be obliged to send and shall send an authorized copy of said register to our Administrator General, if there be one, and if not, then to our officers who reside in the workshop (*fabrica*) of the Mines of Guadalcanal, in order that it may be noted and entered in the book and general registry what he is to have of said Mines, so that all Mines which have been and shall be discovered may be known and an account be taken of them; and that in case such registry be not made in the form and within the time aforesaid, and according to what else is prescribed, any other person whosoever may register said Mine, and have and acquire the right which such discoverer or any other person would have had, if he had caused the registry to be made as aforesaid.

Art. 17. Also, whereas, previous to the publication of these, our ordinances, there have been discovered and registered many mines which were occupied and possessed (*embarazadas*) without being worked or reduced, and without any full account of them being taken, and the registries of them having been differently made:—we ordain and command that all persons who, before the publication of these our ordinances, shall have discovered and registered Mines, shall be obliged within two months to renew and make again the said registries, according to and in the form prescribed and ordered by the preceding ordinance, with respect to those which may hereafter be discovered and registered; and that within the further period of seventy days, they shall be obliged to send and shall send such registries to our said Administrator General, or to our said officers of Guadalcanal, as aforesaid, under the penalty in said ordinance contained.

Art. 18. Also, we ordain and command, that our said officers of Mines, who reside in the *fabrica* of Guadalcanal, shall keep a book in which shall be entered all the registries which may be made of all Mines discovered, and which shall be discovered, taken and sold, or in any other manner contracted for; and that they shall send to our *Contaduria Mayor* a report, signed with their names, of the condition of the Mines of these our kingdoms, and what is done with them; and that after having sent the first report, they shall every two months send one of what shall have taken place or been done with respect to them.

Art. 19. Also, we ordain and command, that no person shall presume to register, or enter in his register, a Mine which is not his own, under penalty of a thousand ducats, to be imposed on the person so offending, one-half to be applied to our exchequer (*camara,*) and the other half to the informer and the Judge who shall pass sentence, and that he shall forfeit the right which he may have to such Mine.

Art. 20. Also, we ordain and command, that when any person shall register a Mine or Mines, which may not be wholly his own, he shall be obliged to declare what share or shares he may hold in them; and, if he hold them in partnership, what share his partner or partners may hold in the said Mine or Mines, under penalty, if he fail to do so, of forfeiting the share or shares which he may have, which shall go to the partner or partners whose share or shares he has omitted to set forth.

Art. 21. Also, we ordain and command, that the person who shall first find and discover a Mine, shall, as first finder and discoverer, make the first registry and enjoy the extent (*medida*) which as such he is entitled to have and stake out, as to him may seem to be or may be best, even though he include and take within his stakes the trial-pit or trial-pits, which others who came after him may have made; provided that, first of all, he shall establish a fixed stake, which he cannot and shall not leave in staking out or altering his boundaries, as he may desire to stake out and alter them; and that the others who come after him, shall proceed in their order to stake out and alter their boundaries, setting up a fixed stake as aforesaid; and if two or more shall come together, it shall be determined briefly and summarily which was the first finder and discoverer; and he who shall be determined to have been first, shall be preferred, saving his right to the other party, who shall nevertheless claim to have been the first finder.

Art. 22. Also, we ordain and command, that any person who shall have discovered, or who shall discover, a new Mine, and shall have made registry as directed in the preceding ordinance, shall enjoy a space of one hundred and twenty *varas* in length upon the vein, and sixty in width; and if he should wish to measure out the said space of one hundred and twenty *varas*, and sixty *varas*, across the vein, he shall be at liberty to do so, and he may do it in such manner as to him may appear expedient; provided that he does not leave the fixed stake, and it shall be without injury to any other person, or persons, who may be on either side of him, and who may hold Mines opened and registered before him; and that, at the stakes of each first discoverer, one Mine be left for us of the same dimensions as his own; and provided, that in the staking out and altering of boundaries, and in all else, there be observed in the Mine, or Mines, which they shall leave, and mark out to us, all the provisions of these ordinances, in the same way and manner as they are to be observed and complied with by all persons who shall take and shall have Mines; and those who, after the first discoverer, shall have taken Mines, or who shall hereafter take them, respecting such Mine which for us is to be left at the stakes of said first discoverer, may proceed to take and work their Mines; and each of the Mines taken after the said first discoverer has taken his, and ours has been taken, shall be one hundred *varas* long and fifty wide, which space they may take across the vein, or as to them shall seem best, provided they do not leave the fixed stake, and do not prejudice a third party.

Art. 23. Also, we ordain and command that, if any person shall apply to have boundary stakes set out for the first discoverer, or for any others who have to set out boundary stakes after having registered their Mines, as well in those already discovered as in those which shall hereafter be discovered, such discoverer, and such other persons, shall be held and obliged to have their boundary stakes set out within eight days from the day on which they are requested to do so, if they be at the mines; and if they shall not set them out, then, the said term having expired, the said courts, taking with them persons who understand staking out Mines, and who have been sworn for that purpose, shall set out said boundary stakes; and if the persons, who are so requested, be not found in the Mines, but being in the vicinity within ten leagues of said Mines, they shall be obliged to set them out within ten days, and if they should fail so to do, at the expiration of the said ten days the said court shall do it, as aforesaid; and if they be not in the vicinity of said mines nor within ten leagues, notice shall be given to their steward (*mayordomo*,) or the person who may have charge of the working and reduction of their Mines, or be in their house if they have one; public notice shall be given on the first holiday (*dià de fiesta*,) and the term of said ten days shall run from the day of the notification which shall have been given to the said steward, or person, or in their house; and the said notice (*pregon*) shall be posted on the door of the church of the said Mines; and if there be no church in the said Mines, of that of the nearest *Pueblo;* and the said term having expired, the said court will establish the said boundary stakes, as aforesaid: and attention must be given, in setting out the said stakes, that there must always be a fixed stake, which must be adhered to, and must not be abandoned, in staking out or altering boundaries.

Art. 24. Also, we ordain and command, that if two or more persons who hold Mines on any side of the Mine for which they ask that stakes be set out, shall come together at the same time to ask stakes to be set out for such first discoverer, or for others who have to set out boundary stakes, in such case it shall be ascertained from the registries which of them is entitled to have the stakes set out first, and which second, and thus going on successively to set out the stakes, observing, with respect to the dimensions and all else, what is contained in these ordinances.

Art. 25. Also, we ordain and command, that when and as often as application shall be made to set out boundary stakes, and they shall be set out as aforesaid, care shall be taken, in setting them out, to make an oblong with right angles, and that in said oblong, and not outside of it, the said fixed stake shall be left, each one taking the number of *varas* to which he is entitled, wherever he may wish or deem best, as aforesaid.

Art. 26. Also, inasmuch as it may happen that, when fixed stakes have been set between two or more persons, one of them, with a view to his own advantage, may take up from its place a stake or stakes as he may think fit, and remove them to another place more desirable to him, thus giving rise to diverse law-suits;—we declare and command that when any person shall apply to have stakes set out for another and it be done, or shall wish to stake out his own Mine without being required to do so, he shall be obliged to make, at the places where he shall set the said fixed stakes on the side of his neighbors, a pit for each one of said stakes, two *varas* in depth and one in width; and in the centre of each one of these pits he shall set up the stake, and he cannot remove it, except in the cases in which, conformably to these ordinances, the boundaries are allowed to be altered: and the stake or stakes, which are thus set up, shall be regarded as *pertenensias* between him who set them up and his aforesaid neighbors; all of which shall be done and observed, under penalty of forfeiting the right which they may have to such mine, and of any other person being at liberty to ask for it, and to register it as his own.

Art. 27. Also, we declare and command, that when any person has already been requested to set out his boundary stakes and has done so, and another person shall again request him to set out boundary stakes on some other side of his Mine, he shall be at liberty to alter them, as between him and the person who thus requests him again to set out said stakes, it being done without prejudice to the stakes already set out, and provided he does not leave his fixed stake outside.

Art. 28. Also, we ordain and command, that, although a person has already set out his boundary stakes between himself and another on one side of his Mine, if he shall wish to alter the boundaries of his Mine before any other person or persons shall request him to set out stakes on some other side, where he has not set them out, he shall be at liberty to do so, provided he go before the Court to manifest his new boundary stakes and the alteration which he makes in his said Mine; and the said Court shall admit such alteration, and it shall be noted in the margin of the registry which has been made of such Mine; provided that it be done without prejudice to third parties, as aforesaid, and the fixed stake be left within his *pertenencia;* and the vacant spaces (*demasias,*) which are left between his Mine and that of his neighbor with whom he has boundary stakes set out, shall be given to the first who shall ask for them; and if the neighbor should ask first he shall be at liberty to take them, provided that, with the spaces he thus takes in, he has only the proper extent of a Mine and does not go outside of his fixed stake; and that he also manifests to the said Court the said alteration, in order that it may be entered in the aforesaid registry.

Art. 29. Also, inasmuch as some person or persons may happen to take Mines by the sides of other Mines which are designated, taken, and staked out, because it appears to them that the vein inclines and may appear on the said sides: we ordain and command, that when this may happen, and the owner of such Mine or Mines at the sides shall apply to the Court, said Court shall protect him in his Mine, and shall not permit the person to whom belongs the Mine from which the said ore inclines, to enter and work it or to take more metal from it: provided that, if the said owner of the Mine from which the said ore inclines shall not have set out boundary stakes with the persons who have Mines on the sides, they may be changed agreeably to these ordinances; and if he shall have set out boundary stakes with said persons, or with any one of them, and he shall have entered into their *pertenencia* while following said ore, he shall restore all the ore which he may have taken out, to the owner thereof, after deducting expenses.

Art. 30. Also, we ordain and command that, if the said ore should incline, as stated in the preceding ordinance, and at the place where it inclines, there be no Mine staked out, or, if there be one, no stakes have been by the owner asked to be set out for the Mine from which the ore inclines, the owner of such Mine shall be at liberty to continue to follow the said ore, working, reducing and enjoying it, although he may go outside of his *pertenencia;* and if any person shall come to establish a Mine in the same place where the ore happens to be and to run, or if he shall have set out stakes without having asked for them, and shall ask for them, then the owner of the Mine from which the said ore inclines, may alter his boundary at the place where the said ore goes, taking for it the dimensions of his Mine, which are 120 *varas* long and 60 wide, he being the first discoverer, and 100 *varas* long and 50 wide, he being any one of those after the first discoverer, or of those who shall derive from them; provided that, they cannot leave and shall not leave the fixed stake which they have set out, and the space be an oblong and at right angles; which, it is to be understood, they can do in any place or places where the ore may run, even though they may have altered the boundaries one or more times; and the vacant spaces which they may leave between their Mine and that of the neighbor, between whom and themselves they shall have set out fixed stakes, shall be given to those who may ask for them, agreeably to the preceding ordinance.

Art. 31. Also, we ordain and command that no person, whether he be the first discoverer or any of the subsequent ones, shall be at liberty to take more than two Mines on one vein; and these two Mines which he may thus take, he shall not be at liberty to take together, but there shall be at least sufficient space of land between for three mines; except in case he shall have purchased, or shall purchase them, because by purchasing them he may have two or more, even though there be many, and all lying together.

Art. 32. Also, we ordain and command, that if two persons shall be in partnership, they may take two Mines at one stake, and also may take other two Mines at one stake in the same vein; provided that, between the first two Mines and the other two, there be left three *pertenencias*, as directed in the preceding ordinance, and if there be more partners, they may take the said Mines in partnership in the same order; and if they have been purchased, they may hold them, as is stated in the said ordinance.

Art. 33. Also, we ordain and command, that no person, of whatsoever condition he may be, shall take a Mine for another, unless he have a power of attorney, or be a servant on wages, of the person for whom he shall take such Mine; and, in default of either of these requisites, the Mine shall be considered as forfeited, and may become the property of the person who shall denounce it, and the Judge shall give immediate possession of it to such denouncer; without allowing any appeal (*recurso*) either to the person in whose name the said Mine was taken, or to him who took it for him.

Art. 34. Also, we ordain and command, that no steward (*Mayordomo*) who may be engaged in the working and reduction of such Mines, nor any other person who may live with the owner of Mines, and may be engaged in the management of them, shall be at liberty to take or hold a Mine or part of a Mine himself, or through the interposition of any other person, directly or indirectly, even though he may have purchased it, or acquired it in any other manner, either in the Mines where he shall exercise his office or serve, or within the space of one league about them, either during the whole time he shall hold the office or serve, or for two years afterwards, except it be for his master (*amo*,) when he can hold them agreeably to these ordinances, or he being in partnership with his master or with some one else by his master's permission; under penalty of forfeiting, and he shall forfeit, the said Mines or parts of them, and they shall be the property of his said master if he can hold them as aforesaid, and if he cannot hold them, then, they shall go to our exchequer (*camara;*) and besides forfeiting the said Mines or parts of Mines, he shall be exiled from the establishments and from such Mines, and from the space of three leagues around them, for the full term of three years, and, if he should break this term, under penalty of serving it over again at the oar in the galleys; which said punishment shall be incurred by any person whatsoever who shall participate in the above offense: and the Mine or Mines, which the said steward, or the said persons who may live with the said owners of Mines and are engaged in managing them, or the slaves of the said owners of Mines, may take, shall belong to the said owners, the same as if they themselves had taken them, carrying out, concerning them, all the proceedings conformably to these ordinances; and no person may or shall enter into them for the purpose of taking them, under penalty of two hundred ducats, to be applied as aforesaid; and by reason of thus entering into them by his own authority, he shall not be able to acquire, nor shall he acquire, either possession or any other right, but on the contrary he shall forfeit any right which he may have or pretend to have.

Art. 35. Also, we ordain and command, that no steward who may be engaged in the working and reduction of the said Mines, nor any other person who may live with the owner of Mines, although he may have his Mines and workmen in charge, shall be at liberty to move the stakes which his employer (*amo*) may have set up, without his permission and authority, even though he may have been applied to, to set up such stakes; and if he shall move them or set them out anew, it shall avail nothing, and shall work no prejudice to the person to whom the Mine belongs.

Art. 36. Also, we ordain and command, that when such steward who may have charge of any Mines and *Hacienda*, shall take or discover a Mine, such steward shall be at liberty to stake out the Mine or Mines which he may so take, and to set out boundary stakes on the side of any party who may apply for that purpose, until such time as his employer shall visit such Mines; but that after his said employer, the owner of such Mine or Mines, shall have arrived, he shall neither apply for, nor set out any more boundary stakes; and the said steward, without the authority of his employer, shall not be allowed to move those which his said employer shall have set out or left fixed.

Art. 37. Also, we ordain and command, that all persons whatsoever who shall hold and take Mines, as well the Mines already discovered as those which may hereafter be discovered, shall be obliged, within three months, counting from the day on which he registered the said Mines, to sink one of the trial-pits (*catas*) which they shall have made in them, three *estados*, each *estado* being seven thirds of a *vara* long, under the penalty, if they shall not sink them or have them sunk the said three *estados*, at the expiration of the said three months, of forfeiting them, and they shall forfeit them, and they shall be for any person who shall denounce them, and the Court shall immediately put the said denouncer into possession, subject to the same obligation of sinking the aforesaid three *estados* within the term aforesaid, notwithstanding any appeal, nullity or revision (*agravio*) which may be interposed.

Art. 38. Also, forasmuch as, in the preceding ordinance and in some others of these our ordinances, it is provided and commanded that such persons as shall take and hold Mines, or shall purchase them, or in any other manner acquire them, shall be obliged to deepen such Mines according to the provisions in said ordinances contained;

3

and whereas it is our intention and will to prevent lawsuits and disputes, and to obviate mal-intents:—we declare, ordain and command, that it shall be understood that they are obliged to deepen the said Mines and to incur the penalties of the said ordinances, if they can deepen them; but that if, by any accidental cause, or because it is more expedient to proceed by following the ore on account of its inclining in some other direction, as often happens, and not by any fault of their own, they shall fail to deepen them, and shall continue to work them as may be most convenient and advantageous, they shall not become liable to, nor incur, the aforesaid penalties; provided, that when this shall be the case, they shall be obliged to give notice of it to our officers who shall reside in said Mines of Guadalcanal, in order that an investigation of said case may be made, to ascertain whether the failure to comply with the provisions of the aforesaid ordinances resulted by reason of following the said ore, and not by their own fault, whereupon, after such inquiry has been made, our said officers shall provide what may be expedient, in such manner that, the inconvenience having ceased, the Mines aforesaid may be deepened according to what by the said ordinances is commanded.

Art. 39. Also, whereas it may happen that, contrary to the provisions of these our ordinances, some persons may take more Mines, of those which they can take, than amounts to two Mines in each vein, having three pertenencias of Mines from one to the other, and four Mines in partnership, according to the provisions in said ordinances cnotained, and whereas, to have many Mines in each vein, gives rise to frauds and concealments, whence may result notable injury to us, and to those who engage in working and reducing said Mines, and to these our kingdoms; we ordain and command, that any one having more Mines than the number which as aforesaid he can have, any other person may apply for the surplus Mine or Mines which he may have over and above those which he can have, whether he has had them registered, or whether they are yet to be registered; and that in order to determine what are surplus, reference shall be had to the registries, and by them the first which shall have been registered shall be declared theirs, and the remainder as surplus; and they shall be adjudicated to him who shall first have asked or shall first ask for them, provided that he shall have them registered, and shall observe all the provisions of these our ordinances, and shall sink said Mine or Mines three *estados* deeper than they were before, for which purpose they shall be measured, under the penalty of forfeiting them, and they shall be for him who shall denounce them, under the same obligation; which is to be understood only with respect to such Mines as each one may take, because any one may freely hold those which they may have acquired by purchase, exchange, or in any other lawful manner.

Art. 40. Also, whereas, it often happens that certain persons hold many Mines which they have taken, or purchased, or acquired in some other way, and do not work or reduce them, either because they cannot, or because they are working others which they consider better, and thus fail to deepen those which they do not work, and to discover and take out ores from them, although sometimes better than the ores taken out of the Mines which they do work; and whereas, the said Mines which they fail to work become filled with water and cause injury to other neighboring and surrounding Mines which are worked and sunk deeper than they are;—therefore, in order to obviate these inconveniences and others which follow or may follow from the Mines not being worked, we ordain and command, that all persons shall be obliged to have their Mines occupied (*pobladas*) by at least four persons to each Mine, whether they be entire owners of such Mines or hold them in partnership, for, in whichever way it may be, by having four persons in each Mine, it is sufficiently shown that the said Mines are occupied (*pobladas*); which said four persons shall be engaged in working the Mines which they occupy, under penalty that, if any Mine whatever shall not be occupied (*poblada*) with said four persons, as aforesaid, for four consecutive months, it shall *ipso facto* be forfeited, and the person who owned it shall forfeit it, and in future shall have no right to it, unless by making a new registry of it and by going through the other proceedings conformably to these ordinances; and said Mine shall be adjudged to any person who shall denounce it as abandoned (*despoblada*), provided he go through the proceedings aforesaid.

Art. 41. Also, we ordain and command, that, in order that any Mine may be pronounced and declared insufficiently occupied (*despoblada*), the person who shall come to denounce it, shall appear before the court and make the denouncement; declaring in it the hill or place where the said Mine is situate, and next to whose boundary stakes, if any there be, to what depth it has been sunk, and whether or not it has any metal; and within forty days, the party being summoned, if possible, in person, or at his house if he have one at the Mines in question or in the vicinity, and, if it can conveniently be done, mentioning it or making it known to his wife, or children, or servants, or his nearest neighbor or neighbors, in such a manner that it may come to his knowledge, and, if he cannot be summoned, not having a house, as aforesaid, then by means of edicts and notices (*pregones*), in the manner hereafter described, it shall be ascertained whether the said Mine has been left unoccupied (*despoblada*) for the four months aforesaid, and within forty days, computed from the day on which the said denouncement was made, both parties shall be at liberty to allege and prove such matters as they deem expedient, and the cause shall be determined upon what is done within said term, without any other postponement (*conclusion*) or adjournment (*prorogacion*); and if such Mine shall be decided to have been unoccupied (*despoblada*), then it shall, as such, be adjudged to the said denouncer, and possession thereof shall be immediately given to him, notwithstanding any appeal, nullity or revision (*agravio*), which may be interposed against the decision; provided, that the person to whom said mine shall be adjudicated, shall be obliged, within three months, to deepen the trial-pit which he may think proper, and sink it three *estados* deeper than it was at the time of his making the said denouncement, and for that purpose it shall be measured; all which he shall do and perform, under penalty of forfeiting it, and of its being adjudged to any person who may denounce it, on the same condition, and subject to the same penalty, provided that he keep an account and reckoning in a book, with the day, month and year, of the silver or metal which may be taken out of said Mine, and of the costs and expenses which may be incurred in working and reducing it; and provided that he give security, in one thousand ducats that, if he lose his case on appeal, and be ordered to render an account with payment, he will do so; and if any one of the parties shall consider himself aggrieved, he may appeal within three days; and upon what, within sixty days, both parties shall state, allege and prove, without any other postponement or adjournment, the case shall be determined and justice rendered; and what shall be thus determined, shall be observed and executed, without there being allowed any supplication, nullity, revision or any other remedy.

Art. 42. Also, we ordain and command that, if any Mine should happen to be denounced as not occupied (*despoblada,*) which shall not appear to have any owner, or having one, he shall be absent and it be not known where he is, or it be known that he is in a place where notice cannot be given to him according to the provisions of the preceding ordinance, the said Court, on some Sunday when coming from mass at the principal church of such Mines, or, if there be no church in them, from that of the nearest Pueblo, where there shall be at least eight persons present, shall give public notice of said denouncement, in order that it may be known and that information thereof may be given to the owner, or to some one who may answer for him, if he please, and make his defence; and such proclamation (*pregon*) being made, a copy thereof shall be affixed to the principal door of such church, where it may be publicly seen; and such proclamation shall be made on two other subsequent Sundays, so that altogether there shall be three proclamations, and copies thereof shall be affixed as aforesaid; which shall be held a sufficient summons in like manner as if it had been made personally; and if during the time of the said three proclamations, or within forty days from the time of the first of them, being made, the owner, or some person competent to oppose such denouncement, shall not appear, then the parties being heard conformably to the preceding ordinance, justice shall be done; and if no appearance be made within such term of forty days, the proclamations being made, the denouncer shall give evidence of the mine aforesaid having not been sufficiently occupied (*despoblada*) for the said term of two months; and this being proved, and the said forty days having expired, the Mine shall be pronounced as such, and adjudged to the said denouncer, and the possession of it shall be given to him, on condition that he shall be obliged to sink it three *estados*, conformably to said ordinance and under the penalty which it prescribes; and if, after the said forty days have expired and within the three days allowed for appeal, the owner, or any one having his authority, shall appear, he shall be at liberty to appeal; and conformably to the said ordinance, justice shall be done.

Art. 43. Also, whereas it may happen that some Mines may be flooded by the water flowing in from the adjoining and neighboring Mines which are of less depth, by which means the working and reduction of such deeper Mines may be stopped, and the owners of them thereby suffer damage;—we command our Administrator General, and the person by him appointed, to take particular care to visit said Mines and to give orders that they shall all be kept clean, drained, worked, and reduced; and if any Mine shall be damaged by the waters of another or of others, the said Administrator General or the said person, upon the request of the party, shall cause two persons, appointed by the parties and sworn in his presence and approved by him, to inspect and appraise the damage, and the expense necessary to clean and drain such Mine, and what shall be thus determined, the Court shall order to be paid immediately; so that the said Mine may be cleaned and drained, in order that it may be worked and reduced.

Art. 44. Also, we ordain and command, that all persons who shall hold, work or reduce any Mine or Mines, shall be obliged to keep them clear, and timbered, in such a manner that they may not fall in or become choked up, leaving in those which have a *ley* of a mark and a half per quintal of lead-silver or under, such bridges and strengthenings as may appear expedient for their security; and those which have a higher *ley*, shall be thoroughly lined and secured with good timbers, in place of the said bridges and strengthenings; and in case this be not done, the said

Court will have it done at the expense of the Mine: and in order that this may be thus done and performed, our Administrator General, or the person by him appointed, is to take and shall take special care to visit such Mines and cause them to be inspected by persons who shall understand the matter, and determine what may be necessary, according to what is prescribed in the preceding ordinance.

Art. 45. Also, whereas it may happen that some persons, who take Mines without working them or ascertaining whether they contain ore, may sell and contract for them, and then proceed to take others for the same purpose, whence divers inconveniences may result;—therefore, in order to avoid these evils, we command that no one shall sell, contract for, or purchase any Mine, if it has not been sunk to the depth of at least three *estados*, under penalty of forfeiting what may have been given to him for it, to be applied as has been stated above, and moreover of the Mine itself being forfeited, and given to the denouncer, subject to the same obligation of sinking it to the aforesaid depth of three *estados;* and if the Mine, which shall be so sold or contracted for, shall have been sunk to the said depth of three *estados*, then, in order that the said sale or contract may be freely made, the purchaser shall be obliged to give notice thereof to the said Court, in order that it may be entered in the book of registries; and an authenticated copy of it shall be sent to the said Administrator, or to the person by him appointed, in order that it may be entered in the general book, and that it may be known from whom the duty is to be collected; which shall be done and performed, under the penalty aforesaid.

Art. 46. Also, we ordain and command, that when two or more persons shall hold a Mine in partnership in order to work it and to take ore from it, if any one of the partners shall apply to have the others put on workmen, they shall be obliged to put on in all twelve persons, if there be ore enough for the purpose, and so many can work conveniently; and if not, as many as can work at once, according to the ore it may have and the condition of the Mine: and he who, on being required to do so, shall not put on his proportion, shall receive no ore all the time that he does not put on workmen, and it shall belong to the other partners who do put on said workmen; and no requisition having been made to any one of said partners, or the steward or person who for him may have charge of said Mine, to put on workmen, they shall give him his part the same as if he had put them on; and if they should not give it to him, the said Court will cause to be given to him the part which belongs to him, without deducting anything for the expenses which shall have been incurred.

Art. 47. Also, we ordain and command, that, if any one of the partners shall wish to put on more workmen than said twelve persons, to work the said Mine, he may do so, provided that he give notice thereof to his partner or partners, in order that, if they shall desire, they may put on more workmen; and if he shall not give such notice, he shall forfeit the ore which he may take out and it shall go to the said partners; and if, having given them notice, they shall not wish to put on more people, they shall not be obliged to do so, because the law is complied with by putting on the said twelve persons; and if, nevertheless the partner shall still desire to put on more people, giving notice as aforesaid, he shall be obliged to give them their proportion of the ore taken out, the same as though the additional people he may have put on and by whom the said ore may be taken out, had been put on by all the partners; and the said Court shall compel him to do so, as is prescribed in the preceding ordinance.

Art. 48. Also, we ordain and command, that if any person shall have a partnership in a Mine which has no ore, and shall desire to work it and sink it, and the partner or partners shall not desire to put on more people than they are obliged to in order to keep said mine occupied (*poblada*,) agreeably to these ordinances, he who shall so desire to work shall be obliged to make requisition on the partner or partners, or on their steward or persons who may have charge of said Mine, to put on more people in proportion to the number which he shall put on, until the ore is reached, and if they shall not so put on, he shall be at liberty to put on; provided that he shall be obliged, on reaching the ore which is to be followed and reduced, to give notice thereof to the partner or partners within another day, in order that they may put on people, agreeably to the preceding ordinance; and within two days the said partner or partners shall elect, whether they desire to pay the part of the laborers (*peones*) which corresponds to them in the working of said Mine, or whether they desire that he who has worked it and discovered the ore, shall himself take out ore, for that purpose, with as many other laborers; and if they shall desire him to take out the said ore, the number of laborers which he has put on shall be determined by his oath, or that of his steward, or the person who had charge of the said Mine; and he may take out with them the said ore for himself; and if the partner or partners shall prefer to pay the laborers' wages (*las peonadas*), they shall pay for each day's work, to each person who shall have worked the proportion which corresponded to them, four *reales;* and until they have so paid them, they cannot and shall not take any ore out of the said Mine in which they have a part; and the partner who shall have deepened it may go on working it, and the ore which he shall take out shall be his alone, until the said wages shall have been paid: and if he who shall discover the ore shall not give notice thereof to the said partners, or to their said steward or person or persons in said charge, within the said term, then they shall not be obliged to pay him the wages (*peonadas*), nor shall he have any of the ore with which to pay himself for them, but he shall give to each one the proportion of the ore corresponding to him from the time he discovered it, and thenceforth all shall be obliged to put on the said twelve persons, as directed in the said ordinance.

Art. 49 Also, that, if they shall not wish to smelt the ore which is thus taken out, all together in partnership, they shall divide it equally according to the share which each one has in the said Mine, and by the measure or weight, and at the time which for that purpose they shall appoint; and that, until it be so divided, it shall be kept together in a secure place; and no one shall dare to take anything from it, under penalty of forfeiting his share, which shall go to the other partner or partners, and, in addition thereto, the amount of the value of the said share, one-half of which shall go to our exchequer (*camara*) and the other half to the denouncer and Judge; and if they shall smelt it in partnership, it shall also be refined in partnership, so that each one may receive therefrom the share belonging to him, under the penalty incurred by those who do not take to be refined the metal which they have smelted, and, without refining, sell and exchange it.

Art. 50. Also, we ordain and command, that no person, in order to work or excavate his Mine, shall be allowed to throw the earth, which he shall take therefrom, into the Mine or *pertenencia* of another, under penalty of ten ducats for each offence, to be applied as aforesaid; and the Court, immediately on its being asked by the party, shall cause such earth to be removed and cleared away from such *pertenencia*, at the expense of the party who put it there or caused it to be put there, notwithstanding any appeal, nullity, or revision (*agravio*) that may be interposed, but any one is permitted to remove the earth of his Mine through any *pertenencia*, provided the said earth be taken entirely outside of such *pertenencia.*

Art. 51. Also, we ordain and command that the washing places (*lavaderos*) which may be required for washing the ore of said Mines, shall be taken wherever most convenient to the miners, provided that if they are prejudicial to any Pueblo or to the cattle, and cannot be made without producing such injury, the water shall be taken from the river or creek to pools where such ores may be washed, and shall be so drained off as not to return to said river or creek; and if this cannot be done, enclosures shall be made, at the expense of those who make such washing places; and for the better provision and performance of this, the Court, in whose jurisdiction such washing places are made, shall cause the above to be complied with, so that the injury may be prevented; and, in taking the said watering places, they shall be staked out in the same order as the said Mines, and the measurement of each shall be sixty feet long, each foot being one-third of a *vara*, and twelve feet wide.

Art. 52. Also, we ordain and command, that no person shall presume to enter, search for, take away, or reduce ore from the refuse-heap (*terrero*,) or washing place (*lavadero*,) or slag-heap (*escorial*) of another, under penalty of ten ducats for the first offence, and for the second, twenty, to be applied as aforesaid; and for the third offence, over and above the said twenty ducats to be applied as aforesaid, that he shall be banished for the full term of three years from the Mines of that *Partido:* and that he do not return within that time, under penalty of banishment for double that period, and, moreover, that all which he may have taken out, or shall take out, shall be for the owner of such refuse-heap, washing-place or slag-heap.

Art. 53. Also, we ordain and command, that for the purpose of working said Mines and of smelting and refining the metals; and for the purpose of lining and preserving them, and of constructing machines, buildings and huts, and all other things required for them, their owners and the persons engaged and working in them, may use and shall be permitted to make use of all forests, timber and trunks of trees, and all else, the same as the residents of each place may do, observing our Royal Letter (*Carta Real*) which on this matter we have issued; which they shall thus do, notwithstanding what has been provided with respect to fire-wood, and timber, and charcoal, in the first Edict (*pragmatica*) which was issued with respect to Mines.

Art. 54. Also, we ordain and command, that all the said owners of Mines and persons engaged in working and in reducing them, may freely drive into said pastures, meadows, outgrounds (*egidos*), or public or municipal forests which may be near the establishments of said Mines, all their beasts and those of their servants which are required for the working of said Mines, either for the machinery, or for draft, or for the saddle, and oxen for the wagons used to bring provisions or timber or other things to the said establishments (*asientos*) and mining works, in the same way and manner as the residents of the boundaries adjacent to said Mines and *Asientos* may do; and if there be pastures (*dehesas*), they shall pay for the grass and pasturage the same as are paid for other cattle; and those who are engaged in making trial-pits, or are traveling for the purpose of finding such Mines, shall be at liberty to take with them each one animal, without paying anything for the grass which may be eaten; but in other matters the provisions of this ordinance must be observed, when the Mines begin to be worked advantageously.

Art. 55. Also, we ordain and command, that all the owners of such Mines, and their servants and the persons engaged in working the Mines and in reducing their ores, shall be permitted to hunt and fish freely, within three leagues around the place where the *Asientos* of Mines in which they reside are situate, observing the edicts of these our kingdoms upon that subject.

Art. 56. Also, we ordain and command, that in any parts or places whatsoever wherein Mines have been or shall hereafter be discovered, the owners of such Mines shall make the establishments (*asientos*), smelting works, furnaces, and other things necessary for their working and reduction, collected and congregated together as much as possible; and that our Administrator General, or person by him appointed, shall take special care that it be thus done; and each one of said owners of mines shall have a stamp of iron, with which they shall stamp and mark the bars of lead-silver and of whatever else shall be taken from their Mine; and without the said stamp they shall not be taken to be refined, nor shall they be refined.

Art. 57. Also, we ordain and command, that no person shall presume to smelt any ore except in furnaces which are his own: and if he shall desire to smelt in any other furnace, he shall signify the same to our Administrator General, or the person by him appointed, and with his license he shall be at liberty to smelt it, and not otherwise, under penalty of forfeiting such lead-silver, one-half to our exchequer (*camara*) and the other half to the denouncer and Judge; and of forfeiting the said Mine, which shall be for the denouncer aforesaid.

Art. 58. Also, we ordain and command, that when in smelting the ore of a Mine it shall happen to be expedient, in order to facilitate the smelting, to mix with it the ore of another Mine, this may be done; provided that it do not exceed the *ley* of the ore with which it is desired to make such mixture by a mark per quintal of lead-silver; and if it should so exceed, then it cannot be done nor shall it be done, without the license of our said officers who reside at Guadalcanal, under penalty of forfeiting the ores which shall be so mixed, and that which shall result therefrom, with as much more, the half to our exchequer (*camara*), and the other half to the denouncer and Judge who shall pass sentence: and we command our said officers that, when such shall be the case, they shall inspect and assay the said ores of said Mines in order that, conformably thereto, the duty belonging to us may be paid; and having done so, and having examined into it as is proper in a thing of such great importance, and having determined the proportion which we are to have, they will give said license, because the said mixing is very advantageous for good smelting.

Art. 59. Also, we ordain and command, that in each of the said *Asientos* there shall be and shall be made, at our expense, a refining-house (*casa de afinacion*) with different kinds of furnaces (*buitrones, y fuslinas*), bellows and tools required for the refining of the lead-silver which shall be smelted in all the Mines of that *Partido;* to which said refining-house all persons shall bring to be refined, and it shall be there refined, all the lead-silver which from said Mines shall be taken, and from all that neighborhood (*comarca*), and no person shall presume to refine such lead-silver, in large or small quantities, in any other place than said refining-house, nor to sell or contract for it, until it shall have been refined, under penalty of forfeiting, and they shall forfeit, what they so refine, sell or contract for, with four times the value, to be applied, one-half to our exchequer (*camara*), and the other half to the person who shall denounce it and to the Judge who shall pass the sentence; which said penalty shall also be incurred by any person whatsoever who shall participate in the above; and where such house cannot be conveniently established, from there being no works erected and no Mines sufficient to render it necessary, the said officers shall make such provisions, and give such orders as may be expedient and necessary, for the smelting of the said ores which may be there; and that the lead-silver, which may be taken out there, shall be carried to the refining-house most convenient, and having reached there, the same order shall be observed in regard to the refining thereof and in all other matters, as is provided in respect to ingots of lead-silver which are ordinarily to be refined in said house.

Art. 60. Also, we ordain and command, that in each of such refining-houses of each *Partido*, there shall be the requisite number of refiners appointed by our said officers who reside in Guadalcanal, which refiners shall, at the expense of the parties, who shall furnish the necessary charcoal, refine the lead-silver of that *Partido* and *Comarca*, and no other persons shall make such refinings, without the license of our said Administrator or the person by him appointed, under penalty of one hundred lashes and of serving three years at the oar in the galleys; and the said officer shall assess the rate which said refiners are to be paid, and the charcoal which they shall expend.

Art. 61. Also, we ordain and command, that in each *Asiento* of Mines where there may be such a refining-house, there shall be, at our expense, a trusty person (*Fiel*) to weigh the lead-silver which is brought to be refined, which person shall, when he enters upon his office, take an oath to well and faithfully discharge its duties, and a Notary (*escribano*) who shall certify to the parcels of lead-silver which are delivered to the refiners, and all the parcels of lead-silver which shall be brought to be refined shall be delivered to the refiner who shall have been selected by our officers to refine it; and the said officers shall keep a book wherein shall be entered all such parcels, and the said Notary shall keep another book for the same purpose; each of which said books shall have an alphabetical index, with a separate account for each one of the persons who shall bring such lead-silver to be refined, and on a separate leaf: the said Trusty (*Fiel*) shall weigh the said ingots and they shall be delivered to the said refiner: and in said book shall be entered, with the day, month, and year, the weight and number of ingots, the persons who brought them to be refined, and their stamp, and the Mine or Mines from which they came, and the refiner to whom they are delivered, in such a manner that there shall be a particular account and reckoning of the whole; and our said officers, if before them it be done, or the person by them appointed, and the said Notary, and the party, if he be able to write, and if not, another for him, shall sign both the said books: and, after all the aforesaid has been done, the said refiner shall refine the said parcel, without permitting the lead-silver of one Mine to be mixed or mingled with that of another, under the penalty, to the party so mixing them, of forfeiting said lead and silver, with four times as much more, to be applied as aforesaid, and if the said refiner shall so mix them, he shall receive one hundred lashes and shall serve at the oar as a prisoner for three years in our galleys: and we charge our said officers to observe, and cause to be observed, particular diligence and care, that such refinings are faithfully performed, so that no fraud may be practiced against our rights, and no injustice be done to the parties.

Art. 62. Also, we ordain and command, that when what is above directed has been done, and the silver refined and taken out, in presence of our said officers or of the person by them appointed, and of the Notary, the Trusty (*Fiel*) shall weigh it, and there shall be taken from it the proportion which, according to these ordinances, belongs to us and we are entitled to have, and it shall be delivered to our Treasurer, and of that which is delivered to him he shall have charge, it being entered in the said books, and in the other book which our said Treasurer shall keep, with the day, and month and year, stating from what Mine or Mines the said silver comes, and the owner of the parcel, and the person who brought it to be refined, and the weight of the silver in such parcel, and the proportion of it which belonged to us and was delivered to said Treasurer: and all the aforesaid persons shall sign all said three books, in order that the said Treasurer may give an account thereof when he shall be ordered to do so; and the remaining silver shall be delivered to the owner, after stamping it in one, two, or more places on each ingot, as each one may require, with the stamp of our Royal Arms, without which said stamp no one shall venture to sell or contract for said silver, which from said Mines is taken, under penalty of forfeiting said silver and, what may be purchased therewith, and the half of all his property, to be applied as aforesaid, and besides this, he shall be banished from said Mines, and the space of ten leagues around, for the full term of six years; and he shall not break said banishment, under penalty of serving the same period in the Galleys or wherever he may be sent; which penalty aforesaid shall be incurred by the purchaser or the person to whom the said silver shall be contracted.

Art. 63. Also, whereas many ores of silver are worked and reduced by quicksilver at less expense and with more profit, and it may happen that some persons shall desire to work and reduce said ores with quicksilver, and therefore cannot observe what is provided and commanded with respect to ores which are worked and reduced by smelting and refining; in order that said silver which is extracted by said quicksilver, may pay us the duty which belongs to us and which we are to receive, conformably to these ordinances, without any fraud therein;—we ordain and command, that any person who shall desire to work and reduce such ores by quicksilver, shall be obliged to give notice thereof to our said Officers and to declare to them the Mine or Mines which he shall desire to work and reduce by quicksilver as aforesaid, in order that the same may be entered, and that it may be known that such Mine or Mines are worked and reduced by quicksilver: and that, during all such time as they shall wish to work and reduce them by quicksilver, they shall not be at liberty to work, nor shall they work or reduce, the same in any other manner, unless upon giving notice thereof, when they shall desire to do so, to the said Officers, in order that the same may be noted, and that it may be known that they no longer work or reduce the said Mines by quicksilver as aforesaid; and if they shall, in any other manner, work and reduce the said Mines, they shall forfeit the silver and ores, and the one-half shall go to our exchequer and the other half to the denouncer and to the Judge who shall pass sentence; and the said Mine or Mines shall be considered as forfeited, and shall be for the said denouncer.

Art. 64. Also, we ordain and command, that, upon all the silver which shall be extracted by quicksilver as aforesaid, there shall be impressed a stamp different from that impressed upon lead-silver, and that it shall be immediately taken before our said Administrator or before the person by him appointed, in order that the said Trusty may weigh it, so that there may be taken from it the duty which we are to have and which to us belongs, and be delivered to our said Treasurer, as aforesaid; and the above mentioned being done, the remainder shall be delivered to the owner; and upon each ingot shall be impressed our Royal stamp, as above directed, and without having our said Royal stamp, the said silver shall not be sold or contracted for in any manner, under the penalty above prescribed to the owner of said silver, and to the purchaser, or person who may contract for it.

Art. 65. Also, we ordain and command, that the poor lead which shall be smelted and shall not be worth refining, from being so poor in silver as not to contain over four *reales* per quintal, shall be brought to the refining-house which may be in the *Asientos* where the Mines are from which the lead is extracted, or those nearest, in order that the person, who shall reside there, appointed by our Administrator, may stamp it and deduct the duty which ought to be taken out of it; and that no lead, even though it be made from litharge, shall be conveyed from one place to another, without such stamp, under penalty, to any person otherwise conveying it, of forfeiting the same, one-half to go to the party who shall denounce it, and the other half to him who shall pass sentence, and likewise four times the value to go to our exchequer; and the same with regard to copper, which, being first assayed, shall be stamped in order that there shall be paid to us our proportion of the silver and gold it may contain; and this is to be understood as applying to places without the limits of the grants (*Mercedes*) which have been made.

Art. 66. Also, we ordain and command, that all those who extract antimony shall pay us our duty on it in the veins where it is extracted, and that, until this is paid, they shall not be at liberty to move it, nor to sell it in any other place, without the license of our said officers or of the person by them appointed, who shall be in the *Asiento* of Mines nearest to the Mine from which said antimony is taken; and that, after such license is obtained, no person shall be at liberty to carry it away without an order (*cedula*) from said person, and that the said seller shall be obliged to advise the purchaser thereof in order that said *cedula* may be obtained, and he shall so advise him, under the penalty of forfeiting the value of said antimony, with four times the amount, to be applied as aforesaid, and to the purchaser who shall in any other manner convey it away, of having it taken from him as contraband, with four times the amount, to be applied as aforesaid; which is to be understood as applicable to places where there are no grants (*Mercedes*) made.

Art. 67. Also, forasmuch as it has been found by experience, that, when suits and differences are raised concerning the possession of Mines, the working and reduction of such Mines cease, and they are ordered to be closed, until it be determined which party has the better right, so that they frequently are one, two or more years without being worked or reduced, which, besides the damage received by the parties, causes notable injury to our Royal Treasury and to all the Republic;—therefore, in order that all the above may cease, and that said Mines may not be left unworked or unreduced for so long a time, we ordain and command, that when and every time such suits shall arise, within forty days, (for which said term and no longer the Mine about which there is litigation may be closed,) the parties shall, before the Court, state and allege their claims, and present the writings and securities which they may have, and they may produce as many as twelve witnesses each, and no more, and upon what is stated, alleged, and proved within the said term, without any other conclusion or adjournment, the said Court shall consider and decide, saving to the party against whom it shall give judgment, the liberty to pursue his claim to the possession and ownership, as he shall think expedient, and shall immediately give the occupancy (*tenencia*) of said Mine to the party in whose favor it shall give sentence, which party shall work and reduce it, keeping an account and reckoning in a book, with the day, month, and year, of the ore which he shall take out, and of the costs and expenses he may incur in the working and reduction, giving security in the sum of one thousand ducats for rendering an account and paying over the above, if upon appeal he should be beaten and should be ordered so to do: which shall be done and performed as aforesaid, notwithstanding any appeal, nullity or revision (*agravio*) which may be interposed against what shall have been determined and executed; and if the party against whom the sentence is given should consider himself aggrieved, he shall be at liberty to appeal within three days, and within sixty days, in case of appeal, nullity, or revision, both parties shall prosecute their case, and present their writings, securities, and witnesses, and they shall be admitted as far as the law allows, as aforesaid, and on what, within said term, without any other conclusion or postponement, they shall state, allege, and prove, the justice of the case shall be determined; and if the decision shall be in confirmation of the former judgment, the party in whose favor it shall be given shall nevertheless keep an account and reckoning of the said ore which is taken out, and of the said costs and expenses, as aforesaid, in order that he may render the same with payment, if in the ordinary proceeding he shall be beaten and adjudged to make such payment; and without having any other grade of appeal, or nullity, or any other remedy, the cause is to be submitted to said Court, in order that justice may be done with respect to possession and ownership; and from the sentence or sentences which are given thereon, an appeal may be taken according to the laws of our Kingdoms, in the grade of appeal, or of supplication to prosecute their case, as to them shall seem expedient: and if the said sentence given in the grade of appeal from the first sentence, shall be revocatory, that such decision shall be carried into full and due execution, and the party in whose favor it shall be given, shall be placed in the occupancy (*tenencia*) of said Mine, to whom the other party, in whose favor the first decision was given, shall render an account, with payment of all that has been taken from and produced by said Mine up to the day it is taken from him, deducting the costs and expenses which have been incurred in working and reducing the same, and this second holder (*tenedor*) shall also keep an account and reckoning in a book, as aforesaid, of what shall be taken out of said Mine, and of the costs and expenses, and shall give said bonds for one thousand ducats to render an account and payment, if in said claim of possession or ownership he shall be beaten and adjudged to pay it; and the said cause with respect to possession and ownership, shall be remitted to the said Court, in order that the parties may seek justice in the ordinary mode of proceeding.

Art. 68. Also, we ordain and command, that when and so often as any person shall ask for a Mine of which another has quiet and peaceful possession, and shall also ask that such Mine be closed, then, inasmuch as the principal object sought for in such a case, is the ores which are taken from such Mines, and in order that the working and reduction may not be discontinued, on account of the damages which would thereby ensue;—the said court shall command, that, within the peremptory term of thirty days, the party be cited and give evidence of the right which he may have; and that the other party, if he think proper, may give evidence to the contrary, or to such purport as he may deem expedient; and immediately on the expiration of the thirty days, if it shall appear that the plaintiff is in the right, the party in possession shall be ordered to henceforth keep an account and reckoning of the ore and silver which may be taken from the said Mine and of the costs and expenses which shall be incurred, as is mentioned in the preceding ordinance, in order that he may render the same with payment, if he should be beaten: all which shall be observed, fulfilled and executed, notwithstanding any appeal, nullity or revision which may be interposed; and this being done, the said cause shall be proceeded in, without allowing any delay or malicious procrastination (*dilaciones de malicia*), and justice shall be done.

Art. 69. Also, we ordain and command, that when and as often as cases shall arise in which arbitrators shall be appointed by the parties or by said court, such arbitrators shall, first of all, take an oath that they will well and truly give and declare their opinion; and that if such arbitrators shall not agree, then in case of disagreement another arbitrator shall be appointed, and if this one shall coincide with either of the aforesaid arbitrators, such opinion shall be observed and carried into effect; and if they should not agree, but should all differ in opinion in whole or in part, then other arbitrators shall be successively appointed until two shall agree, and having so agreed, what they shall give and declare shall be observed and carried into effect.

Art. 70. Also, we ordain and command, that all thefts committed in said Mines and in their *assientos* and limits, of gold, silver, lead, and ores of any class or description, and of any other things belonging to or connected with the working and reduction of said Mines, shall be punished with the utmost rigor; and that any person who shall steal any of the things aforesaid, shall, besides restoring and paying to the party the thing stolen, be condemned in seven times the amount, which we apply, one-half to our exchequer and the other half to the informer and the judge who shall pass sentence; and if he shall not have wherewith to pay the said penalty of sevenfold the amount, it shall be commuted to some corporal punishment, or to banishment, according to the gravity of the offense.

Art. 71. Also, we ordain and command, that our said officers who reside at Guadalcanal, and the persons appointed by them and by their successors, to attend alone in any of their places, and our treasurers, *contadores*, *factores*, who are or may be in said Mines or in any *partidos* thereof, and the courts, and notaries, which by us are appointed or shall hereafter be appointed, to use and exercise their offices therein, shall not be at liberty to have or hold any Mine or part of a Mine in any *Partido* of the kingdom, either by themselves or through any person, directly or indirectly, during all such time as they shall hold the said offices, under penalty of being forever deprived of such offices and of forfeiting the Mine or Mines which they may hold, which shall go to the denouncer, and under the further penalty of forfeiting half of their property to our exchequer, which penalty of forfeiture of property and Mines shall be incurred by any person who shall participate in the above offense.

Art. 72. Also, we ordain and command, that no other persons who, by the appointment of our said officers or by persons whom they shall appoint, shall be engaged in the working and reducing of said Mines, or who shall in any manner receive salary or pay from us for that purpose, shall be at liberty to hold Mines or shares in Mines, either by themselves or through any person acting for them, directly or indirectly, in the *Partidos* where they shall be employed and shall work, or within two leagues around; and if they shall take or shall have any Mine or Mines or shares thereof, whilst they receive our salary or pay as aforesaid, they shall forfeit such Mine or Mines or shares thereof, and they shall go to the person who shall denounce the same;

and besides this, they shall be banished from said Mines and from the space of six leagues around for the full term of three years, and they shall not break that term under the penalty, if they be of noble rank, of the said term of banishment being doubled; and, if of lower rank of being forced to serve said three years at the oar in the galleys.

Art. 73. Also, we ordain and command, that our said officers shall be at liberty to assign to the persons whom they may appoint in the *Partidos*, and all others whom they may deem necessary for the benefit and collection of our Royal dues, the salaries which they shall deem proper; and to give an order to each one for what he is to have, in order that he may be paid by the Treasurers and persons who collect from what is produced in said *Partidos*; and he may also make the other expenditures which for the above purpose they may determine to be expedient: of all which a book shall be kept with the amount and reckoning of what is issued and expended, and they shall send accounts, signed with their names, to our *Contaduria Mayor* every three months, in order that they may there be examined, and if anything should require changing, it may be done as deemed expedient.

ORDINANCES FOR MINES OF GOLD.

Art. 74. Also, we ordain and command, that all persons who shall search for, find and take Mines of gold, whether as first discoverers or as others, shall, in taking, registering and staking out the boundaries of said Mines, observe what is contained in these ordinances with respect to taking, registering, and staking out Mines of silver, and under the penalties in them contained: and that, in conformity to the said ordinances, and under the penalties therein contained, they shall be obliged to send the registries to our officers who reside at Guadalcanal: and these shall keep books of registries of Mines of gold, as is prescribed with respect to those of silver.

Art. 75. Also, we ordain and command, that the first discoverers of said Mines of gold shall take and hold fifty *varas* in length and twenty-five in width, and that the Mine which is to be left for us shall have the same dimensions; which dimensions they shall be at liberty to take as may best suit them; and in all else they shall observe what is contained in said ordinances with respect to silver, and under the penalties therein contained.

Art. 76. Also, we ordain and command, that all those who have Mines of gold shall be obliged to keep them occupied (*pobladas*) as is commanded with respect to keeping Mines of silver occupied: and that they shall also observe with respect to surplus Mines (*Minas demasiadas*,) what is prescribed with respect to such Mines of silver, under the same penalties as in all the foregoing.

Art. 77. Also, we ordain and command, that no person shall presume to treat for, contract, sell or purchase gold in dust, or bars, or ingots (*rieles*), without being stamped with our Royal stamp; which we command to be kept by the person who shall be appointed in our name to collect the proportion which to us may belong: and that there shall also be a smelter who shall smelt and make bars (*vergas*) of the gold which shall be taken out, and who shall also act as the Trusty to weigh it (*Fièl del Peso*), and he shall smelt, weigh and stamp the same with our Royal stamp aforesaid, in the presence of our said Administrator, or in the presence of the person appointed by him, and the proportion which shall belong to us shall be given and delivered to our Treasurer who for that purpose shall have been designated, and the remainder shall be delivered to the owner; and the said Treasurer shall keep a book wherein he shall enter the said parcels, with the day, month and year, and also whose gold it is, from what Mine it came, and how much there is, and what proportion belongs to us, of what the said Treasurer has taken charge, and what has been given to the owner of said parcel, which shall be signed by the said Treasurer, and by the said party, if he know how to write, and if not, then by some other person for him, and by the smelter and the Notary before whom it is passed; which said Notary shall keep a book in which the same shall be entered, and he shall sign it, as aforesaid; and no person shall be at liberty to sell or contract for said gold, unless it shall have been smelted and marked as aforesaid, under the penalty prescribed in the ordinance with regard to silver which relates to this subject, and the like penalty shall be incurred by him who shall purchase or contract for it, as is prescribed in said ordinance in regard to silver.

Art. 78. Also, forasmuch as it may sometimes happen that the servants of said owners of Mines or other persons, without the knowledge of said owners, sell or contract for gold or silver, without its being stamped with our Royal stamp, against the provisions of these ordinances: we ordain and command, that any servant or person who, without the knowledge or fault of said owners, shall sell or contract for gold or silver, without its being stamped with our Royal stamp, as aforesaid, and any one who shall purchase or contract for it, besides restoring and paying for what is so sold or contracted for, to the owner, shall forfeit all their property; and it shall go, one-half to our exchequer, and the other half to the informer and the Judge who shall pass sentence, and they shall be compelled to serve for ten years at the oar in the galleys.

August 10th, 1564.

Don Philip II, at Madrid, August 10th, 1564.

[From Book IX, Title VIII, Law XIX, Recopilacion de Castilla. Vide also Book IX, Title XIX, Law I, Novisima Recopilacion.]

Incorporation in the Crown and Royal Patrimony of all the Salines of the Kingdom; and prohibition to make salt outside of them.

Whereas we have been several times informed that, on account of the limits and *gias* (*limites y guias*) which our salines (*salinas*) have, and on account of certain noblemen and individuals who are in possession and claim to have title and privilege, and of the prohibitions and restrictions, and the pains and penalties which, with respect to them, are prescribed by the laws of our books, letters and provisions, which on that subject have been issued, the cities, *villas* and places, and their citizens, included within said limits, have received and receive great injuries, troubles, vexations, and damages; and that many of the said cities and *villas*, being very far off and distant from the salines within the limits of which they are situate, and being able to use and have salt nearer and cheaper, are obliged and compelled, by reason of said limits, to use that of said salines, at great cost and labor; and that, besides this, the lessees and collectors, and other persons who are engaged in this, with examinations, and inquiries, and trials and other pretexts, cause them many extortions and vexations: and that some of the said cities and places compromise and combine with the said lessees and pay them large sums of money (*maravedis*) in order to be permitted to use salt from other places: and ourself desiring, on this matter, as one of such great importance, to provide for the welfare and public good of these our kingdoms and of our subjects and vassals, we have commanded certain investigations and proceedings to be had, which being had, we have commanded some of the persons of our Council to examine them, and to consider and discuss the matter, so that they might recommend a better order and proper remedy; which persons having at different times considered and discussed the matter, and consulted with us, it has been resolved that we ought to command and ordain, and, by these presents, we do command and ordain, that now and hereafter, so long as it may be our will, and without prejudice to the duty (*derecho*) of the said limits and *gias*, all the cities, *villas* and places of these our kingdoms, and the citizens and residents thereof, as well those included in said limits and bounds, as those not so included, shall be at liberty to purchase and use the salt of the salines, salt-mines, and deposits (*salinas y saleros y alfolies*) which by my order and command are worked, made and provided, freely, and of whichsoever may be to each one of them most near and convenient; without being obliged to purchase or use that from one place rather than from another, notwithstanding said limits and bounds, and prohibitions and restrictions, penalties and ordinances, which, respecting the above, have been prescribed and enacted; which, so far as relates to the above, and so long as it shall be our will, as aforesaid, and without prejudice to our said duty, we annul and make void. And whereas, besides the said salines which we hold and possess and which have the said *gias* and limits, there are, as aforesaid, other salines held and possessed by noblemen and private persons, who hold title and privilege for the said *gias*; and, in order that the grant and benefit which we make and confer upon these our kingdoms and upon the subjects and natives thereof, may, from this cause, meet with no impediment or difficulty, we have ordered to be taken and incorporated, and we take and incorporate, in our patrimony all the said salines of *gias* and limits, which the said noblemen and private persons hold; and we have commanded to be given to them, and we have given to them, just recompense, forasmuch as all said salines remaining, as they do remain, in our hands and power, they may freely use and enjoy said grant and benefit which to our said kingdoms and subjects are made. And forasmuch as, said limits and *gias* being revoked, and a new order in this matter of salt being given, as we now give it, it is proper and necessary in these our kingdoms that so much salt be worked and made as shall be necessary for their supply, and that it be provided in other parts where it is wanting, so that there may be in them an abundance and full provision and supply of salt; for this purpose we have commanded that orders be given, that in all said salines which have limits and *gias* said salt be worked and made, in the same manner as it was previously made, and even in greater quantities, according as it may be required: and in addition to this, we have ordained and commanded, that, persons of practice and experience being sent for that purpose, search be made in other parts of these kingdoms for salt wells, springs and waters; and that, in parts and places where it may be suitable to work and make salt, and it may appear to be expedient to do so, said salt shall be made and worked, in order that the cities, *villas*, and places which are distant from the salines may have it nearer and purchase it at less cost and trouble. And, moreover, with respect to the cities, *villas* and places where there is no such manufacturers, or from which they are very distant, we command that it be provided and ordered that there be deposits and custom houses, in which the said salt may be provided and brought when expedient; having, as we do have, in all, the object and intent that our subjects and natives shall receive, as much as possible, benefit and favor, and that they shall have and use said salt with the greatest convenience and least expense. And whereas, the said limits and *gias* being revoked and repealed, and a new order being given, and provision being made by us for abundant and sufficient supply of said salt, there would result great inconvenience and confusion, and much injury to us, if said salt should not be made, worked, or provided in other salines, and in other parts and places, besides those which, by our order and command, or by us, shall be made, worked, and provided:—we ordain and command, that, in these our said kingdoms, no salt shall be worked or made in salinas or wells, except in those which, by our command, order and hand and license, shall be worked and made; nor shall any be provided or taken away from them, except that which by us is taken for the said deposits and salt magazines, which shall be declared and assigned, under the penalties contained in the laws and edicts of these kingdoms against those who smuggle salt away from them. And in what relates to the salt of Andalusia and the kingdom of Granada, in which we now make no change, they will observe the order and form which ought to be followed: provided, that thence no salt shall be taken into other parts of these our kingdoms, except that which by our command shall be carried and taken there; since we, as aforesaid, have to provide so that there shall be an abundant and sufficient supply.

May 8th, 1572.

Royal Cedula, dated at Aranjuez May 8th, 1572.

[From the "Comentarios de Gamboa," cap. II. § 51.]

Authorizing the sale of Quicksilver to the Miners of New Spain on credit, and prohibiting its sale to merchants for the purpose of trade.

To Don Martin Enriquez, our Viceroy, Governor and Captain General of New Spain, and President of our Royal Audiencia thereof, *know ye:*—

That both from what you and our officers of that country have, at different times, written to us, and by the report, petition, and proceedings which, on the part of the Miners thereof and of the Province of New Galicia, have been made to us, with respect to the sale and disposal of the quicksilver taken to that country on our account, and from the Provinces of Peru, we have been advised of the inconveniences which have resulted, with the order which has been had in the sale of it, with respect to the necessity of said Miners, and the small *ley* of the ore which is worked, and the silver which for this reason is left unworked, and the injury which they receive from obtaining it through merchants who get it at first sale, and the injury which results to our Treasury and to the duties of one-tenth, and to the trade and commerce of that country, and the other inconveniences and points which on this matter you have referred to us: respecting which matters and others that have been represented to us, and because we desire the advancement of that country, and wish to give every aid to the said Miners, we have commanded that some remedy be devised and discussed which may provide therefor; and what has appeared possible and necessary now to be done, is, that all the quicksilver which is conveyed to that country, as well from these kingdoms on our account, as from the Provinces of Peru, be deposited in our store houses in that country; that, a list being made of all the Miners in your government of that New Spain and New Galicia, the quicksilver be issued to them one-half on credit, so that the proceeds thereof may be brought to these kingdoms by the same fleet (*Flota*) by which it is carried out, and the other half by the second fleet, good securities being taken, and that the price which is to be given for the same shall be what you and our officers of that country may determine upon, as most advantageous to our Royal Treasury. Wherefore, I command you to provide for the observance and fulfillment of the aforesaid order respecting the sale and disposal of the said quicksilver; and for the better execution and observance of the same, you will prohibit in our name, as we by these presents do prohibit, every merchant and other person from purchasing quicksilver in that country for the purpose of resale, under penalty of forfeiting it, with twice the value; and we forbid and condemn henceforth to the said penalty those who shall act to the contrary hereof.

Dated at Aranjuez May 8th 1572 and countersigned by Antonio de Erasso.

PART II.

MINING LAWS OF SPAIN AND MEXICO,

FROM 1584 TO 1783.

PART II.

MINING LAWS OF SPAIN AND MEXICO,

FROM 1584 TO 1783.

August 22d, 1584.

Don Philip II, at San Lorenzo, August 22d, 1584.

[From Book VI, Title XIII, Law IX, Recopilacion de Castilla.

Vide also Book IX, Title XVIII, Law IV, Novisima Recopilacion.]

ORDINANCES,

Which are to be observed in the discovery, working, and reduction of Mines of Gold, Silver, Quicksilver, and other metals; (denominated by Gamboa, The New Ordinances, or New Code of Mines.)

Art. 1. In the first place, we revoke, annul, and make void the edicts (*pragmaticas*), and orders (*ordenamientos*) issued at Valladolid and at Madrid, which are the 4th and 5th laws of this Title, and all laws of the *Ordenamiento* and *Partidas*, and all other Laws and edicts, and privileges (*fueros*) and customs, so far as they are in opposition to the provisions of this Law: and we will and command, that they shall, in that respect, be without force or authority, leaving only in force and effect the 4th law of this Title, so far as it treats of the incorporation in our Royal patrimony of the Mines of gold, silver and quicksilver of these our kingdoms, of which gift (*merced*) had been made to private persons, by Departments (*Partidos*), Bishoprics, and Provinces: in conformity with which, and with these our laws and ordinances, exclusive of all others, we will and command, that the said Mines be worked, and reduced, and that all suits and disputes be decided and determined, which may in any manner arise respecting the said Mines, or any matter annexed to, touching, or concerning them.

Art. 2. And in order to benefit and favor our subjects, and natives, and all other persons whatsoever, even though foreigners to these our kingdoms, who may work or discover any Mines of silver, already discovered or to be discovered, we will and command, that they shall have them and that they shall be their property, in possession and ownership (*propiedad*), and that they may do with them as with any thing their own, observing, as well in regard to what they have to pay by way of duty to us, as in all else, what is prescribed and ordered in this Edict (*pragmatica*) in the following manner:

Art. 3. If the metals, which are taken from the said Mines, shall yield at the rate of a mark and a half, or twelve ounces per quintal of lead-silver (*Plomo-Plata*), or under, they shall pay to us the tenth part of the silver which is taken from said Mine and its ores, without deduction for expenses or for anything else, for all these shall be at the charge of the said persons who work, discover, and reduce said Mines; and all that remains after deducting the said tenth part of said silver, they may have and retain to themselves.

Art. 4. In Mines which yield more than a mark and a half per quintal of lead-silver, and up to four marks, there shall be paid to us the fifth part of the silver which is extracted, without deducting expenses; and the persons who work and reduce the said Mines and ores shall have the remainder, as aforesaid.

Art. 5. In Mines which yield more than four marks per quintal of lead-silver and up to six marks, there shall be paid to us the fourth part of the silver which is extracted, without deducting expenses; and the said persons shall have the remainder, as aforesaid.

Art. 6. In Mines which yield more than six marks per quintal of lead-silver, whatever their goodness, quality or richness may be or become, whether expected or not expected, there shall be paid to us the half of the silver which is extracted, without deducting expenses, and the said persons shall have the remainder, as aforesaid.

Art. 7. In Mines of gold, whatsoever their lay, quality, quantity, or richness may be or may become, there shall be paid to us the half of the gold which is extracted from them, without deducting any expenses; and the persons who discover and work them shall have the other half. And this is to be understood of every description of mining for gold, whether it be procured from Mines proper, or from sources in rivers, or in any other way.

Art. 8. And inasmuch as there are some old Mines in these our kingdoms, which, before the publication of our said Edict (*pragmatica*), issued on the tenth day of January, 1559, were worked and reduced, but at present are not worked or reduced by their owners, and, in fact, were not worked at the time said edict was issued; and inasmuch, also, as some other Mines have been discovered and worked since then, and from both classes heaps of rubbish (*terreros*) and slag have been taken, we command that such persons as may desire to work said Mines, and to reduce said rubbish-heaps and slag, without prejudice to the right which their owners may have in them, may do so; and that out of the ores which are taken from such Mines there shall be paid as follows:

Art. 9. In Mines which, before the publication of the said edict, were abandoned and had ceased to be worked, and have since then been discovered and worked; from those which have been sunk to the depth of ten *estados* or more, the ores taken out, which yield two marks or less per quintal of lead-silver, shall pay to us, of the silver extracted, one-twelfth part; and if they yield more than said two marks per quintal they shall pay in the proportion above declared for newly discovered Mines, without deducting any expenses. But it is declared that any Mines, old or new, which are less than ten *estados* in depth, are to be considered as new Mines, and shall pay duties as such, in the same manner and form as is declared in the ordinances which treat of said new Mines.

Art. 10. And if the rubbish-heaps and slag belonging to the Mines referred to in the preceding ordinance, shall be smelted alone and unmixed with other ores, which have been or shall be taken from the Mines after the accumulation of such rubbish-heaps and slag, there shall be paid to us the tenth part of the silver produced by said rubbish-heaps and slag, smelted, as aforesaid, alone. But if they be mixed with other ores, there shall be paid to us of the silver which they produce, in proportion to the duties paid by other Mines, having regard to the kind of ore mixed with them.

Art. 11. And the lead, chalk, ashes, and the impregnated cupels and sweepings and all else that may result from the refinings, after extracting the silver, of which there is to be paid to us the proportions declared above, free of all expenses, may and shall remain to the owners of the said Mines; without that, of the said lead, chalk, ashes, impregnated cupels, there be paid to us anything; nor shall there be imposed any prohibition or restriction thereon.

Art. 12. And whereas, poor-lead which is not worth refining as containing little or no silver, and also antimony and copper, are required for the reduction of the produce of Mines of silver: we command that Mines of such lead, antimony and copper, which have been or may be found in places where there have been made no grants of Mines and ores, may be searched for and worked by all the aforesaid persons, and there shall be paid to us, of copper one-thirtieth part, and of antimony the one-tenth part, and of poor-lead, which is to be understood such as will not yield more than four *reales* of silver per quintal, the one-twentieth part; all clear of expenses: provided, that if such copper should contain gold, there shall be paid to us one sixth part of the gold in addition to the duty on the copper; and if it should contain silver, there shall be paid to us one-half of the duty above directed to be paid on ores of silver, according to the number of marks it may yield per quintal, and in addition, the duty on the copper, as aforesaid.

Art. 13. It is to be understood that all above mentioned proportions which we are to receive out of the produce of the said Mines, new and old, and refuse-heaps and slag, must be paid to us in silver, at the refining houses and smelting works which we are to have for the said refinings, and not in ore, nor in lead-silver; and those out of the poor-lead and copper, in sheets, and those out of the antimony, in metal, all of the same kind and quality as the proportions which remain to the owners, and clear of all expenses.

Art. 14. And whereas, according to the said Edict of 1559, issued on the 10th of January of that year, those who had grants of Mines were to enjoy all that was not gold, or silver, or quicksilver, according to their privileges, and also were to enjoy Mines of gold, and silver which had been begun to be worked and were actually worked by them, or by other persons in their name, prior to said Edict; and whereas some doubts have arisen respecting those words, it being said that it might happen that the Mines had been discovered and begun to be worked, one, two, or more years prior to the said Edict, and the working had been discontinued some time before it was issued, and that such Mines are consequently excluded by the said Edict, as not being actually worked at the time it was issued; therefore, it is declared that said Mines of gold and silver, which are to be enjoyed by the owners of such privileges, shall be such as were actually worked and producing at the time said Edict was issued or within four months previous, and no others.

Art. 15. And whereas, in the said Edict of 1559 we prohibit and command, that no person shall be at liberty to search for or discover Mines within one league around the Mine of Guadalcanal, or within a quarter of a league around those of Cazalla, Galaroca, and Aracena respectively; and whereas it has since been found to be expedient for our service to enlarge the said limits of said quarter of a league, and to declare from what point they are to run; we therefore command, that in said three places and in that of Guadalcanal, and in each one of them, no person shall take or hold Mines within the space of one league around each one of the said places, such leagues to be computed and measured in the following manner; that of Guadalcanal from the house which is built there for the workshop of the said Mines; and that of Cazalla from the house which is over the Mine of Pedro Candil; and that of Aracena from the house which is built at the Mine of El Cerro de los Azores; and that of Galaroca from the first Mine which was discovered there, and which is near the town (*Lugar*); and that the said leagues be of the legal length of 15,000 feet, each foot being one-third of a *vara*, land measure; and all the Mines which shall be found within those limits shall be for us; but if previous to the day of the promulgation of this our Letter (*carta*) any Mines shall have been found outside of the said quarters of a league, and within the spaces of a league which are now designated, the discoverers thereof shall enjoy them in accordance with said first Edict (*Pragmatica.*)

Art. 16. Also, we ordain and command, that all persons whosoever, even though they be foreigners, may freely search for Mines of gold and silver, and such others as by these our ordinances are declared, and make trial-pits (*catar*), and do all things necessary for the discovery of the said metals in all our said kingdoms and lordships (*señorios*) of the crown of castile, (outside of the places excepted,) in the fields, woods, vacant places, and outside lands (*egidos*), and pastures (*dehesas*), whether belonging to us, to Pueblos or to private persons, and in any inheritances whatsoever, without that any owners of such pasture grounds and inheritances, or any person whosoever, can impose any impediment or prohibition to their so doing. And that if it should be necessary to dig or sink pits in said pasture grounds or inheritances, they may do so; provided that if any damage be done, the court of Mines will appoint two persons of trust to appraise the damage, who shall inspect it and declare on oath what it is; and if they should not agree in their declaration, the said court shall appoint a third one, or other sworn persons, until there is an agreement; and what the majority shall agree upon declaring, shall be ordered to be paid, and shall be paid for it. And that if ore shall be found which shall appear to them worth following out, and they shall make an establishment (*Asiento*) and workshop, and other things necessary for the working of the Mine or Mines and the reduction of the ore, the aforesaid two persons shall inspect the damage which by reason thereof such pasture ground or inheritance may have received or may receive, and, upon a due consideration of the whole matter (under their said oaths,) shall appraise such damage, which the court shall order to be paid as aforesaid.

Art. 17. Also, we ordain and command, that whoever may discover a Mine of gold or silver or other metal, shall, within twenty days after he has discovered or found the ore, be obliged to register it before the Mining court (*Justicia de Minas*) within whose jurisdiction such Mine may be situate, in the presence of a Notary, producing the ore which he shall have found; and in the registry shall be stated the person who discovered and registered it, and the place where it is situate, and where was found the ore presented: and that within sixty more days from the date of such registration, he who made it shall be obliged to send, and shall send, an authenticated copy of said registry to our Administrator General, if there be one in the district (*comarca*), and if not, then to the Administrator of the *Partido* within the district of which the said Mine may be situate, in order that there may be noted and entered in the book and registry what part each one may have in said Mines, so that all the Mines which have been or shall be discovered may be known and an account be taken of them; and that in case such registry be not made in the form and within the time aforesaid, and according to what else is prescribed, any other person whosoever may register said Mine, and have and acquire the right which such discoverer or any other person would have had, if he had caused the registry to be made as aforesaid.

Art. 18. Also, whereas, previous to the publication of these our ordinances there have been discovered and registered many new and old Mines which are occupied and possessed (*embarasadas*), without being worked or their ores reduced, and without any full account of them being taken, the registries of them having been differently made:—We ordain and command, that all persons, who, before the publication of these our ordinances, shall have discovered and registered old or newly discovered Mines, shall be obliged, within two months, to renew and make again the said registries, according to and in the form prescribed by the preceding ordinance with respect to those which may be hereafter discovered: and that within the further period of sixty days they shall be bound to send, and they shall send, such registries to our said Administrator General, if there be one within the district (*comarca*), and if not, to the Administrator of the *Partida* within whose district the said Mine may be situate; and that if they shall fail to do and comply with this and to procure an authenticated copy of such registry, they shall be deemed to have forfeited and shall forfeit the right which may belong to them, or which they may claim, to the said Mine, and any other may have it, who shall institute the proper proceedings, agreeably to this our Edict (*pragmatica.*)

Art. 19. Also, we ordain and command, that the Administrators of Mines of each *Partido* shall keep a book in which shall be entered all the registries which may be made in each district, of Mines discovered and which shall be discovered, taken, sold, or in any manner whatsoever contracted for; and that the said Administrators shall send to our *Contaduria Mayor* a report, signed with their names, of the condition of the Mines of these our kingdoms, and of what is done with them: and that, after having sent the first report, they shall, every six months, send a like report of what shall have taken place or been done with respect to them.

Art. 20. Also, we ordain and command, that no person shall presume to register, or enter in his register, a Mine which is not his own, under penalty of a thousand ducats, to be imposed upon the person so offending, one-half to be applied to our exchequer (*camara*,) and the other half to the informer and to the Judge who shall pass the sentence; and, moreover, that he shall forfeit the right which he may have acquired to such Mine.

Art. 21. Also, we ordain and command, that when any person shall register a Mine or Mines which may not be wholly his own, he shall be obliged to declare what share or shares he may hold in them: and if he hold them in partnership, what share his partner or partners may hold in the said Mine or Mines, under penalty, if he fail to do so, of forfeiting the share or shares which he may have, which shall go to the partner or partners whose share or shares he has omitted to set forth.

Art. 22. Also, we ordain and command, that the person who shall first find and discover a Mine, shall, as first finder and discoverer, make the first registry and enjoy all the *pertenencias de Minas* which he shall stake out, or be pleased to stake out, at the Mines and veins which he may discover or may have discovered, provided, that he shall, within ten natural days from the time of his making registry of such Mine, stake out, declare, and mark the *pertenencias* which he may desire, and that he shall enjoy the extent (*medida*) which to each portion staked out (*estacada*) properly pertains, throughout all the *pertenencias* staked out (*de estacada*) which he shall designate as such discoverer;

and that he shall be obliged, within the said ten days, to stake out all the *pertenencias* which, as aforesaid, he may desire, and in the manner which shall be or shall appear to him best; even though he should take and include within his boundaries the trial-pit, or trial-pits, which others who came after him may have made or may make, provided that, first of all things, he has set up a fixed stake (*estaca fixa*) in each *pertenencia* of those which he shall thus mark out and take, which fixed stakes he cannot and shall not leave, when he shall stake out or alter his boundaries, howsoever he may desire to stake out and alter them; and that the others who come after him shall proceed, in their order, to stake out and alter their boundaries, as they, from time to time, discover ore. And that when they shall have made registry, as they are obliged to do, they shall proceed to set up a fixed stake in all the *pertenencias* which they, may desire to take and to mark out, within the said term of ten days, after the expiration of the first ten days which are allowed to the first discoverer: for those who stake out a Mine shall always have ten days to view the Mine, and to take in it all the *pertenencias* which they desire, and to set up a fixed stake which they cannot leave, nor enter in the *pertenencias* which shall have been staked out qefore, because all the *pertenencias* and boundaries which have been taken and marked out, are always to be preserved to those who first staked them out. And if two or more persons shall apply to have their stakes set, it shall be determined in a short and summary manner which of them applied first; and he who shall be decided to have been first, shall be preferred to the others, saving the right of the other party if he shall, nevertheless, claim to have applied first to have such stakes set out.

Art. 23. Also, we ordain and command, that any person who shall have discovered or shall discover a new Mine, and shall have made registry, as directed in the preceding ordinance, shall enjoy a space of 160 *varas* in length upon the vein, and 80 in width; and if he shall wish to measure out the said space of 160 *varas* and 80 *varas* across the vein, he shall be at liberty to do so, and may do it in such manner as to him may appear expedient. And it is declared, that after the first discoverer of a Mine shall have marked out, within the said ten days which are allowed to him for that purpose, the *pertenencias* which he shall have taken, no other person can ask to have boundaries staked out, or stake them out, until the expiration of a further period of ten days, in order that there may be determined the *pertenencias* which he shall desire to take as first discoverer, provided that he does not leave his fixed stake, or prejudice any other person or persons who may be on either side of him, and who may hold Mines opened and registered before him: and those who, after the first discoverer, shall have taken Mines, or shall hereafter take them, may proceed to take and work their Mines and *pertenencias;* and each of the Mines taken after the said discoverer has taken his, shall be 120 *varas* long and 60 wide, which space may be taken across the vein, or as it shall seem best, provided it does not leave the fixed stake, and is without prejudice to a third party.

Art. 24. Also, we ordain and command, that if any person shall apply to have boundary stakes set out for the first discoverer, or for any others who have to set out boundary stakes, after having registered their Mines, as well in Mines already discovered as in those which shall hereafter be discovered; such first discoverer and such other persons shall be held and obliged to have their boundary stakes set out, within ten days from the day on which they are requested to do so, if they be at the Mines; and if they shall not set them out, then, the said term having expired, the court of Mines which has cognizance of these things conformably to these our ordinances, taking with it persons who understand staking out Mines, and who have been sworn for that purpose, shall set out such boundary stakes; and if the person of whom it is asked be not found in the Mines, but being in the vicinity within ten leagues of said Mines, he shall be obliged to set them out within fifteen days; and if he shall fail so to do, at the expiration of the said fifteen days, the said court shall do it, as aforesaid; and if he be not in the vicinity of said Mines nor within ten leagues, notice shall be given to his steward (*Mayordomo*) or the person who may have charge of the working and reduction of his Mine, or left in his house, if he have one; and public notice (*pregon*) shall be given on the first holiday (*dia de Fiésta*) which shall occur, and the term of fifteen days shall run from the day that the notification was given to said steward, or person, or was left in his house; and the said notice (*pregon*) shall be posted on the door of the church of such Mines, or if there be no church in them, of that of the nearest *Pueblo:* and the said term of fifteen days having expired, the said court will establish the said boundary stakes, as aforesaid; observing in setting them out, that there must always be a fixed stake, which must be adhered to, and must not be abandoned, in staking out or altering boundaries.

Art. 25. Also, we ordain and command, that if two or more persons make application, at the same time, to have the boundary stakes set out by such first discoverer, or by other persons who have to set boundary stakes, they, (the first mentioned persons), holding Mines on any side of the Mine with respect to which the application to set out boundary stakes has been made; in such case, it shall be ascertained from the registries which of them is entitled to have the stakes set out first, and which second; and thus going on successively to set out the stakes, observing, with respect to the dimensions and all else, what is contained in these our ordinances.

Art. 26. Also, we ordain and command that, when and as often as application shall be made to set out boundary stakes, and they shall be set out as aforesaid, care shall be taken, in setting out the stakes, to make an oblong with right angles: and that in said oblong, and not outside of it, the said fixed stake shall be left; each one taking the number of *varas* which he is entitled to, wherever he may wish or deem best, in the form before set forth and declared.

Art. 27. Also, inasmuch as it may happen, that, when fixed stakes have been set between two or more persons, one of them, with a view to his own advantage, may take up from its place a stake or stakes, as he may think fit, and remove them to another place more desirable to him, thus giving rise to divers lawsuits:—We declare and command, that when any person shall apply to have another set out his boundary stakes, and it be done, or shall wish to stake out his own Mine without being required to do so, he shall be obliged to make, at the places where he shall set the said fixed stakes on the side of his neighbors, a pit for each one of said stakes, two *varas* in depth and one in width: and in the centre of each one of these pits, he shall set up the stake, and he cannot remove it, except in the cases in which, conformably to these ordinances, the boundaries are allowed to be altered: and the stake or stakes, which are thus set up, shall be regarded as *pertenencias* between him who sets them up and his aforesaid neighbors; all of which shall be done and observed, under penalty of forfeiting the right which they may have to such Mine, and of any other person being at liberty to ask for it, and to register it as his own.

Art. 28. Also, we declare and command, that when any person has already been asked to set out his boundary stakes, and has done so, and some other person shall again ask him to set out boundary stakes on some other side of his Mine, he shall be at liberty to alter them, as between him and the person who thus again asks him to set out said stakes, it being done without prejudice to the stakes already set out, and provided he does not leave his fixed stake outside.

Art. 29. Also, we ordain and command, that, although a person has already set his boundary stakes between himself and another on one side of his Mine, if he shall wish to alter the boundaries of his Mine, before any other person or persons shall ask him to set out stakes on some other side where he has not set them out, he shall be at liberty to do so, provided he go before the court having cognizance of such matters, and manifest his new boundary stakes, and the alteration which he makes in his said Mine: and the said court shall admit such alteration, and it shall be noted in the margin of the registry which has been made of such Mine, provided that it be done without prejudice to third parties, as aforesaid, and the fixed stake be left within his *pertenencia;* and the vacant spaces which are left between his Mine and that of his neighbor with whom he has boundary stakes set out, shall be given to the first who shall ask for them; and if the neighbor should ask first, he shall be at liberty to take them, pro vided that, with the spaces he thus takes in, he has only the proper extent of a Mine, and does not go outside of his fixed stake; and that he also manifest to the said court the said alteration, in order that it may be entered in the aforesaid registry.

Art. 30. Also, we ordain and command, that, if any Mine shall extend beyond the boundary stakes or limits

6*

which belong to it, conformably to these Edicts (*Pragmaticas*), either in length or in width, and the ore therein contained is joined to the ore of the Mine of some other party, and both Mines shall in depth become one, the Miner who shall have first sunk and entered into the Mine of the other, shall and may enjoy the ore which he shall take out, until the owner of the other Mine shall carry on his works so as to meet him, and then the latter may demand that, he who anticipated him, shall set out his boundary stakes; and it being found to be within the *pertenencia* and stakes of the other, he shall withdraw, relinquish the occupation, and leave the vein to the Miner within whose *pertenencia* he shall have entered; and all the ore which he shall have taken from the other's *pertenencia* prior to that time, shall belong to him who took it out, without any obligation to give it to another, inasmuch as he has acquired and gained it by his diligence and care in working deeper than his neighbor. But if any person shall have staked out *pertenencias* next to the Mine of another, either on the side of its length or of its breadth, and shall have no vein, or a vein which contains no ore, or no appearance of ore, and shall work it solely with the intent of profiting by the ore of his neighbor when he shall get within his boundaries:—We command that such person shall not be able to acquire, and shall not acquire, any right, even though his neighbor's ore should take its course inside of his *pertenencia:* and that our judges and courts of Mines shall so decide, and that they shall not allow or permit such Mines without a vein or ore, to be worked.

Art. 31. Also, we ordain and command, that the first finder and discoverer of such Mines may take all the stakes and *pertenencias* which he wishes, observing, in respect to the same, what is contained in these ordinances on that subject; and that he may also hold and possess as many

Mines and *pertenencias* as he shall purchase, or inherit, or as shall belong to him by any title or cause whatsoever.

Art. 32. Also, we ordain and command, that no person, of whatsoever condition he may be, shall take a Mine for another, unless he have a power of attorney, or be a servant on wages, of the person for whom he shall take such Mine: and in default of either of these requisites, the Mine shall be considered forfeited, and may become the property of the person who shall denounce it, and the Judge shall give immediate possession of it to such denouncer; without allowing any appeal (*recurso*) either to the person in whose name the said Mine was taken, or to him who took it for him.

Art. 33. Also, we ordain and command, that no steward (*Mayordomo*) who may be engaged in the working and reduction of such Mines, nor any other person who may live with the owner of Mines, even though he have charge of his Mines and workmen, shall be at liberty to move the stakes which his employer (*Amo*) may have set up, without his permission and authority, even though he may have been applied to to set out such stakes; and if he shall move them or set them out anew, it shall avail nothing, and shall work no prejudice to the person to whom the Mine belongs.

Art. 34. Also, we ordain and command, that when any such steward who may have charge of any Mines or *Hacienda*, shall take or discover a Mine, such steward shall be at liberty to stake out the Mine or Mines which he may so take, and to set out boundary stakes on the side of any party who may apply for that purpose, until such time as his employer shall visit such Mines. But that after his said employer, the owner of such Mine or Mines, shall have arrived, he shall neither apply for, nor set out any more boundary stakes; and the said steward or servant, without the authority of his employer, shall not be allowed to move those which his said employer shall have set out or left fixed.

Art. 35. Also, we ordain and command, that all persons whatsoever, who shall hold, take, or acquire Mines, whether already discovered or hereafter to be discovered, shall be obliged, within three months, reckoning from the day on which they shall register such Mines, to sink in such Mine, if new, one of the trial-pits (*catas*) which they have made, and, if an old one, one of the pits (*pozos*) which shall have a vein or ore, three *estados*, each *estado* being seven-thirds of a *vara* long, under the penalty, if they shall not sink them or have them sunk to the depth of said three *estados*, at the expiration of the said three months, of forfeiting them, and they shall forfeit them, and they shall be for any person who shall denounce them, and the Court of our Mines shall immediately put the said denouncer into the possession, subject to the same obligation of sinking the aforesaid three *estados* within the term aforesaid, notwithstanding any appeal, nullity, or revision (*agravio*) which may be interposed.

Art. 36. Also, forasmuch as in the preceding one, and in some others of these our ordinances, it is provided and commanded, that such persons as shall take and hold Mines, or shall purchase them, or in any other manner acquire them, shall be obliged to deepen such Mines according to the provisions in said ordinances contained; and whereas, it is our intention and will to prevent law suits and disputes, and to obviate mal-intents:—We declare and command, that it shall be understood that they are obliged to deepen, if they can, the aforesaid trial pits and wells (*catas y pozos*), or otherwise incur the penalties of these ordinances. But that if, from any accidental cause, or from its being more expedient to proceed by following the ore, on account of its taking some other direction, sa often happens, and not by any fault of their own, they shall fail to deepen them, and shall continue to work them as may be most convenient and advantageous, they shall not become liable to, nor incur, the aforesaid penalties; provided, that when this shall be the case they shall be obliged to give notice of it to the Administrator of the *Partido*, in whose district such Mine may be, in order that an investigation may be made of the said case, to ascertain whether the failure to comply with the provisions of the aforesaid ordinances, resulted by reason of following the said ore, and not by their own fault: whereupon, after such inquiry has been made, the said Administrator shall declare and provide as may be proper, in such a manner that, when the inconvenience ceases, the Mines aforesaid may be deepened according to the provisions of the said ordinances.

Art. 37. Also, whereas it often happens, that certain persons hold many Mines which they have taken, discovered, or purchased, or acquired in some other manner, and do not work or reduce them, either because they cannot, or because they are working others which they consider better, and thus fail to deepen, and to discover and take out ores from those which they do not work, although sometimes better than the ores taken from the Mines which they do work: and whereas, the Mines aforesaid which they fail to workbe come filled with water, and thus cause injury to other neighboring or surrounding Mines, which are worked and sunk deeper than they are: Therefore, in order to obviate these inconveniences and others which follow, or may follow, from the Mines not being worked, we ordain and command, that all persons shall be obliged to have their Mines occupied (*pobladas*) by at least four persons to each Mine or *pertenencia*, whether they be entire owners of such Mines, or hold them in partnership; for, in whichever way it may be, by having four persons in each Mine in every *pertenencia* of it, it is sufficiently shown that the said Mines are occupied (*pobladas*); which said four persons shall be engaged in working the Mines which they occupy, in raising water or ore, or in doing some other thing for its improvement, either within or without the Mine, under penalty that, if any Mine whatsoever shall not be occupied and worked by said four persons, as aforesaid, for four consecutive months, it shall *ipso facto* (*por el mismo caso*) be forfeited, and the person who owned it shall forfeit it, and in future shall have no right to it, unless by registering it anew, and by going through the other proceedings, conformably to these ordinances: and said Mine shall be adjudged to any person who shall denounce it as unoccupied (*despoblada*), provided he go through the proceedings aforesaid. But that if, on account of any reasonable impediment, such as war, pestilence, or famine, occurring in the part or place within the jurisdiction of which, or within twenty leagues around, the said Mine is situate, it cannot be kept occupied by the aforesaid four men, in such cases the term of said four months shall not run. But if such impediments shall exist outside of the jurisdiction in which such Mine is situate, and beyond said twenty leagues around, this shall not be admitted as an excuse for not keeping the Mine occupied, according to, and under the penalties prescribed in, this our ordinance.

Art. 38. Also, we ordain and command, that in order that any Mine may be pronounced and declared insufficiently occupied (*despoblada*), the person who shall come to denounce it, shall appear before the court of Mines and make the denouncement, setting forth therein the Mine, the hill, or place where it is situate, and next to whose boundary stakes (if any there be), its condition as to depth and whether or not it has any ore ; and within forty days it shall be ascertained, whether the said Mine has been left unoccupied (*despoblada*) for the four months aforesaid, the party being summoned, if possible, in person, or at his house, if he have one at the Mines in question or in the vicinity, and, if it can conveniently be done, mentioning it or making it known to his wife, or servants, or nearest neighbor or neighbors, in such a manner that it may come to his knowledge: but if he cannot be summoned in the vicinity, not having any house there, as aforesaid, then by means of edicts and notices in the manner hereafter described: and within forty days computed from the day on which the said denouncement was made, both parties shall be at liberty to allege and prove such matters as they may think proper, and the cause shall be determined upon what is done within said term, without any other postponement (*conclusion*) or adjournment; and if such Mine shall be pronounced to have been unoccupied (*despoblada*), then it shall, as such, be adjudged to the said denouncer, and possession thereof shall be immediately given to him, notwithstanding any appeal, nullity, or revision (*agravio*), which may be interposed against the decision: provided, that the person to whom said Mine shall be adjudged, shall be obliged within three months to deepen the trial-pit (*cata*) or well (*Pozo*), whichever he may think proper, and sink it three *estados* deeper than it was at the time of his making the aforesaid denouncement, and for that purpose it shall be measured in the presence of our Judge of Mines: all which he shall do and perform, under penalty of forfeiting it, and of its being adjudged to any person who may denounce it on the same condition, and subject to the same penalty, and provided, that he keep an account and reckoning, entered in a book, with the day, month and year, of the ore and silver which may be taken out of the said Mine, and of the costs and expenses which may be incurred in working it and in reducing the ores; and provided, that he give security in one thousand ducats that, if he lose his case on appeal and be ordered to render an account with payment, he will have the means to do so, and will do it; and if any one of the parties shall consider himself aggrieved, he may appeal within three days, and the case shall be determined and justice rendered, within sixty days computed from the day of pronouncing the decree, upon what both parties shall state, allege and prove, without any other postponement or adjournment; and what shall be thus determined shall be observed, and executed, without there being allowed or admitted any appeal, supplication, nullity, revision, or other remedy.

Art. 39. Also, we ordain and command that, if any Mine should happen to be denounced as not occupied (*despoblada*), which shall not appear to have any owner, or having one, he shall be absent, and it be not known where he is, or it be known that he is in a place where notice cannot be given to him according to the provisions of the preceding ordinance, the said court, on some Sunday, when coming from mass at the church of such Mines, or, if there be no church in them, from that of the nearest Pueblo, where there shall be at least eight persons present, shall give public notice of said denouncement, in order that it may be known, and that information thereof may be given to the owner, or to some one who may answer for him, if he please, and make his defense. And such proclamation being made, a copy thereof shall be affixed to the principal door of such church, where it may be publicly seen; and such proclamation shall be made on two other subsequent Sundays; so that altogether there shall be three proclamations, on three Sundays, and copies thereof shall be affiixed as aforesaid; which shall be held and accounted a sufficient summons, in like manner as if it had been made personally; and if during the time of the said three proclamations, or within forty days from the time of the first of them being made, the owner, or some person competent to oppose such denouncement, shall appear, then, the parties being heard conformably to the foregoing ordinance, justice shall be done: and if no appearance be made within such term of forty days, the proclamations being made, the said denouncer shall give evidence that the Mine aforesaid has not been sufficiently occupied (*despoblada*) for the said term of four months; and this being proved, and the said forty days having expired, the Mine shall be pronounced as such, and adjudged to the said denouncer, and possession of it shall be given to him, on condition that he shall be obliged to sink it three *estados*, conformably to the said ordinances, and under the penalty which they prescribe; and if, after the said forty days have expired, and within the three days allowed for appeal, the owner, or any one having his authority, shall appear, he shall be at liberty to appeal, and justice shall be done conformably to said ordinance.

Art. 40. Also, whereas, it may happen that some Mines may be flooded by the water flowing in from the adjoining and neighboring Mines which are of less depth, by which means the working and reduction of such deeper Mines may be stopped, and the owners of them thereby suffer damage: — We command our Administrator General, and the Administrator of the *Partido*, and each and every one of them, to take especial care to visit said Mines, and to give orders that they shall all be kept clean, drained, worked and reduced; and if any Mine shall be damaged by the waters of another or of others, the said Administrator General, or Administrator of the *Partido*, upon the request of the party, shall inspect it, and cause two persons, appointed by the parties, sworn in his presence and approved by him, to inspect and appraise the damage, and estimate the expense necessary to clear out and drain such Mine: and what shall be thus determined, the court of Mines shall order to be paid, so that the injury may cease, and the working and reduction be proceeded with, and indemnity be made to the person who has received the injury.

Art. 41. Also, we ordain and command, that all persons who shall hold, work, or reduce any Mine or Mines, shall be obliged to keep them clear, and timbered, in such a manner that they may not fall in or become choked up, leaving in those which have a *ley* of a mark and a half per quintal of lead-silver or under, such bridges, strengthenings and supports as may seem expedient for their security and permanence; and those which have a higher *ley* shall, in addition to the above, be thoroughly lined and secured with good timbers; and, in case this is not done, the court of said Mine shall have it done at their expense. And in order that this may be done and performed as aforesaid, our Administrator General, or the Administrator of the *Partido*, is to take, and shall take, special care to visit such Mines and cause them to be inspected, taking with him persons who understand the matter, in order that he may provide what shall be necessary, according to what is prescribed in this and the preceding ordinance.

Art. 42. Also, whereas, it may happen that some persons who take Mines, without working them, or even ascertaining whether they contain ore, may sell them or contract for them, and then proceed to take others for the same purpose, whence divers inconveniences result; therefore, in order to avoid these evils, we command that no one shall sell, contract for, or purchase any Mine if it has not been sunk to the depth of at least three *estados*, under penalty of forfeiting what may have been given to him for it, to be applied as has already been mentioned, and, moreover, of the Mine itself being forfeited and given to the denouncer, subject to the same obligation of sinking it to the aforesaid depth of three *estados*. And if the Mine which shall be so sold or contracted for, shall have been sunk to the said depth of three *estados*, then, in order that said sale or contract may be freely made, the purchaser shall be obliged to give notice thereof to the aforesaid court, so that it may be entered in the book of registries, and an authenticated copy of it shall be sent to the said Administrator of the *Partido*, in order that it may be entered in the book, and be known from whom the duty is to be collected: which shall be done and performed, under the penalty aforesaid; and the same shall be done, if, from any other cause, there shall be a change of the ownership of the said Mine.

Art. 43. Also, we ordain and command, that when two or more persons shall hold a Mine in partnership, in order to work it and to take ore therefrom, if any one of the partners shall apply to have the others put on workmen, they shall be obliged to put on, in all, twelve persons, if there be ore enough for the purpose, and so many can work conveniently; and if not, as many as can work at once, according to the condition of the Mine and the ore it may contain: and if any one of them, on being required so to do, shall not put on his proportion of workmen, then the Judge of the Mine shall cause to be examined, and shall examine, the condition of said Mine, and shall put on, at the expense of the owners of the Mine, such number of workmen as the partner was bound to put on, so as to make up his proportion of the twelve persons: in order that, by reason of these disputes, the working of said Mines may not cease.

Art. 44. Also, we declare and command, that, if any of the partners shall wish to put on more workmen than said twelve persons to work the said Mine, they may do so, provided, they give notice thereof to their partner or partners, in order that, if they shall wish more workmen to be put on, it may be done: and if he shall not give such notice, he shall forfeit the ore which he may take out, and it shall belong to the said partners. And if, when he shall have given them notice, they shall not wish to put on more people, they shall not be obliged to do so, because the law is complied with by putting on said twelve persons between all the partners; and if, nevertheless, any of the partners shall still wish to put on more people, giving notice as aforesaid, he shall be obliged to give the other partners their proportion of the ore taken out, the same as though the additional people he may have put on, and by whom the said ore may be taken out, had been put on by all the partners; and the said Court shall compel him to do so.

Art. 45. Also, with respect to the ore which may be taken out of the Mines held in partnership, that, if they shall not desire to smelt it in partnership and to divide it between them when smelted and refined, according to the share which each one has in the Mine, they shall divide it in ore, in like proportion to their said respective shares; and that, until it be so divided, it shall be kept together in a secure place, and no one shall dare to take anything from it, under penalty of forfeiting his share, which shall go to the other partner or partners; and, in addition thereto, the amount of the value of the said share, one-half of which shall go to our exchequer (*camara*) and the other half to the denouncer and Judge; and if they shall smelt it in partnership, it shall also be refined in partnership, so that each one may receive therefrom the share belonging to him, under the penalty incurred by those who do not take to be refined the metal which they have smelted, and, without refining, shall sell and exchange it.

Art. 46. Also, we ordain and command, that no person, in order to work and excavate his Mine, shall be allowed to throw the earth which he shall take therefrom into the Mine or *pertenencia*, of another, under penalty of ten ducats for each offense, to be applied as aforesaid. And the court of Mines, immediately on its being asked by the party, shall cause such earth to be removed and cleared away from such *pertenencia* at the expense of the party who put it there, or caused it to be put there, notwithstanding any appeal, nullity or revision that may be interposed; but any one is permitted to remove the earth of his Mine through any *pertenencia*, provided that the said earth be taken entirely outside of such *pertenencia.*

Art. 47. Also, we ordain and command, that the washing places (*lavaderos*) which may be required for washing the ores of said Mines, shall be taken wherever most convenient to the Miners; provided that, if they are prejudicial to any Pueblo, or to the cattle, and cannot be made without causing such injury, the water shall be taken from the river or creek to pools where such ores may be washed, and shall be so drained off as not to return to said river or creek: and if this cannot be done, enclosures or *corrales* shall be constructed at the expense of those who make such washing places; and for the better provision and performance of the above, the court of the Mine in whose district such washing places are made, shall cause the same to be complied with, so that injury may be prevented; and, in taking said washing places, they shall be staked out in the same order as the said Mines, and the measurement of each washing place shall be sixty feet long and twelve feet wide, each foot being one-third of a *vara;* but if the washing places shall be made with water taken from the

Mines, and not from any river or creek, there shall be no obligation to perform any of the things above specified, but the parties may make them wherever they please, near the Mine or works where the ores shall be smelted.

Art. 48. Also, we ordain and command, that no person shall presume to enter, search for, take away, or reduce ore from the refuse-heap (*terrero*), washing place, or slag-heap (*escorial*) of another, the owner thereof being known, under the penalty of ten ducats for the first offense, and twenty for the second, to be applied as aforesaid; and for the third offense, over and above the twenty ducats to be paid and applied as aforesaid, that he be banished, for the full term of three years, from the Mines of that *Partido*, and that he do not return within that time, under penalty of banishment for double that period. And moreover, that all which he may have so taken or shall so take, shall be given to the owner of such refuse-heap, washing place, or slag-heap: but we grant, that the old slag-heaps, which have resulted from ores of silver, copper, iron and other metals, and have no owners, from having been made a long time ago, of which slag-heaps there are many in these our kingdoms, may be made use of by persons who work Mines, because we are advised that such slag-heaps are useful and necessary for the smelting of ores; and we command that any Miners whatsoever, and from whatsoever places they may be, shall be at liberty to take them away and make use of them, without any person having the power to hinder them, under pretense that they are on his pastures or grounds, or that he has registered them, or for any other cause or reason whatsoever, if he be not the owner who made them.

Art. 49. Also, we ordain and command, that for the purpose of working said Mines, and of lining and preserving them, and of constructing machines, buildings and huts, and all other things required for working and maintaining them, their owners and the persons employed about them, may use and shall be permitted to make use of all forests, common grounds, municipal grounds, and vacant lands nearest to said Mines, and of their wood timber and trunks, and to cut down by the root the dry trees, without paying anything therefor. And that they shall likewise be at liberty, for the purposes aforesaid, to make use of the wood, timber and trunks of trees and to cut down the dry trees to the roots, in the pastures of private persons and municipal bodies, which shall be nearest to the aforesaid Mines, paying for what they may so cut down in such pastures, the just value thereof, which shall be appraised by the Judge of Mines of the *Partido*, citing the person or municipal body to whom such pasture shall belong. And with respect to green timber and wood, they shall also be at liberty to cut, in the public and municipal forests aforesaid, so much as may be necessary for the buildings and machines, and to line and sustain said Mines, without paying anything therefor, first obtaining a license for that purpose from the Administrator of the Mines of that *Partido*, and not otherwise. And if there shall not be found in such public and municipal forests, such green wood as may be necessary for the aforesaid purpose, they shall be at liberty to cut the same in the pastures of individuals and municipal bodies, as aforesaid, first obtaining a license for that purpose, as aforesaid, from the Administrator, and, first of all, summoning the municipal body or person to whom such pastures may belong, or under whose care they may be, in order that they may be present when the order is given for this cutting. And the said Administrator will take particular care not to give such licenses, except for so much as may be necessary for working and keeping up said Mines, and no more, and that as little injury and damage as possible may be done to such forests and pastures. And although we have commanded that the parties shall be summoned for the cutting of said green timber, the said Administrator may enforce the cutting of what he may think ought to be cut, notwithstanding any opposition thereto, on account of the great damage which might ensue to the works and buildings of the said Mines from any delay being occasioned.

Art. 50. Also, we ordain and command, that all the said owners of Mines and persons engaged in working them and in reducing their ores, shall be at liberty freely to drive into such pastures, meadows, out-grounds (*egidos*), commons, or public and municipal forests, which shall be near said Mines and their establishments, all their oxen and animals, and those belonging to their servants, which are required for the working of said Mines, either for the machinery or for draft, burthen or riding, and the oxen for wagons used to bring provisions, or timber, or other things to said Mines, establishments (*Asientos*) and buildings; provided, that, if the pastures belong to municipal bodies or private persons, they shall pay for the grass and pasture, the same as is paid for other cattle; and those who shall be engaged in searching and trying for Mines, or shall be on their way to search for them, shall be at liberty to take with them one beast each, without paying anything for the grass which shall be eaten by such beasts.

Art. 51. Also, we ordain and command, that all the owners of such Mines, and their servants and persons occupied in working said Mines and reducing their ores, shall be at liberty to hunt and fish freely, within three leagues around the place where the *Asientos* of Mines in which they reside are siutuate, in the same manner as they might do if they were inhabitants of the places (*Lugares*) which are situate within the said three leagues, observing the laws and edicts (*Pragmaticas*) of these our kingdoms upon that subject.

Art. 52. Also, we ordain and command, that, in any parts or places whatsoever, wherein Mines have been or may hereafter be discovered, the owners thereof shall be at liberty to make, and may construct establishments (*Asientos*), houses, smelting works, different kinds of furnaces (*Hornos*, *Buitrones*, *Fuslines*), and all other things necessary for the working, reduction, smelting and refining of said Mines and ores, where, how, and in such form and manner as they may think proper, even though at a different place from that of the Mines; provided, that, if all the owners of a Mine shall desire and shall be able to make such structures together and in the same place, the Administrator General, or the Administrator of the *Partido*, shall take particular care that it be so done and performed, if without damage or injury to the owners of said Mines and the ores which may be produced. And if, for the better smelting and refining of the ores, the owners of the Mines or any of them should wish to make their establishments and smelting furnaces and refineries, in places where there are rivers or creeks, for the purpose of working the bellows with water, they shall be at liberty to do so, and to use freely, for that purpose, the said rivers and creeks, in such part or place as may be most convenient and least expensive, so that no injury accrue to a third party, and that they pay for the place which they occupy, what shall be estimated and appraised by two persons whom the Judge of Mines of the *Partido* shall appoint. And in order that there may be no fraud in regard to the lead produced from the smelting works, we command that each one of the said owners of Mines shall keep an iron stamp, with which he shall stamp and mark the bars of lead-silver (*Plomo-Plata*), and all others whatsoever which are produced from his

7*

Mine and ores, and that without such stamp he shall not take them to be refined, nor shall they be refined.

Art. 53. Also, we ordain and command that, no person shall presume to smelt any ore except in furnaces which are his own, unless he has had them erected in partnership; and if any person, not having a furnace of his own, shall wish to smelt in that of some other person, he shall signify the same to our Administrator of the *Partido*, and with his (the Administrator's) license he shall be at liberty to smelt it, but not otherwise, under the penalty of forfeiting such lead-silver, one-half to our exchequer (*camara*) and the other half to the denouncer and to the Judge; and of forfeiting such Mine, which shall be for the denouncer aforesaid.

Art. 54. Also, we ordain and command, that when, in smelting the ore of any Mine, it shall happen to be expedient, in order to facilitate the smelting, to mix with it the ore of another Mine, this may be done with the license of the Administrator of the *Partido*, provided, that the quality of the ore into which such mixture is introduced shall not exceed that of the ore which may be so mixed and combined with it: and if the former should exceed the latter in richness of *ley*, it cannot and shall not be done, under the penalty of forfeiting the ores which are so mixed, and the produce thereof, with as much more, one-half to our exchequer, and the other half to the denouncer and to the Judge who shall pass sentence. And we command our Administrator in each district, that, in order that what is contained in this our Edict may not be contravened, he shall take particular care to inspect and assay the ores of said Mines, which it may be wished to mix together, in order that what may belong to us shall be paid accordingly. And having done so and made inspection as is proper in a matter of such great importance, and having determined what proportion we are entitled to have according to the quality of such ores, he will give the said license, because the aforesaid mixture is very advantageous for good smelting.

Art. 55. Also, we ordain and command that, in each of the said *Asientos* or works of Mines, there shall be erected at our expense a refining house (*casa de Afinacion*) with different kinds of furnaces (*Hornos, Buitrones y Fuslines*) as may be most expedient, which shall have bellows, tools, and such other things as are required for the refining of the lead-silver, which may be smelted in each *Asiento* of Mines, to which refining house all persons shall be obliged to bring all the lead-silver which may be produced by and smelted from such Mine or Mines, to be refined, and it shall be there refined. And no person shall presume to refine such lead-silver, whether in large or small quantities, in any other place than our said refining house, nor to sell, give away, or contract for the same, until it shall have been refined, under penalty of forfeiting, and he shall forfeit, what he may so refine, sell, give away, or contract for in any other way, with four times the amount, one-half to be applied to our exchequer, and the other half to the denouncer and to the Jugde who shall pass sentence; which penalty aforesaid shall also be incurred by any person whatsoever who shall participate in the above; and where such refining house cannot be conveniently established, from there being no works constructed, and the Mines not being sufficient to render it necessary, the said Administrator of the *Partido* shall make such provisions and give such orders as may be expedient and necessary for the refining of such of the said leads as may be found there. And that the lead-silver, which shall be taken out there shall be carried to the most convenient refining house; and having reached there, the same order shall be observed in regard to the refining thereof and in all other matters, as is provided with respect to ingots of lead-silver which are ordinarily to be refined in said refining house. But it is our grace and will, that such owners of Mines shall be spared as much expense as possible in transporting such lead, which for said cause cannot be refined in said Mines.

Art. 56. Also, we ordain and command, that in each of such refining-houses, of each Mine and its *Asiento*, there shall be the requisite number of refiners appointed by our Administrator of the *Partido*, to the satisfaction of the owners of the Mines, who shall refine the lead-silver which in said *Asiento* or Mine may be produced, at the expense of the parties, they furnishing the necessary charcoal; and that no other person shall intermeddle in such refining, under penalty of one hundred lashes, and of serving three years at the oar in our galleys without pay; and the said Administrator shall fix the rate which such refiners are to be paid for each quintal which they may refine.

Art. 57. Also, we ordain and command, that in each *Asiento* of Mines where there may be such a refining-house, or other place where one shall be established by the order of our said Administrator, there shall be a trusty person (*Fiel*) to weigh the lead-silver which is brought to be refined, which person shall take an oath, when he enters upon his office, to faithfully discharge its duties; and also a Notary (*escribano*), who shall certify to the parcels of lead-silver which are delivered to the refiners; and all the parcels of lead-silver, which shall be brought to be refined, shall be delivered to the refiner who shall have been selected, by the said Administrator of the *Partido*, to refine it. And the said Administrator shall keep a book, wherein shall be entered all such parcels, and the said Notary shall keep another book for the same purpose; which said books shall have each an alphabeticai Index, with a separate account for each of the persons who shall bring such lead-silver to be refined; and on a separate leaf the said Trusty shall note the weights of said ingots, and they shall be delivered to the said refiner; and in said book shall be entered the day, month and year, the weight and number of ingots, the persons who brought them to be refined, and their stamp, and the Mine or Mines from which they came, and the refiner to whom they are delivered; in such a manner that there shall be a particular account and reckoning of the whole; and the said Administrator of the *Partido*, or the person appointed by him, and the said Notary, and the party, if he be able to write, and if not, another for him, shall sign both the said books; and after all the above mentioned has been done, the said refiner shall refine the said parcel; but the lead-silver of one Mine shall not be mixed with that of another, under the penalty, to the party so mixing them, of forfeiting such lead-silver, with four times as much more, to be applied as aforesaid. And if the said refiner shall so mix them, he shall receive one hundred lashes, and shall serve at the oar as a prisoner for three years in our galleys. We charge our said Administrator to observe, and cause to be observed, particular care and diligence that such refinings are faithfully performed; so that no fraud be practiced against our rights, and no injustice be done to the parties.

Art. 58. Also, we ordain and command, that, when what is above directed has been done, and the silver refined and taken out in presence of our said Administrator of the *Partido*, or of the person appointed by him, and of the said Notary, the Trusty (*Fiel*) shall weigh it, and there shall be taken from it the proportion which, according to these ordinances, belongs to us and we are entitled to have, and it shall be delivered to the person whom we shall direct to be appointed for that purpose; and the person to whom it is so delivered shall take charge of it, noting it in said books, and also in that which our said Administrator is to keep, with the day, month and year, stating from what Mine or Mines such silver comes, the owner of the parcel, the persons who brought it to be refined, the weight of the silver in such parcel, and the proportion which belonged to us, and was delivered to said Administrator; and all the aforesaid persons, and the party himself, shall sign all said three books, in order that the said Administrator may give an account thereof, when he shall be ordered to do so; and the remaining silver, (after taking out our proportion as aforesaid), shall be delivered to the owner, after stamping it in one, two, or more places on each ingot, (as each may require,) with the stamp of our Royal Arms, without which said stamp no one shall venture to sell, purchase, or contract for, said silver, which from said Mines is taken, under penalty of forfeiting said silver, and what may be purchased therewith, and the half of all his property, to be applied as aforesaid. And besides this, he shall be banished from said Mines and the space of ten leagues around, for the term of full six years; and he shall not break said banishment, under penalty of serving the same period in the galleys or wherever he may be sent: which penalty aforesaid shall be incurred by the purchaser or the person to whom such silver shall be contracted.

Art. 59. Also, whereas, many ores of silver are worked and reduced by quicksilver, at less expense and with more profit, and it may happen that some persons will desire to work with quicksilver certain ores adapted to that method, and therefore cannot observe what is provided and commanded with respect to ores worked and reduced by smelting and refining; in order that said silver, which is extracted by said quicksilver, may pay us the duty which belongs to us and which we are to receive, conformably to these ordinances, without any fraud therein;—we ordain and command, that any person who shall desire to work and reduce such ores by quicksilver, shall be obliged to give notice thereof to our said Administrator, and to declare to him the Mine or Mines which he shall desire to work and reduce by quicksilver as aforesaid, in order that the same may be entered, and that it may be known that such Mine or Mines are worked and reduced by quicksilver; and that during all such time as they shall wish to work and reduce them by quicksilver, they shall not be at liberty to work, nor shall they work or reduce the same, in any other manner, unless upon giving notice thereof, when they desire to do so, to the said Administrator, that the same may be noted, and that it may be known that they no longer work or reduce the said Mine or Mines by quicksilver, as aforesaid. And if they shall work and reduce the said Mines in any other manner, they shall forfeit the silver and metal, and the one-half shall go to our exchequer, and the other half to the denouncer and to the Judge who shall pass sentence; and the said Mine or Mines shall be considered as forfeited and shall be for the denouncer: and the part or duty which we may be entitled to have, agreeably to these our ordinances, shall be determined by weighing the quintals of metal which may be mixed with the quicksilver, in the presence of the Trusty, and Notary, and our Administrator; and when the quicksilver shall have been driven off from the amalgam (*pellas*) obtained, so that the silver shall remain fine, it shall also be weighed, that it may be known and ascertained how much silver has been produced from the quintals of metal mixed with the quicksilver; and our duties shall be levied according to the produce, in conformity with these ordinances, as aforesaid, keeping thereof the same books, account, and reckoning, in the order, and form, and manner, as is required to be observed with respect to silver to which the refining process is applied, and under the same penalties, to be applied as aforesaid.

Art. 60. Also, we ordain and command, that the silver shall not be removed from the place where it may have been put for the purpose of having the quicksilver driven off, except in the presence of our Administrator of the *Partido*, or of the person whom he shall appoint, in order that, before him, and the Trusty, and the Notary, it may be weighed and the duty which we may be entitled to and which may belong to us, shall be taken from it, and be delivered to the person whom we shall command to be appointed for that purpose, and the same account and reckoning shall be kept thereof, as of that which shall be refined by fire; and the silver which shall remain shall be delivered to the owner, and upon each ingot there shall be impressed our Royal Stamp, as aforesaid: and without having our said Royal Stamp, said silver cannot be sold or contracted for in any manner, under the penalty above prescribed to the owner of said silver, and to the purchaser or the person who shall contract for it.

Art. 61. Also, we ordain and command, that the proportion belonging to us of the poor lead which may be smelted, and which will not bear refining, from its not containing over four *reales* of silver per quintal, shall be stamped, in the part or place where it is smelted, by the Administrator of the *Partido*, or by the person whom he shall appoint; and upon its being found by assay to be poor-lead, the person whom we shall have appointed for that purpose, shall receive the duty which may belong to us, agreeably to our ordinances: and that no lead, even though it be made from litharge, shall be conveyed from one place to another, without such stamp, under penalty, to any person otherwise conveying it, of forfeiting the same, one half to go to the party who shall denounce it, and the other half to the Judge who shall pass sentence, and likewise four times the value, to go to our exchequer;

and the same with regard to copper, which, being first assayed, shall be stamped, in order that our proportion of it, and of the silver and gold it may contain, shall be paid: but this with respect to poor-lead and copper, is to be understood as applying to places without the limits of the grants (*Mercedes*) which have been made.

Art. 62. Also, we ordain and command, that all those who shall convey antimony out of the *Partido*, where no grant thereof has been made, shall pay us the duty thereon at the Mines or veins from which it shall be taken, and that, until this be paid, it shall not be removed nor sold in any other place, without the license of our Administrator of the *Partido*, or of the person by him appointed, who may be in the *Asiento* of Mines nearest to the Mine from which said antimony is taken; and that after such license is obtained, no person shall be at liberty to carry or convey it without an order (*cedula*) from said Administrator, or from the person whom he may have appointed; and that such seller shall be obliged to advise the purchaser thereof, in order that said *cedula* may be obtained, and he shall so advise him, under the penalty of forfeiting the value of said antimony, with four times the amount, to be applied as aforesaid; and if the purchaser shall otherwise convey it away, it shall be taken from him as contraband, with four times the amount, to be applied as aforesaid, which is to be understood, as aforesaid, to apply to places where there are no grants (*mercedes*) made.

Art. 63. Also, forasmuch as it has been found by experience, that when suits and differences arise concerning the possession of Mines, the working and reduction of them cease, and they are ordered to be closed until it be determined which party has the better right; so that they are frequently one, two or more years without being worked and reduced; wherefore, and in order to prevent the damage that results from such Mines being left without being worked or reduced, for so long a time, we ordain and command, that when, and as often as, such suits shall arise, the term of forty days shall be allowed, and, no longer, during which the Mine in dispute may be closed, and within which the parties shall, before the court of Mines, state and allege their claim, and present such writings and securities as they may have, and shall be at liberty to produce as many as twelve witnesses, each one on every interrogatory, and no more; and upon what is stated, alleged and proved within the said term, and without any other conclusion or adjournment, the said court shall consider and decide, saving to the party against whom it shall give judgment, the liberty of prosecuting his claim to the right of property, as he shall think expedient, before the said court of Mines; and it shall immediately give the occupancy (*tenencia*) and possession of the said Mine, to the party in whose favor it shall give sentence; who shall work and reduce it, keeping an account and reckoning, in a book, with the day, month and year, of the ore which he shall take out, and of the costs and expenses he may incur in the working and reduction, and giving security, in the sum of one thousand ducats, for rendering an account and paying over the proceeds, if upon appeal he should be beaten, and should be ordered so to do: which shall be done and performed in manner aforesaid, notwithstanding any appeal, nullity, or revision (*agravio*) which may be interposed. And if the party against whom the sentence is given, should consider himself aggrieved, he shall be at liberty to appeal, within three days, to our Administrator General of Mines, and within sixty days, in case of appeal, nullity or revision, both parties shall prosecute their case before said Administrator, and shall present writings, securities and witnesses, and they shall be admitted so far as the law allows, as aforesaid. And on what they shall state, allege and prove within said term, without any other conclusion or postponement, the justice of the case shall be determined: and if the decision shall be in confirmation of the former judgment, such suit shall be at an end, so far as the right of possession is concerned, and there can be no appeal from it. But, nevertheless, the party in whose favor the suit is decided, shall keep an account and reckoning of the said ore which shall be taken out, and of the said expenses as aforesaid, in order that he may render the same with payment, if, in the question of property, he shall be beaten and shall be adjudged to make such payment. But if the said decision should not be confirmatory of the former judgment, and the parties should appeal therefrom, the said appeal shall be to the Chief Accountant of Finance (*Contaduria Mayor de Hacienda*), and to no other Tribunal. And if the parties, or either of them, should set up a claim to the right of property of the said Mines, this case shall be brought before the Administrator of the *Partido*, or before the Administrator General of Mines, and not before any other Judge, and he shall hear the parties concerning the same, and from the sentence which he shall give, an appeal may be taken to the said *Contaduria Mayor*, and not to any other Tribunal. And if a writ of execution shall have issued, by virtue of which the possession of such Mine or Mines may have to be restored to some other person, with the produce thereof, we command that the person who shall have had possession, and the sureties which he has been obliged to give, agreeably to this our Letter (*Carta*), shall give a correct and faithful account, with payment, of all that may have been taken from, or produced by, such Mine, up to the day it is taken from him, deducting the costs and expenses of working and reducing the same: of which expenses he shall put in a sworn account, signed with his name, to which full faith and credit shall be given.

Art. 64. Also, we ordain and command, that when, and as often as, any person shall lay claim to a Mine of which another has quiet and peaceful possession, and shall also ask that such Mine be closed; then, inasmuch as, in a case of this kind, the principal object sought for is the ore taken from such Mines, and, in order, that the working and reduction may not be discontinued, on account of the damages which would ensue thereby, the said Court shall command that, within the peremptory term of twenty days, the party be cited, and give evidence of the right which he may have, and that the other party, if he think proper, give evidence to the contrary, or to such purport as he may deem expedient. And immediately on the expiration of the twenty days, if it shall appear that the Plaintiff is in the right, the party in possession shall be ordered to henceforth keep an account and reckoning of the ore and silver which may be produced from said Mine, and of the costs and expenses which he shall incur, as mentioned in the preceding ordinance, in order that he may render the same, with payment, if he should, on trial of the case, be beaten. All which shall be observed, fulfilled and executed, notwithstanding any appeal, nullity or revision which may be interposed; and this being done, the said cause shall be proceeded in, without allowing any delay, or malicious procrastination (*dilaciones de malicia,*) and justice shall be done.

Art. 65. Also, we ordain and command, that when, and as often as, cases shall arise in which arbitrators shall be appointed by the parties, or by the said court of Mines, such arbitrators shall, first of all, take an oath that they will give and declare their opinion, truly and faithfully; and that if such arbitrators shall not agree, then in case of disagreement, another arbitrator shall be appointed, by consent of the parties, or by the court of Mines; and if this one shall coincide in opinion with either of the aforesaid arbitrators, such opinion shall be observed and carried into effect. And if no two should agree, and if they should all differ in opinion, upon the whole case or in part, then other arbitrators shall be successively appointed, until the majority agree in opinion upon the whole matter; and this being the case, what the majority shall say and declare, shall be observed and carried into effect.

Art. 66. Also, we ordain and command, that all thefts, which may be committed in the said Mines, and in the *Asientos*, and in the bounds, and places where there may be Mining works, of gold, silver, lead, or ores of any class or description, and of any thing belonging to said Mines or connected with their working and reduction, shall be punished with the utmost rigor; and that any person who shall steal any of the things aforesaid, shall, besides paying and restoring to the party the thing stolen, be condemned in seven-fold the amount, which we apply, one-half to our exchequer, and the other half to the informer and to the Judge who shall pass sentence. The Administrator of each *Partido* shall have cognizance of such thefts; and from the sentence which he may give, appeal may be taken to the Administrator General. But if he who shall have been so condemned in seven-fold the amount, shall not have property wherewith to pay it, the sentence shall be commuted to some corporal punishment, or to banishment, according to the gravity of the offense: from which commuted sentence there shall be an appeal to our *Contaduria Mayor de Hacienda*, and to no other tribunal, whether said commutation be made by the Administrator of the *Partido*, or by the Administrator General.

Art. 67. Also, we ordain and command, that our Administrator General, and the Administrators of *Partidos*, and the persons appointed by them and their successors, to attend officially at any one of the Mines, and the Courts, and Notaries and Trústies, who have been, or shall be, by us appointed to use or exercise their offices therein, shall not be at liberty to have or hold any Mine or part of a Mine, in any *Partido* of the Kingdom, either by themselves or through any person, directly or indirectly, during all such time as they shall hold the said offices, under penalty of being forever deprived of such offices, and of forfeiting the Mine or Mines which they may hold, which shall go to the denouncer, and under the further penalty of forfeiting a moiety of their property to our exchequer: and that any person who shall participate in the above offense, shall incur the penalty of forfeiture of property and Mines.

Art. 68. Also, we ordain and command, that all persons appointed by us, or by our said Administrator, or by our Administrators of *Partidos*, to attend the working and carrying on of said Mines, or who shall, in any manner, receive a salary or pay from us for that purpose, shall be incapacitated for holding any Mines or any shares in Mines, either by themselves or through any other person acting for them, directly or indirectly, in the *Partidos* where they shall be employed and shall work, or within two leagues around; and if they shall take or have any Mine or Mines, as aforesaid, or any share in them, whilst they receive our salary or pay, as aforesaid, they shall forfeit such Mine or Mines, or shares thereof, which shall go to the person who shall denounce the same; and besides this, they shall be banished from said Mines and from the space of six leagues around, for the full term of three years; and they shall not break that term, under the penalty, if they be of noble rank, of the said term of banishment being doubled; and, if of lower condition, of being forced to serve said three years at the oar in the galleys.

Art. 69. Also, we ordain and command, that all persons who shall search for, find or take Mines, or sources of gold, as well the first discoverers as others, shall, in taking, registering and staking out the boundaries of said Mines, observe the provisions of these ordinances which relate to the taking, registering, and staking out of Mines of silver, and under the same penalties as are therein imposed, and that, agreeably to the said ordinances and under their penalties, they shall be obliged to send the registries to our Administrator General, or to the Administrators of each *Partido*, and these shall keep books of registries of Mines of gold, in like manner as is provided for those of silver.

Art. 70. Also, we ordain and command, that the first discoverers of said Mines or sources of gold, shall take and have a space of eighty *varas* in length and forty in width, which they may take as they shall deem best, and those who come after them shall take and have a space of seventy *varas* long and thirty wide, which they also may take as they shall deem best; and, in every other respect, they shall observe what is contained in the said ordinances with regard to silver, under the penalties prescribed therein.

Art. 71. Also, we ordain and command, that all those who may have Mines or sources of gold, shall be obliged to keep them worked (*pobladas*), as is commanded with respect to silver Mines, under the penalties hereinbefore declared.

Art. 72. Also, we ordain and command, that no person shall presume to treat for, contract, sell, or purchase gold, either in dust, or bars, or ingots (*rieles*), unless stamped with our Royal Stamp, which we command to be kept by the person who shall be appointed, in our name, in each *Partido* to collect the proportion which to us may belong. And there shall also be a smelter, who shall smelt and

8

make bars (*vergas*) of the gold which shall be taken out, and who shall also be the Trusty to weigh it (*Fiel del peso*); and he shall smelt, weigh, and stamp the same with our Royal Stamp aforesaid, in the presence of our said Administrator, or in the presence of the person appointed by him, and the part which shall belong to us shall be given and delivered to the person who shall attend, for that purpose, in the *Partido* where the above shall be performed, and the remainder shall be given to the owner; and our said Administrator shall keep a book, wherein he shall enter the said parcels, with the day, month and year, and also to whom such gold belongs, from what Mine or source it comes, and how much there is, and what proportion belongs to us, of which the said Administrator was charged, which entry the said Administrator shall sign, and also the party, if he shall be able to write, and if not, then some other person for him, and the smelter and the Notary before whom it is passed; which Notary and smelter aforesaid, shall each keep another book, wherein the same shall be noted and signed, as aforesaid, by all; and no person shall be at liberty to sell or contract for said gold, unless it shall have been smelted and stamped as aforesaid, under the penalty prescribed in the ordinance with regard to silver, which relates to this subject, and the like penalty shall be incurred by him who shall purchase or contract for it, as is prescribed in the said ordinance with regard to silver.

Art. 73. Also, inasmuch as it may happen that the servants of said owners of Mines or other persons, without its coming to the knowledge of such owners, may sell or deal in gold, without its being stamped with our Royal Stamp, contrary to the provisions of these ordinances; we ordain and command, that any servant or person who, without the knowledge or fault of the owner thereof, shall sell, or deal in gold or silver not stamped with our Royal Stamp as aforesaid, and any one who shall purchase or contract for the same, shall, besides restoring to the owner or paying him for what shall have been so sold or dealt in, forfeit all his property, one-half of which shall go to our exchequer and the other half to the denouncer and the Judge who shall pass sentence, and shall serve ten years at the oar in the galleys.

Art. 74. Also, whereas we are informed, that from sinking pits (*pozos*) in Mines near together at the surface, and from sinking them continuously without making landing-places, great inconveniences and dangers arise, not only with regard to their permanence, but also with regard to working and draining them ; and in order to remedy this:—We ordain and command, that hereafter, when any new Mine shall be discovered, the pits which are to be continued shall be ten *varas* apart, and each pit shall be fourteen *estados* in depth ; and, if they shall be sunk deeper, an excavation (*Mineta*) shall be formed before sinking any deeper, and from that the next pit shall be formed. But, forasmuch as in many places, circumstances will not admit the adoption of this order, in such cases that mode shall be adopted which shall seem most expedient to the Administrator of the *Partido*, and to the other Miners who understand the matter.

Art. 75. Also, forasmuch as we are informed, that, by not assaying ores for smelting, or rich leads for refining, there is great negligence in the smelters and refiners, whence results injury, not only to our treasury, but also to individuals, and there is morever afforded an opportunity for frauds: therefore, in order to remedy this, we ordain and command, that our Administrator General, and those of the *Partidos*, shall take great care, in providing that, where there are many Mines together, there shall be sufficient sworn assayers, as well for the ores to be smelted, as for the rich lead which is to be refined, so that the smelters and refiners shall make the returns of the smeltings and refinings, which they may perform, correspond with the assays which may have been made.

Art. 76. Also, forasmuch as, when old Mines reach the depth of thirty, forty, or more *estados*, the expense of taking out the water, rubbish and ore, and of putting in the timber and other necessary things, is much greater than in other Mines which have less depth, and frequently amounts to more than the produce, and, under these circumstances, the owners of such Mines cannot pay so high a duty as that which these ordinances have imposed on such old Mines, and as it is reasonable that in such cases there should be moderation: therefore, we ordain and command, that when this shall happen, and it shall be proved to our Administrator General that the old Mine has become, from its depth or from other causes, so expensive as to be of scarcely any profit to the owner, he shall send a particular report of it, with his opinion, to our Council of Finance (*Consejo de Hacienda*), together with the investigations which he shall have made on the above subject, whereupon we order, that the matter be considered and determined in as summary a manner as possible.

Art. 77. Also, whereas we are informed, that one of the causes which impede the good order and working of the Mines already discovered, and which prevent the search for and discovery of new ones, is the suits and disputes which occur in them, and among the people who engage in and work them, and the annoyances and vexations which the courts and other persons cause to the officers and workmen engaged in them, as well on account of the said courts not having the proper practice and experience in mining affairs, as on account of the proceedings in such causes being long and in the ordinary form; by which means, their capital (*haciendas*) is wasted and consumed, in proceedings before these and the other tribunals to which the parties have recourse by way of appeal, thus rendering it impossible for them to prosecute the discovery and working of such Mines, whence follow notable damage and injury to us, and to these our kingdoms, and our subjects therein: therefore, in order to remedy this, as a matter of such great importance, and in order that all may be encouraged to discover, work and reduce said Mines,—we have resolved to appoint, and we will appoint, an Administrator General, and such other Administrators, as may be necessary for the *Partidas* and districts, which may be marked out, who shall be persons of practice and experience in such matters; and who shall have the government and jurisdiction of all said Mines, and all matters relating thereto, and shall be superior to all other persons engaged therein, and they shall take account and reckoning of the Mines, and particular care that all the provisions of these ordinances are kept, observed, and complied with, and they shall execute them, and cause them to be observed and complied with, conformably to the order, and the instructions which we shall command them to give agreeably to these ordinances; and they shall have jurisdiction to take cognizance, and shall take cognizance, in *first instance*, of all suits, causes, and civil and criminal matters, and matters of execution, which in any manner may arise, or occur, or be agitated, in each district, of which they may and ought to take cognizance, agreeably to these ordinances, in manner as follows:—That, of the causes which are thus offered, the Administrator General shall take cognizance, if he be in the district of the *Partido* where they may happen to arise, and if he be not found there, the Administrator of the *Partido* shall take cognizance of them, and the causes, of which cognizance is taken by the Administrator General, if he shall absent himself from the *Partido*, he shall refer, in the condition they happen to be, to the Administrator of the said *Partido*, who shall proceed with and determine them, agreeably to these ordinances; and if the said Administrator General shall return to the *Partido*, and shall find the causes which he so referred, still undetermined, he shall be at liberty to resume the cognizance of the same while he may be there. And we command such Administrator General, and Administrators of *Partidos*, to do and administer justice to the parties, in the causes and matters of which they take cognizance, in a brief and summary manner, agreeably to these ordinances; so that, by reason of said suits, the working and reduction of said Mines may not be impeded or embarrassed. And we command our courts, as well those of general jurisdiction (*ordinarïas*), as those of brotherhoods, and of commission, and all others of these our kingdoms, and those of seigniorage (*señorio*), not to intermeddle in the cognizance of such causes, touching and concerning said Mines, and the persons, beasts, oxen, or wagons which may serve, work, or be employed in them or in reducing them; nor to proceed upon or entertain claims, petitions, complaints, or other matters, either *ex-officio* or upon the request of the parties, in relation to the aforesaid matters or any part thereof: and if any such matters shall be pending before them, they will immediately remit them to the said Administrators of each *Partido*, in order that they, as Judges thereot, may take cognizance, and do justice to the parties. And, by these presents, we prohibit, and hold as prohibited, the said courts, and ordinary Judges, and those appointed by commission, and all others whatsoever, from taking cognizance, in any manner, of such causes and matters touching, proceeding from, or in any way depending on, the said Mines, and the workmen, officials and officers thereof, as aforesaid, notwithstanding any law, edict, or other thing, to the contrary, which (as far as relates thereto) we dispense with, abrogate, annul and make void and of no force or effect, leaving them in force and effect as to all other matters. And with respect to the persons who are to be appointed as Administrators, and Receivers (*Receptores*), and other officers connected with said Mines, it is our will that they be appointed by our Council of Finance (*Consejo de Haciendo*), with our Titles and Edicts (*Titulos y Cedulas*), signed with our hand; and that the same be done with respect to orders and instructions which shall be given to them, for their conduct in their several offices.

Art. 78. Also, we ordain and command, that all persons whatsoever who shall wish to carry provisions, supplies and other things to the said Mines, for the supply and support of those who may be and may work in them, shall be at liberty to take out and carry, and shall take out and carry, the same freely from all the cities, villages and places of these our kingdoms and lordships. And that the courts thereof shall not prevent them from so doing, nor put any obstacle or impediment in their way, nor enhance the price; but that, on the contrary, they shall assist and favor them, in order that the said Mines, and those engaged in them, may be always provided and supplied therewith.

Art. 79. Also, forasmuch as we are advised that many Mines are so situated as to admit of tunnels (*contraminas*) being constructed in them, and that new ones, hereafter to be discovered, may possibly admit of the same disposition, so that the water of the Mines may flow out from the bottom, or be got out at less expense; which is of great importance, as well for the permanence of the Mines, as for the working and reducing of them;—Therefore, we ordain and command, that wherever there shall be an opportunity to construct such *contraminas*, the owners of the Mines shall make them, and each shall contribute to the same, according to the nature and condition of his Mine which can be drained by said *contramina*. And that when the owners thereof shall not agree amongst themselves to construct it, the Administrator General, having inspected and made himself acquainted with the character of the place, and the utility of the work, will arrange with them to construct it. And that in such case, (the said owners having agreed to it,) he shall make such apportionment or apportionments, as may be necessary, amongst the owners of the Mines which are to enjoy the benefit, of what each one has to contribute, according to the advantage which is to result to them, and he shall compel the payment and fulfillment of the said apportionments, which are made for the said purpose. And that the ore which may be taken out, in opening and working the said *contramina*, shall be used for the expenses incurred in constructing it; and the deficiency shall be apportioned according to the arrangement entered into by the owners, or, in default of such agreement, according to that made by the said Administrator.

Art. 80. Also, we ordain and command, that if, in the said *contramina* or *contraminas* which, in conformity with what is aforesaid, are opened, there shall be discovered any new Mines which were not previously found and discovered on the surface, even though they fall within the stakes of other Mines discovered on the surface, those which are so discovered where the said *contramina* is opened, shall belong to the owners who shall have contributed to such *contramina;* and each of them shall have a share of the produce, proportionate to the share of the expense which has been apportioned to him respectively, as aforesaid.

Art. 81. Also, we ordain and command, that if there be any Mines very distant from the place where such *con-*

tramina shall be made, and, for that reason, their owners shall not be willing to contribute to its expense, then, and as often as it shall be ascertained that the water of said distant Mines is drained off, or diminished, by reason of said *contramina*, or that they derive any benefit therefrom, in getting out ore, rubbish or anything else, they shall pay to the owners of said *contramina* so much as shall be rated and assessed by the Administrator General, or by the Administrator of the *Partido*, or of the one nearest thereto, for the benefit which, from said *contramina*, may result to their Mine; having regard to the expenses to which they would have been put, had not such *contramina* been constructed, and which are thereby saved to them.

Art. 82. Also, we ordain and command, that if in any of the *Asientos* of Mines where it may be convenient to make the said *contramina* or *contraminas*, the owners of the Mines are unwilling to incur the expense of working it, and some individual shall desire to undertake the work, the Administrator having approved the expediency of making it, and the commencement of such *contramina* being registered, he shall be at liberty to make it, and he may make it as far as he pleases, without regard to the order of stakes, or to the boundaries measured. And all the ore and the produce thereof, which may be discovered by means of said *contramina*, shall belong to the person who shall have made the same: provided, that he shall not take more of the ore of another's Mine than shall be contained within the cavity of such *contramina*, and that he who constructs said *contramina* shall not be permitted to increase its depth, height, or width beyond the dimensions which shall have been assigned at the commencement of said *contramina*, which are understood to be eight quarters (of a *vara*) in height, and five in width. And that he shall enjoy this preference and the ore, so long only, as there shall be no other Mine deeper, and from which more benefit results to said Mines, for this right belongs to that one which is the deepest.

Art. 83. Also, in order to confer a benefit and favor upon those who hold and work said Mines, and upon their Administrators, assayers, smelters, refiners, accountants, and paymasters:—We ordain and command, that in the parts and places where they shall reside in said Mines, they shall be free and exempt from having guests and baggage trains (*vagages*) quartered on them, and that they shall not be obliged to furnish beds, nor guard beasts (*bestias de guia*), nor wagons. And that, besides this, they shall be at liberty to carry arms in said Mines, at all times of day and night, offensive and defensive, provided they be not such as are prohibited, nor carried in prohibited places; and that our courts shall observe this and not oppose it, so long as they shall be employed in said Mines and in the working thereof.

Art. 84. Also, it is our will and pleasure, and we command, that the incorporation, which we thus ordered to be made into our Royal patrimony, of the Mines of gold, silver and quicksilver, by the said Edict (*Pragmatica*) of the year 1559, be, and shall be, understood to be, without prejudice to the contract and agreement which we ordered to be made with Don Diego de Cordova, our Master of Horse, respecting the Mines which he holds by gift (*de Merced*), signed with our name, on the 15th day of August of the year 1568.

By which said laws and ordinances, and by each one of them, we command that the said Mines, and the things touching, annexed to, and concerning them, shall be directed and governed; and that all the Judges, Courts and *Audiencias*, in their districts and jurisdictions, shall observe them and cause them to be observed, complied with, and executed, in all and by all, as in them, and in each one of them, is provided, and that against their tenor and form, they shall not go or pass, nor consent to go or pass, under the penalties in these our said laws and ordinances contained, and under the penalty of our displeasure, and of ten thousand *maravedis*, to be paid to our exchequer by each one who shall violate the same. And we command our principal Auditors (*Contadores Mayores*) to enter a copy of them in the books of our *Contaduria Mayor*, and to cause them to be printed for the benefit of all.

And, moreover, we command our said *Contadores Mayores* to keep books, accounts, and note of all which, from said Mines shall come to us, and of the reports and transcripts which our said Administrators and officers shall have sent, of the condition of said Mines and of their costs and expenses.

August 18th, 1607.

Don Philip III, in Cedula August 18th, 1607.

[Book VI, Title XIII, Law X, Nueva Recopilacion.]

In which are changed some of the provisions of the ordinances of August 22d, 1584, with respect to the Royal dues, and to the working and the reduction of Mines of gold, and silver, and other metals.

Art. 1. Whereas the King, my lord and father, (who is now in Holy Glory), by his Letter (*Carta*) and Provision, signed with his hand and countersigned by Juan Velasquez de Salazar, his Secretary, issued by some of his Council, dated at San Lorenzo el Real, the 22d of August, 1584, ordered to be made, and made, certain ordinances concerning the Mines and minerals in these kingdoms, discovered and to be discovered, in which (among other things) it is provided what dues are to belong to His Majesty from said Mines and minerals, and the form and order to be had and observed in collecting them, as in said Letter and Provision is at greater length set forth; and whereas experience has shown it to be necessary and expedient, for my service, as well as for the benefit of these kingdoms, and of their subjects and natives, to grant grace and power to the discoverers and workers of the said Mines, in addition to those granted to them by the said ordinances, and to facilitate the payment and collection of the said duties and of other things, it having been discussed and considered in different *Juntas*, as well in my time as in that of His Majesty, and lastly having ordered it to be discussed and treated of in my Council of Treasury and its *Contaduria Mayor*; and having had its opinion, I have granted, resolved and determined, that, although, according to the said ordinance, there belongs to me, of the Mines of silver, whose metals yield at the rate of a mark and a half of silver, (which is twelve ounces per quintal of lead-silver), the tenth part of the silver, without deducting expenses; and of those which yield at the rate of from a mark and a half per quintal of lead-silver to four marks, the fifth part; and of those which yield from four marks to six, one-fourth part; and those above six marks, one-half; and of Mines of gold, of whatsoever *ley* and richness they may be, one-half of the gold, clear of all expenses; and of abandoned Mines, which shall have been sunk to the depth of ten *estadas*, and shall yield at the rate of two marks of silver per quintal of lead and below that, the twelfth part; and if they shall yield more, the same as from new Mines; and of the silver which is taken out of the refuse-heaps and slags of old abandoned Mines, the tenth part, if it be smelted alone; and if it be mixed with other ores, it must pay the same as other Mines: nevertheless, in order to confer a favor, as aforesaid, on the subjects and natives of these kingdoms, I have been pleased to order that, for the term of ten years, counted from the date of this my *Cedula*, there only be paid to us from the Mines of gold and silver, and from the waste-places and slag, one-fifteenth; and after the expiration of the said ten years, one-tenth; all without deducting expenses; with the declaration, that after the expiration of twenty years from the date hereof, orders may be issued to increase the said duties, provided that they do not exceed one-fifth, it being left as a charge to my said council of Treasury and its *Contaduria Mayor*, that, at the end of the said twenty years, agreeably to the condition of the Mines, they shall consult me respecting those on which the duties may be raised, provided that, as aforesaid, it shall not exceed a fifth in any of them.

Art. 2. By the ordinances it is commanded, that no person shall smelt ore, except in his own furnaces, without the license of the Administrator, and that no one shall mix ores in order to smelt them, and that in each one of the *Asientos* of Mines there shall be a refining house at my expense, where all shall have their lead and silver refined, and that it shall not be done elsewhere; and that from places where they cannot have a refining house, it shall be taken to some place where there is one, and that in each refining house there shall be the requisite refiners, appointed by the Administrator, who shall make the refinings at the expense of the parties: and that in each *Asiento* of Mines where there is a refining house, there shall be a Trusty (*Fiel*), and Notary, and that the silver shall be refined and extracted in presence of the Administrator and of a Notary, that the Trusty shall weigh it, and take out the proportion which shall belong to me, and that it shall be delivered to the person whom I shall appoint, and shall be taken charge of by him, it being entered in the books, and in that of the Administrator, with the day, month and year, and stating from what Mine the silver is taken, and the owner of the parcel, and the person who brought it to be refined, and what it weighed in all, and the proportion of it which belonged to me, and, in the three books which are to be kept, they shall all sign, and also the party; and that which shall belong to the owner shall be delivered to him with the stamp of my Royal Arms, without which, no one shall be at liberty to sell or purchase it, under certain penalties; and that those who extract silver by means of quicksilver, shall give notice thereof to the Administrator, and that they shall not take silver from the place where it shall have been deposited to drive off the quicksilver, unless the Administator and Notary are present, and many other proceedings are had: nevertheless, I am pleased to suspend, and I have suspended, so far as relates to the above, the use of the said ordinances, and I direct that, with respect to the Mines which may be worked, and the places where they may be worked, my said commissary of Treasury and its *Contaduria Mayor*, shall give the form which may be deemed proper for all the above, until I shall provide and order something else, particular care being taken to collect my duties: so that the working of the Mines shall not thereby be impeded; insomuch as it can well be done.

May 18th, 1680.
LAWS OF THE INDIES RELATING TO MINES.

[Extracts from the "Recopilacion de Indias," declared of authority by Don Carlos II, at Madrid, May 18th, 1680.]

NOTE BY THE TRANSLATOR.—This *Recopilacion* is a mere digest, or compilation, of the Royal orders, decrees, edicts, &c., issued from time to time for the government of the Spanish American Colonies. In many cases several decrees, edicts, &c., are condensed into a single law, rendering it impossible to arrange these laws in the order of the dates of the original decrees. They are, therefore, placed according to their order in the *Recopilacion*, placing at the head of each law the titles and dates of the original decree or decrees upon which it is founded, and the name of sovereign by whom each decree was made.

Book I, Title XII.
OF THE CLERGY.
Law IV.

Don Philip II, at Vienna de Navarra, November 15th, 1592. Don Philip III, at Madrid, March 29th 1621.

That clerigos and religiosos are not permitted to work Mines.

Whereas, the working of Mines by *clerigos* and *religiosos*, besides being improper in them, may result in scandal and bad example; we charge the prelates not to consent to nor permit it, punishing with severity and exposition those who contravene this law.

Book II, Title I.

OF THE LAWS, PROVISIONS, CEDULAS AND ROYAL ORDINANCES.

Law III.

Don Philip III, at Valladolid, November 26th, 1602.

The Viceroys will cause to be observed in the Indies, so far as expedient, the laws of these kingdoms (Spain) respecting Mines.

The Viceroys of the Indies will consult with intelligent persons who have experience in the laws of these our kingdoms of Castile, respecting matters of Mines; and, where they are found to be suitable, they will cause them to be observed, enforced and executed, in all those kingdoms, where they may not be opposed to what shall have been specially provided for each province, and they will dispose and determine what may be requisite, and in this form, or as may be most expedient, they will send us a particular account of what laws of Mines are not enforced in each province, and why, and also the reasons they may have for ordering those to be observed which they may deem necessary.

Book II, Title XVI.

OF THE PRESIDENTS AND OIDORES OF THE ROYAL AUDIENCIAS AND CHANCILLERIAS OF THE INDIES.

Law LX.

The Emperor Don Carlos, at Barcelona, November 20th, 1542; the Queen of Bohemia, Regent, at Valladolid, April

29th, 1549, and April 26th and May 2d, 1550; Don Philip II, in Ordinance 30 of Audencias, 1563, at Valladolid, May 9th, 1565, and in Ordinance 37 of Audiencias, 1596.

That Officers shall not engage in Armadas, Discoveries or Mines.

The Presidents, Oidores, Alcaldes, and Fiscales shall not engage, nor shall they be permitted to engage, in *armadas*, or discoveries, without our express command, or in Mines, either to a great or small extent, in any district where they reside; and those who contravene the above shall incur the penalties prescribed in the preceding laws.

Book IV, Title III.

OF DISCOVERIES BY LAND.

Law XXIV.

Don Philip II, Ordinances 96 and 97 of *Poblaciones*.

That the settlement being completed the principal settler (poblador) may establish the right of primogeniture to what he may have in it, and enjoy the Mines, paying the fifth.

To him who shall have performed his contract, and have made a settlement (*poblacion*) in conformity to his agreement, we give permission and power to establish the right of primogeniture (*mayorazgo*) to what he may have built, and to what, within the boundaries, may have been granted to him, and to what, within the same, he may have planted and constructed, and also to the Mines of gold and silver, and other minerals, and salts, and pearl fisheries, on condition that, of the gold, silver, pearls and all else which he may get out of said metals and Mines, the founder (*poblador*) and the inhabitants of the settlement, or other person whatsoever, shall give and pay to us, and to our successors, the one-fifth, free from all charges, after the expiration of the first ten years.

Book IV, Title XIX.

OF THE DISCOVERY AND WORKING OF MINES.

Law I.

The emperor, Don Carlos, at Granada, December 19th, 1526. Don Philip II, at Madrid, June 19th, 1568.

Permitting all Spaniards and Indians, vassals of the King, to discover and work Mines.

It is our grace and will, that all persons, of whatsoever state, condition, rank, or dignity, being our vassals, shall be at liberty to take out (*sacar*) gold, silver, quicksilver, and other metals, either personally, or by their servants or slaves, from all Mines which they shall discover, or wherever they shall choose or think proper, and to hold them and work them freely, without any kind of impediment, having given an account to the Governor and to the Royal officers, for the object mentioned in the following law, so that the Mines of gold, silver, and other metals may be common to all, and in all places and limits, provided that there result no injury to the Indians or other third person, and that this permission is not to extend to the Ministers, Governors, Corregidores, Alcaldes Mayors, and their professional (*letrados*) substitutes, to Alcaldes, and Notaries of Mines, nor to those who have been especially prohibited: and with respect to designating and taking Mines, and marking them out, they shall observe the laws and ordinances made in each province, and by us confirmed.

Law II.

The emperor Don Carlos, at Toledo, November 24th, 1525. Don Philip IV, at Madrid, June 19th, 1627. Vide Law 3, Title 5, Book 8, § Han de tener.

That the discoverers of Mines shall swear to render an account of the gold; and in order to discover them and beds of pearl-oysters, a license shall be first obtained.

We command that Miners, and all others who shall collect gold from Mines, rivers, ravines, or any other places, shall appear before the Governor and Royal officers, and take an oath that they will give an account thereof and personally declare it at the smelting: and for discoveries of Mines, and beds of pearl-oysters, they shall obtain a license from the Governor, who shall hold a special meeting (*Junta*) on the subject with the Royal officers, and they shall there agree upon what may be expedient for the proper collection of our Royal revenue.

Law III.

The Emperor Don Carlos, at Zaragoza, March 8th, 1530.

That, of what shall be promised to the discoverer of a Mine, two-thirds shall be paid by the Royal Treasury, and the other third by the parties interested.

Whenever it shall happen that money, or any other reward, shall be promised to Miners who shall discover Mines of gold, silver, quicksilver, or other metal, there shall be paid from our treasury only two-thirds of what was promised, and the other third shall be paid by the persons who shall take out the metal.

Law IV.

Don Philip III, at Madrid, January 19th, 1609.

That the discovery of Mines of quicksilver be secured.

We charge and command the Viceroys, *Audiencias*, and Governors that they take particular care to procure that Mines of quicksilver, of which they shall receive information in any parts of the Indies, be discovered and worked, and that they bestow upon those who shall discover and work them, such advantages as to them may seem, and shall be, just, announcing to them that no allotment (*repartimiento*) of Indias will be made to them for working them.

Law V.

Don Philip IV, at Madrid, June 7th, 1630.

That the ordinances of Mines be observed, and especially that one which directs that those who serve shall register, for their masters, the Mines which they shall discover.

We ordain and command, that the ordinances and particular laws which treat of Mines, shall be observed, complied with, and carried into effect; and that, in carrying them into execution, that one be especially observed, which ordains that those who serve others shall register for their masters the mines which they shall discover, and not in their own names.

Law VI.

Don Philip IV, at Madrid, June 18th, 1629.

That the ordinances be observed in denouncements of Mines, and that the time limited be not extended.

The decline of some *Asientos* of Mines results from the non-observance of our Royal ordinances, and especially those with respect to deserted and abandoned Mines, and in which it was resolved that, the term of four months having passed without their being worked, any persons might denounce them, before the ordinary court, as abandoned (*despobladas*), and the proceedings prescribed in the new code (*cuadernilla*) of Mines being had, they should be adjudged to the denouncer, in order that he might work them, as the real owner, under the conditions which are there declared, so that the Mines should not remain unworked, and new veins be undiscovered. And, whereas, it having been commanded by some of our Royal *Audiencias*, that the ordinances, of Mines, given for this purpose, be observed and carried into effect, the Miners and those interested in abandoned Mines, have gone to the Viceroys or Presidents to ask orders of protection, so that they cannot be denounced, as abandoned, for some length of time, thereby leaving them unoccupied (*despobladas*) and suspending the execution of the ordinances:—We command the Viceroys and Presidents and *Oidores* of our *Audiencias*, to strictly and punctually observe and comply with the said ordinances, and not to extend the prescribed time, which is deemed proper and is our will.

Law VII.

Don Philip III, at San Lorenzo, November 14th, 1603.

That there shall not be wasted, at the Mines, the slag, and refuse stone and earth, and the washings and sweepings.

The refuse matter (*desmontes*) and slag which are taken out by the assays and smeltings, the earthy matters, washings and sweepings, after their owners have derived all benefit from them possible by the machinery in common use, shall be preserved and collected, in order that they may be open for the public good, the benefit of their owners, and the increase of our Royal Treasury.

Law VIII.

Don Philip II, at Madrid, March 5th, 1571, and at Toledo, August 11th, 1596.

That the Asientos of Mines be furnished with provisions, and that no monopoly be allowed.

We command the Viceroys and Courts to cause the towns and *Asientos* of Mines to be furnished with abundance of necessary provisions, and that they be furnished and carried by the native Indians of the vicinity, at fair and moderate prices, and that they shall require and compel the muleteers to carry them, they being paid for the freight, and that they do not permit monopolies of provisions.

Law IX.

Don Philip III, at Aranjuez, August 14th, 1610. Vide Law I, Title XI, Book VIII.

That care be taken of the Mines and their working.

Inasmuch as the discovery, reduction and working of the Mines conduce so much to the prosperity and increase of these kingdoms, and those of the Indies;—we charge and command the Viceroys, Presidents, Governors, and Alcaldes Mayores to take very particular care to observe, and cause to be observed, the orders which have been, or shall be given, with respect to the personal services of the Indians, in the cases where, by the laws of this book, they are allowed.

Law X.

Don Philip III, at El Pardo, November 22d, 1609; Don Carlos II, and the Queen Regent.

That the Viceroys and Presidents shall, in their executive capacity, have cognizance to determine upon the expediency of issuing execution against machines for grinding ores, and the Royal Officers, in judicial suits, with the right of appeal to the Audiencias.

Many inconveniences having been experienced from the renting of the machines for grinding ores, by the practice introduced of Miners contracting many debts to our Royal Treasury, and of the Officers, in coercing the payment to us, being under the necessity of seizing these machines and of afterward, leasing them in order, by this means, to collect the debts:—We declare that if, on the day of payment, our Royal Treasury has to collect any debts, it pertains to the Government and the Administration of the Treasury, to determine whether it be expedient, or inexpedient, that execution be levied on the machines of the Miners. And we ordain that the Royal Officers, before making their attachments and leases, shall communicate it to the Viceroy, or President-Governor of the *Audiencia* of the District, and they shall not be at liberty to proceed in any other way, and that the Viceroy or President shall declare what ought to be observed, as an executive matter; and the result being that execution issue, and levy be made and payment enforced upon the machines, if petitions be presented and defense be made, which are, in law, judicial proceedings on judgments rendered, no recourse or appeal can be had to the Viceroy or President, because, being a judicial matter, it pertains to the *Audiencia*.

Law XI.

Don Philip III, at Madrid, December 22d, 1608.
Don Philip IV, at Madrid, February 12th, 1622.

That the copper of the Mines in Cuba be reduced and transmitted agreeably to this law.

We command that the persons who have under their charge, by our commission, administration, or direction, or in any other form, the Mines of copper in the Island of Cuba, shall cause the ores to be reduced with much care, so that the metal shall become ductile and malleable, by roasting and proper refining, and not so hard and dry as it has heretofore been sent, in order that it may be more suitable for the founderies of Artillery, and that it may be sent to Havana consigned to our Royal officers to be transported to these kingdoms in gallens of armada, and captains and admiral's ships of the line, invoiced and directed to the *casa de contratacion*, all to be duly accounted for to us by the *Junta* of war of the Indies.

Law XII.

Don Philip III, at Ventosilla, October 17th, 1617.

That no one except the owner of Mines can sell the metals.

No Spaniard or Mestizo, who is not owner of Mines, is at liberty to sell, nor shall he sell, any kind of metals, under penalty of forfeiting them, and, for the first offense, also one hundred dollars, all to be applied to our exchequer; and for the second offense, two hundred dollars; and for the third offense, of being forever banished from the Mines and a space of ten leagues around them, and the person who shall purchase them shall incur the same penalty.

Law XIII.

The same Ordinance 14 for personal service, 1601.

That Spaniards, Mestizos, free Negroes and Mulattoes be induced to work in the Mines.

We ordain and command that, in order to secure the reduction and working of the Mines, idle Spaniards who are able to work, Mestizos, free negroes and mulattoes shall be required to work and hire themselves in them, and of this the *Audiencias* and Corregidores shall take particular care, and shall tolerate no idle people in the land.

Law XIV.

The Emperor, Don Carlos, and the Princess Regent, at Madrid, December 17th, 1551. Don Philip II, at Madrid, April 5th, 1563, and March 6th, 1575.

That Indians may hold and work Mines of gold and silver, the same as Spaniards.

We command that no restriction be placed upon the Indians in discovering, holding and occupying Mines of gold, or silver, or other metals, and in working them the same as the Spaniards may do, agreeably to the ordinances of each Province, and that they may extract the metals for their own profit and for the payment of their taxes; and that no Spaniard or *Cacique* shall have part or management of any Mines which the Indians shall have discovered, held or worked.

Law XV.

Don Philip IV, at Madrid, March 28th, 1633; Don Carlos II, and the Queen Regent.

That to Indians who shall discover Mines there shall be secured the rights (preeminencias) *which are declared and granted to Spaniards and Mestizos.*

We ordain and charge the Viceroys, Presidents and Governors, that they exercise particular care and diligence in inquiring and ascertaining if, in their districts, there be any Mines of gold, silver and other metals of which the Indians have, or can obtain, knowledge, and that, after due inquiry and advice they cause to appear before them the Indians of the most reliance, in order that these, personally, and others who may have more skill and intelligence, may give information of the places and positions where it has been understood that there are Mines which are concealed in order that they may not be worked, (although such working really results to their benefit), because they are naturally inclined to idleness; and that they be assured in our name that, for their care and trouble, if successful, there shall be granted to them, and henceforth there are granted to them, many rewards and exemptions; and especially, that henceforth they shall not be assigned to work in any Mines, and that they, and their descendants, shall be forever exempt from personal taxes (*tributo*); and, if they are Spaniards or Mestizos, they shall receive gifts (*Mercedes*) corresponding to their rank.

Law XVI.

Don Philip II, and the Princess Regent, at Valladolid, May 23d, 1559.

That, with respect to staking out the boundaries of Mines, the same rules be observed toward Indians as is observed toward Spaniards.

In some Provinces of the Indies, the practice has been introduced, that, if several Indians discover a vein, one only is selected, who may, as owner, ask the marking out of boundaries of what belongs to him as such; and as we desire that the Indians shall have and enjoy the benefit and profit which may be due from diligence and industry:— We command that, in respect to the staking out the boundaries of Mines which they shall discover, there be observed with respect to them, what is observed with respect to Spaniards, without any difference whatever.

Book IV, Title XX.

CONCERNING MINERS, AND AMALGAMATERS (*AZOGUEROS,*) AND THEIR PRIVILEGES.

Law I.

Don Philip II, at ———, May 18th, 1572; at San Lorenzo, September 12th, 1590; Don Philip III, at Madrid, December 12th, 1619, and March 8th, 1620. Also, Vide Law III, Title XIV, Book V.

That miners shall be favored, and that instruments used in mining shall be exempt from execution.

We order the Viceroys, Presidents, Governors, Alcaldes Mayores of Mines, and Courts of our Indies, that they favor Miners and Amalgamaters, (*azogueros*) and that they observe, and cause to be observed, toward them, all the privileges (*preeminencias*) granted by the Kings, our ancestors, and by us, as having in every respect the force of law; and especially that, for no debts of whatsoever nature they may be, if not due to us, may or shall any execution be issued against the slaves and negroes, the tools, provisions, and other things necessary for the supply, working and provision of the Mines, and the persons who work in them. And we command, that the executions which, conformably to law, may be issued, shall be against the gold or silver which may be or may have been extracted from the Mines, from which the creditors shall be paid in their order and grade, in such manner as not to impede or stop the discovery, use and working of the Mines, and, at the same time, to satisfy the executions.

Law II.

Don Philip III, at Valladolid, November 26th, 1602.

That when Miners are to be imprisoned for debts, it shall be in the Real and Asiento of the Mines.

It is important that the Miners and Amalgamaters should be favored and relieved in every possible way, so that the working of the Mines may not be suspended or stopped. And in order that no inconvenience may result from their absence, it is our will that, when persons are to be imprisoned for any debt, they be imprisoned in the *Asiento* and *Real* of the Mines where they are employed, and that they be not removed from them.

Law III.

Don Philip IV, at Madrid, October 9th, 1635.

That the Miners and Amalgamaters of Potosi be not detained in Lima for debts due the Royal Treasury, they having given security in that town.

It is our will that when any Miners and Amalgamaters of the Imperial Town of Potosi, who are debtors for any amount to our Royal Treasury, shall happen to go to the City of the Kings, and shall give security to present themselves, within the term which is prescribed to them, before the Royal officers of said Imperial Town, they shall not be detained nor molested on that account, nor for any other civil cause, notwithstanding any edicts and ordinances to the contrary.

Law IV.

Don Philip III, at Valladolid, November 26th, 1602.

That the Miners be provided with the necessary materials at fair prices.

In order to benefit the Miners, we command the Viceroys and Governors to favor them, and to cause them to be supplied with corn from our public stores (*tributos*), and all other articles which may be necessary for the supply of their Mines and the reduction of the ores, at fair prices, prohibiting the excessive charges which, in this respect, have been made.

Law V.

Philip III, at Valladolid, November 26th, 1602.

That the law-suits of Miners be dispatched in the Audiencias with brevity.

We charge and command our Royal *Audiencias* that they dispatch, and cause to be dispatched, with great brevity, the causes, law-suits, and business of Miners and Amalgamaters which are pending before them, so that they shall not be perplexed by law-suits, nor subjected to long absences, to the damage and prejudice of the supply of their Mines and *Hacienda*.

Law VI.

Don Philip IV, at Madrid, April 16th, 1635.

That the Miners of the Philipines shall enjoy the privileges granted.

Whereas in the Province of Camarines in the Philipine Islands, distant from the city Manilla more than sixty leagues, there have been discovered Mines of gold of rich specimens, which extend from north to south nine leagues, of which assay has been made by washing and quicksilver, and whereas they have gone on discovering others, and different persons have begun to work and reduce them: it is our will that the Miners of said Islands shall enjoy all the privileges which are directed and established by the laws and ordinances. And we command the governors and captains-general to take particular care that they be observed, and that the Mines be worked and reduced in such manner as may be most expedient for our service, the increase of our Royal Treasury, and the good of our vassals.

Law VII.

Don Philip IV, at Madrid, April 16th, 1635.

That the Miners and Amalgamaters of Potosi may be chosen corregidores and to other public offices.

Notwithstanding the provisions of Laws XVII and XLII, Title II, Book III, we permit Miners and Amalgamaters of the Imperial Town of Potosi to be appointed *corre-*

gidores, and to hold other public and Municipal offices, although they may be debtors to our Royal Treasury to some amount, on account of quicksilver entrusted to them, or in any sum not resulting from the office for which they are candidates, or from another which they hold or the jurisdiction of which does not extend over the place where they are debtors: and we grant to them, that, if they have been *capitulares*, they may vote in the elections of public officers, except when any one may desire to vote in virtue of an office which he may have purchased and not paid for, if the period within which he ought to have paid the price, or a part of it, shall have expired.

Book IV, Title XXI.

CONCERNING ALCALDES MAYORES AND NOTARIES OF MINES.

Law I.

Don Philip III, at Valladolid, November 26th, 1602; at San Lorenzo, September 5th, 1620; Don Philip IV, at Madrid, February 28th, 1637; Don Carlos II, and the Queen Regent.

That the Alcaldes Mayores of Mines shall have the qualifications specified, and shall neither trade nor make contracts.

Inasmuch as it is very expedient that the Alcaldes Mayores of Mines should be persons capable and practiced in working them, and should possess the qualifications required for such offices:—We command the Viceroys and Presidents to whom this matter pertains, that they cause to be selected and appointed persons suitable and fitted for the discharge and performance of the office which they are to administer, and that they do not permit them to trade nor to contract with Miners, under pretext of supplies (*avio*), or any other pretence, nor with any other persons; which we hereby prohibit and forbid. And inasmuch as, it has been claimed that some *corregimientos* of land and extensions be added to them, thus giving them greater jurisdiction and boundaries;—We order the Viceroys and President-Governors to consult on the matter with intelligent persons, and to decide what may be most expedient for our Royal service, the administration of justice, the supply and working of the Mines.

Law II.

Don Philip III, at Madrid, June 9th, 1618.

That the Alcaldes Mayores of Mines shall neither purchase nor exchange silver.

We command the Alcaldes Mayores of Mines that, neither personally nor through the interposition of others, shall they barter or purchase of the Miners, any gold, silver or other metals, either by way of advances or by payment of full price down, nor shall they enter into any such, or other, understandings or contracts with the Miners, under penalty that said Alcaldes Mayores be deprived of their offices and condemned to pay fourfold, and that the Miners be banished at the discretion of the Judge who shall have cognizance of the cause, and also be subject to a fine of the amount of the contract, if they shall not make it known; and if there shall be proof of the contract, one-half of the fine shall go to the Miner who shall make it known.

Law III.

Don Philip II, and the Princess Regent, at Valladolid, May 23d, 1559.

That no Alcalde Mayor, Judge or Notary of Mines, shall enter into partnership with an owner of Mines, or discover them.

We prohibit and forbid all Alcaldes Mayores, Judges, and Notaries of Mines from forming any partnership with any owner of Mines, and from instituting any proceedings for the discovery of Mines, during the term of their offices, either personally or through the interposition of others, under penalty, for said offense, of having forfeited and of forfeiting their offices, and one thousand dollars of gold to our exchequer and treasury.

Law IV.

Don Philip II, at Madrid, January 15th, 1575.

That the salaries of the Alcaldes Mayores and Inspectors of Mines are to be paid out of the profits derived from them.

The salaries which the Alcaldes Mayores and Inspectors of Mines are entitled to receive, shall be assigned and paid to them from the produce of the Mines, and shall be taken from those Mines which are under their administration, and not from our treasury or from any other source.

Book VI, Title III.

OF THE REDUCTIONS AND PUEBLOS OF INDIANS.

Law X.

Don Philip III, at Valladolid, November 24th, 1601; Ordinance 21 of personal service.

That Pueblos of Indians be established near where there may be Mines.

To assign, for the working and reduction of Mines, Indians who are drawn from Pueblos and provinces very distant, will cause damage and injury. And as we wish that this be avoided as much as possible, we charge and command our Viceroys and President-Governors, that they cause to be made and established, in the vicinity of such Mines, selecting suitable and healthy places, towns of Indians, where they may be collected and live in regular Pueblos, and have religious instruction, hospitals and all other necessary things, where the sick may be attended to, and where they may apply themselves more willingly on account of the interest which they will derive from their work; provided, that it shall not be necessary to bring others by assignment from a greater distance. And as the working and preservation of the Mines is a matter of such great importance that, under no circumstance, ought it to be diminished, and as it is proper that it always be increased,—we will and command, that if, while the towns are being established, or after they have been established, there be a deficiency in the number of Indians necessary for each *Asiento*, they be brought from the nearest places, in order that the Mines may be supplied, and the change may not be made from a cold to a hot climate, or the reverse; and in all there must be observed what will be ordered with respect to the *Cerro de Potosi* in Law XVI, Title XVI, of this Book, providing and ordering what may be expedient for its execution and observance, and the good treatment and pay of the Indians.

Book VI, Title XV.

OF THE SERVICE OF INDIANS IN MINES.

Law I.

Don Philip II, at Madrid, January 10th, 1589, cap. 46.

That Indians may be assigned to Mines according to the provisions of this law.

We declare that Indians may be ordered to go to the Mines, provided there be no change of climate to injure their health, they having religious instruction and courts to protect them, provisions to support them, good pay for their labor, and a hospital where those who get sick may be attended to and taken care of, and that the labor be moderate, and there be an inspector to take care that the foregoing be observed: and with respect to the payments for religious instruction and for courts, they shall be at the expenses of the Miners, as the distribution of the Indians is made for their benefit; and they shall also pay the necessary expenses of taking care of the sick.

Law II.

The Emperor Don Carlos, at Inspurg, December 25th, 1551; Don Philip II, at Pardo, December 1st, 1573.

Indians who wish to do so may work in Mines.

We grant to Indians, who may wish to do so and who are paid a just price, the liberty to go to and work in Mines of gold, silver and quicksilver; provided, that no one who is intrusted with their care shall take his own Indians; but we give permission, that the Indians of one trustee may go to work in the Mines of any other of such trustees (*encomenderos*).

Law III.

Philip II, at Madrid, January 24th, 1594; at San Lorenzo, August 26th, 1595; at Madrid, February 22d, 1597; Don Philip III, ordinance 15 ot Personal service, 1601; and December 10th, 1618; Don Philip IV, at Madrid, January 13th, 1627.

That the Indians of Mita and volunteers shall be paid, and the courts shall enforce it, and the quicksilver of the King shall be delivered to Miners at cost.

The wages of the Indians shall be reasonable and proportioned to their work, and to the other circumstances which constitute the just value of things, and there shall be paid to them the road allowance for going and coming, as is provided in Law III Title XII of this Book, computing it at the rate of five leagues per diem, to which matter the Viceroys and President-Governors will give much attention and care, in order to enforce the part which has respect to Miners; and considering the high price it bears, we command that the quicksilver, which may be sold on our account, be delivered to them at its cost and charges, placed in Potosi and in the other *Asientos* of Mines, and there shall be introduced into the pay and wages of the Indians, the equality and justice which is so desirable, even though, from this cause, the gain of Miners, and owners of farms (*chacras*), cattle and sowings, may be diminished; but even if the road allowance, and the increase of day-wages, should so raise the price as to ruin the Mines, farms and stock-raising, at least there should be given to the poor and miserable Indians, in this matter, the equivalent and pay which, within these limits, may be considered practicable: and supposing the Indians of the manufactories (*obrajes*) should be volunteers, the preceding Law II will be observed, and a just price be fixed for their daily wages, without respect or attention to what we say above about the workings (*labores*): and the daily wages which may be so fixed, shall be paid in money in hand every day, or at the end of the week, as they may prefer, with intervention of the court, or of the protector. And inasmuch as, there are none of our officers in some working places which are unpeopled, nor any persons who can act in defense of the Indians, and therefore this proceeding and precaution cannot be observed: we command all courts of the Pueblos, which have cognizance of Indians of Mita and distribution, to take special care to inquire, by means of public proclamations or otherwise, if any Indians, who return from serving their assignment, have not been paid for their work and occupation, and, finding any one to whom a part of his wages is due, to immediately take measures to have it punctually paid. And we command that, to any one who shall violate the provisions of this law, no more Indians be assigned for any purpose: and the Judge who shall have been remiss or negligent in its execution and observance, shall be deprived of his office, and shall pay, from his own property, what may be due and cannot be collected from the debtors. And inasmuch as, it is expedient to remedy the inequality of payment which the Miners have to make for the expenses of the going and returning of the Indians, because some Mines are more distant than others:—We command that they be distributed among all the Miners, according to the amount of what may be reduced and be shown in their account, making the distribution, for this purpose, with perfect equality.

Law IV.

Don Philip II, at Madrid, December 29th, 1593; and January 21st, 1594.

That Indians of Mita shall not be distributed to any one who is not an owner of Mines, machines and workings.

In many provinces of the Indies, there are made distributions of Indians of *Mita*, for Mines and other works, to persons who do not have such works, they procuring this grant from the Governors and courts by favors and other illicit means, in order to enrich themselves by the large sums paid by the owners of machines, Mines and works, for the labor of the Indians; and whereas this is an oppressive species of servitude to the Indians, and a practice equally injurious to the owners of Mines, machines, and works, which it is in no manner expedient to permit: we command the Viceroys, Presidents, and Governors, and all other courts which it may concern, to neither consent nor permit Indians of *Mita*, destined for this service, to be distributed to persons who are not owners of Mines, machines and works, and who do not, with their own capital, work Mines or grind ores; and in no case shall they be assigned or distributed to others, nor to those who are in partnership with the owners of machines or Mines, unless it be shown that they really have an interest therein, so that in no case, way, or manner, shall this pass through the hands of a third person; and the distribution shall be made equally, according to the quality of the *Haciendas* of each, under penalty that the Judges and distributors incur the loss of their offices, which the Viceroys, Presidents and Audiencias will irremissibly execute. And those who shall sell the labor of the Indians, and shall not use them for the purpose for which they were assigned to them, shall forfeit all their property and be banished from the Indies, and it shall be thus executed.

Law V.

Don Philip III, ordinance 18th of personal service, 1601.

That Indians shall be distributed to owners and lessees of Mines, and shall not be employed in other occupations.

To any one who shall not have Mines of his own in the

Cerro of Potosi, or in some other place, and who shall not actually work them on his own account, Indians of no quality or condition shall be assigned; but we permit that, to those who lease Mines, whether belonging to us or to any other persons or communities, Indians may be given, the same as to owners of Mines, having consideration and respect to their condition and produce, for the time that the lease, working, or reduction, is to continue. Also we command that, to those who shall have, and shall work, Mines of their own, or Mines under lease, no more Indians shall be given or distributed than shall be requisite and necessary, according to the quantity and quality of the Mines, which they shall have, work, and actually reduce, so that they shall employ them in the working and reduction thereof, and not for any other purpose or business, and if they shall do so, the Indians shall be immediately taken from them, and shall not be again given to them.

Law VI.

Don Philip III, December 10th, 1618.

That the Indians, who shall be assigned to Mines, shall not be required to supply or pay for absentees, refugees, and the dead.

Because of the oppression and injustice of charging the Indians of *Mita* with the obligations and payment of the absentees, refugees and dead, and of the expediency of remedying it: we command that, in no case, it be permitted that, under title of service or any other pretense, they be charged for the absent, escaped, or dead:

and that, the time and obligation of their service having expired, they may return, and shall return, freely and without impediment, to the place from whence they were taken.

Law VII.

Philip III, at Madrid, July 15th, 1620.

That proceedings be had against the Miners who shall receive money from the Indians of Mita, to excuse them from work.

Many Indians assigned to work Mines fail to work in them, because the Miners to whom they are assigned relieve them, and collect every week a certain sum of money from each Indian, saying that, with this money, they hire others: and although it be true that some do this, yet in general they keep the money and no one works; by which violation of conscience and justice our Royal fifths are diminished: we command the Viceroys and *Audiencias* to proceed against those who are thus delinquent: and not only shall they be deprived of the Indians, but they shall be condemned to the corporal and pecuniary punishments which shall appear just. And we command that it be a cause of accusation (*capitulo de residencia*) against the *Corregidores* of Potosi and other *Asientos* and *Reales* of Mines, if they cloak or consent to such excess; and the *Fiscales* of our Royal *Audiencias* will appear in these causes and demand what is expedient against the guilty parties.

Law VIII.

Don Philip III, at Aranjuez, May 26th, 1609, capitulo 21.

That Indians be not assigned to poor Mines, and only to those persons who have Mines or machines.

In the distribution to Mines, particular attention shall be given to the amount and quantity of the ores, and to their value and yield, in order that Indians be not distributed to poor Mines and to those of little utility, and that they be distributed only to those which afford occupation to each Miner in the business thereof: and in no case shall an assignment be made to persons who shall desire to sell them to the owners of Mines and of machines for grinding ores, nor shall Indians be given, except to those who actually, and on their own account, work machines or Mines which they own or hold by lease, and the same shall be observed with respect to other *Haciendas*.

Law IX.

Philip III, ordinance 15 of personal service, 1601; at Aranjuez, April 20th, 1608.

That the Indians and laborers in Mines be paid punctually every Saturday afternoon.

We command that all the Indians of *Mita*, and volunteers, and other persons who, agreeably to what is provided, shall work in Mines, be paid reasonable wages, in proportion to their work and occupation, every Saturday in the afternoon, into their own hands, so that they may take recreation and rest on Sunday or any other day, as they may wish: and the officers shall take particular care of their health, and good treatment in spiritual and temporal matters, and that the sick be well cared for.

Law X.

The Emperor Don Carlos, at Toledo, December 4th, 1528; Don Carlos II, and the Queen Regent.

That Indians and slaves in the Mines receive religious instruction.

The Indians and slaves who work in the Mines, shall have *clerigos* or *religiosos* to administer to them the holy sacraments, and to teach them the christian doctrine, and the parties interested in the Mines shall pay the stipend: and the diocesan prelate, observing the *patronazgo* in making the nominations and appointments, will cause mass to be said and doctrine to be taught on Sundays and feast-days.

Law XI.

Don Philip III, at Aranjuez, May 26th, 1609, capitulo 18.

Mines shall not be worked in unhealthy places, and the Indians shall be induced to engage, of their own free will, in certain occupations.

Mines shall not be worked in places which are dangerous

to the health and lives of the Indians; and those who are engaged in reducing with quicksilver, shall be so distributed in the different employments, as to equally participate in those which are more or less onerous, in which employments they should be induced to engage voluntarily, by giving to them the privilege of exemptions and other corresponding advantages; and in case these motives shall be insufficient to incline and attract them to the particular work and employment, then the requisite number of Indians will be allotted, according to what is provided, and the wages will be increased to such a price that, besides what may be requisite for their daily support, they may obtain out of their earnings sufficient to pay the tributes (*tributos*) to their *encomenderos*, if they have not already earned more by their work than will be sufficient for that purpose.

Law XII.

Don Philip III, ordinance 23 and 26 of personal service.

That Mines be not drained by Indians, even though they be volunteers.

The labor which the Indians endure in the draining of Mines is very severe, and by continuing in it they are made sick: and whereas it is our will that they be relieved as much as possible; we ordain that draining be not done by the Indians, even though they may wish to do it of their own will, but by negroes or other classes of people, and we so charge the Viceroys, and we command that they take special care to provide and order that it be enforced and observed as far as possible, and as may conduce to the greater benefit, security and alleviation, and to the less vexation, of the Indians; provided, that for this reason the working of the Mines shall not cease.

(NOTE BY TRANSLATOR.—The other laws of this title to law XXI have respect to the service of Indians in particular Mines, and are therefore of little general interest.)

Law XXI.

Don Philip III, in ordinance of personal service, 1601.

That Indians be settled near Mines of quicksilver, and that they be favored.

It has been found by experience that it is not possible to reduce silver ores without quicksilver, and that it is expedient to prosecute and continue the working and reduction of these Mines; and inasmuch as this cannot be done without the industry and labor of the Indians; we command the Viceroys, Presidents, and Governors to induce them to reside near these Mines, in order that, when necessary, a distribution of them may be made, and if possible, that Indians be not brought from other places, the work being distributed so that it shall be tolerable, and being divided equally among all, so that the same persons shall not be always occupied in extracting the metal. And we ordain, that in the liberty, good treatment and pay of the Indians who shall work in the Mines and reduction of quicksilver, the same rules be observed as in all the others.

Book VIII, Title V.

OF NOTARIES OF MINES AND REGISTRIES.

Law I.

Don Carlos II and the Queen Regent.

That the Notaries of Mines and Registries be examined.

We ordain and command, that the Notaries of Mines and Registries be examined by the *Audiencias* of their districts, before entering upon and exercising their offices, and that they have the general qualifications prescribed for other Notaries in Law III, Title VIII, Book V.

Law II.

[Don Philip II, at Toledo, March 10th, 1561; at the Escorial, July 9th, 1565; at Madrid, May 29th, 1594; Don Philip IV, at the same place, June 1st, 1623.]

That the Notaries of Registries assist at the auctions, collection of fifths, and smeltings.

In some parts of the Indies the Notary of Registries does not personally assist at the auctions (*almonedas*), collection of fifths (*quintos*), and smeltings of gold, or the introduction of silver into the exchequers, or inspect the weighings, but there is substituted in his place an assistant who is not a Royal Notary, from which there may result inconveniences and nullities: we therefore command that the principles (*proprietarios*) assist in person at all the acts above mentioned, under penalty of losing their offices, except in case of sickness or very necessary cause, in which cases we permit any one to substitute another who is a Royal Notary.

Law III.

Don Philip II, at the Escorial, July 9th, 1565.

Instruction for the Chief Notaries of Mines and Registries.

In the creation of the office of Chief Notary (*Escribano Mayor*) of Mines and Registries, an instruction is given by the Emperor Don Carlos, May 4th, 1534, ratified by Don Philip II, our predecessors, the 9th of July, 1565, with different heads, for the use and exercise of that office, which, it is our will, shall be observed by all those who, in the provinces of the Indies, fill and exercise the office; it is as follows:

First, to the Chief Notaries of Mines, of Registries, and of Royal Treasury, there shall be given, by our officers, an account of all the *Haciendas*, revenues, houses, cattle, and other agricultural property (*grangerias*), which we may have in the province and territory, and of all else which may belong to us, or which may have been, by custom, applied to our Royal possession, in order that they may have an account of their capital and produce, and what addition and increase may be given to our Royal Treasury.

There shall be given to them, and they shall have, a statement of all the gifts, offices and salaries to be paid from our Royal exchequer, at which payments they shall assist with the rolls which our *contadores* shall have of the drafts, and with all other special provisions, for payments from the Royal exchequer, in order that they have an account and record of everything.

They must keep a book and record of the names of the persons to whom licenses are given to collect gold and silver and other metals, with the oath, and the day, month and year they were given, in order that they may have record and knowledge of the persons who ought to come and give an account and report of their license, and of the gold, silver and metals, which, in virtue of such license, they shall have collected, with a statement thereof, and they shall show these to the Governor and Royal officers, in order that they may take the proper measures to have them looked after or punished, conformably to justice and to what is commanded in Law II, Title XIX, Book IV.

The Notaries of Mines and Royal Treasury will reside in the smelting houses and refineries, as well to take an account and make a record of the permits (*cedulas*) which shall be issued for the extraction of gold and silver or other metals, as to keep a book in which shall be entered the metals brought to be smelted, and the persons who bring them, and by what means they were collected, and the part paid to us, and how it is turned over to the treasurer; and at the end of every smelting, our officers will compare their books, and will sign them with their names.

If pearls and precious stones shall have to be assessed in order to collect the fifth which belongs to us, the Notaries of Mines and Royal Treasury will be called, and they shall be present and take account and record of what the treasurer shall receive therefrom, and when the days of the week shall have been designated on which the fifths are to be paid, the Notary shall be informed what those days are, so that, without being called, it shall be his duty to go and be present at the paying of the fifths, and at their delivery to the treasurer; and this may be done on the days assigned and on no others; and if, from any necessity, it shall be done on other unusual days, the Notary shall be called, and he shall sign with his name the charge which is thus made to the treasurer, both in the book of the Notary and in that of the *contador*, the one being compared with the other; and, as it is thus done in all private matters, it is right that it be observed in our Royal Treasury for its proper collection, account and record.

When any gold or silver shall come from abroad to be delivered to, and taken charge of by the treasurer, it shall be delivered in the smelting house on the days appointed and not on others; and if it should be necessary to do it on another day, the Notary of our treasury will be called, and will make record of it, and the treasurer will sign it in his book, as is directed.

(NOTE BY TRANSLATOR.—The remainder of this law, and the other laws of this title are omitted, as having but little relation to matters of Mining.)

Book VIII, Title XI.

CONCERNING THE ADMINISTRATION OF MINES, AND THE REMISSION OF COPPER TO THESE KINGDOMS, AND MINES OF SULPHUR.

Law I.

Don Philip II, in Instructions to Viceroys, 1595, and 1596; and Don Philip IV, in Instruction to Viceroys, 1628.

The Viceroys and others directed to encourage the discovery and working of Mines.

We order and command the Viceroys, Presidents, and Governors to attach great importance, and give special attention to the working and reduction of Mines already discovered, and to apply all their diligence to the searching for, discovering and working of other new ones; because the richness and abundance of silver and gold is the main nerve from which springs the wealth of both those and these kingdoms; observing, with respect to personal services, Law IX, Title XIX, Book IV, and all other prohibitions.

Law II.

Don Philip II, at Madrid, May 26th, 1573; at El Pardo, October 17th, 1575; Don Philip III, at Madrid, February 6th, 1613.

That the Mines of the King may be worked, leased, or sold, as may be most expedient.

We grant to the Viceroys and Prætorial Presidents

power and authority, if they consider that any of our Mines of silver, gold or quicksilver, discovered in their districts, cannot be conveniently worked on our account, and that it will be more advantageous expedient and profitable to lease or sell them, to so lease or sell them, as may be most advantageous and beneficial to our Royal treasury. And inasmuch as there are other Mines which belong to us and are not worked, although very rich, and from which, if they were leased or sold, we might derive profit, and as it will be well that some suitable means be adopted for this purpose:—We command the Viceroys and Presidents, that, having informed themselves of the quality and richness of each one, they proceed to work, lease or sell them, as may best promote the increase of our Royal Treasury, and that they render an account of the whole to the Council of the Indies.

Law III.

Don Philip IV, at Madrid, August 10th, 1628.

That the Royal officers of Terra Firma compel the masters of the armada to transport the copper which may be delivered to them.

The officers of our Royal treasury of the Province of *Terra Firma* will give the proper orders for the masters of galleons to receive the copper which may be delivered to them, and to transport it, signing the bills of lading, and the officers will compel them to do so, with the utmost rigor. And we order the Captain General of the said *armada* to place no impediment in their way, but, on the contrary, to give them all the aid and assistance which may be necessary for the execution of this order.

Law IV.

Don Philip III, at Madrid, April 14th, 1609.

That the copper which is brought from Havana and other parts shall not be disposed of without the order of the Junta of war of the Indies.

The copper from the Mines of Santiago de Cuba is brought to these kingdoms for casting the necessary artillery to arm the forts of the Indies, and the galleons and vessels which are built to guard their trade and coasts. And as it is thus deemed expedient, we command our Captain General of Artillery of Spain, to in no manner dispose of, for any other purpose than our Royal service, the copper from those Mines, and from the city of San Cristobal de la Havana, and from other parts of the Indies, which has been or shall be brought to the *casa de contratacion* of Seville, without the order of the *Junta* of war of the Indies, which, it is our will, shall have the direction of everything relating to this matter.

Law V.

Don Philip II, 1571.

That Mines of sulphur be taken for the King, and that some of them be worked for munitions.

We order that the Mines of sulphur of all the Provinces of the Indies be taken for us, and that our officers manage them; and that, without our express license or that of the Governor, no one shall take anything from them, and those which shall be, or appear to be, necessary for munitions, shall be worked and reduced.

Book VIII, Title XXIII.

OF ESTANCOS (GOVERNMENT MONOPOLIES).

Law I.

The Princess Regent, at Valladolid, March 4th, 1559; Don Philip II, at Aranjuez, May 8th, 1572; at Madrid, May 26th, 1573, and April 27th, 1574, and May 8th, 1577; Don Philip IV, at Madrid, February 28th, 1637. Vide also Law LXII, Title VI, Book IX.

Quicksilver is not to be carried to the Indies, or made an article of trade there, except on the account of the King, and its re-sale is prohibited.

We ordain and command that no person, of whatsoever rank and condition, shall be permitted to carry, from these kingdoms to the Indies, or from those of Peru to New Spain, or from New Spain to Peru, any quicksilver, even in small quantities, publicly or secretly, or to receive it in the Indies, or in the provinces, parts, or ports thereof, except on account of our Treasury, under penalty of its being forfeited, with double its value, which shall be applied, one-third to the informer, and two-thirds to our exchequer and treasury, and the same penalty shall be incurred by the merchant or person who shall purchase it in said kingdoms and provinces, in order to sell it again, even though it be what has been divided out and distributed on our account: and the same shall be observed with respect to the quicksilver which may be taken from Peru to Guatemala and Honduras, and which the Viceroy of New Spain may transmit to New Galicia, and to all other parts where Mines of silver are worked and it is necessary to use this metal. And forasmuch as it is understood that there is great abuse resulting from the Miners re-selling the quicksilver, which is sent on our account and distributed among them for the supply of their Mines:—We command the Viceroys, Presidents, Governors, and courts to proceed to the investigation and punishment of these offenses according to law, considering the quicksilver, with double the amount, as forfeited, applying it as aforesaid, and imposing such other penalties as the offense may seem to deserve.

Law II.

Don Philip II, at Madrid, May 25th, 1573.

That the Royal officers receive and deliver quicksilver in conformity with this law.

The Royal officers of the ports of the Indies, by delivering the quicksilver, which, on our account, they shall receive, to other officers of the places where it shall have to be delivered, comply with their obligations and are free from responsibility; and, consequently, if these latter shall have it delivered to others, where it is ordered to be sent and consigned, they also are free from responsibility, each taking the proper receipts from the other. And those to whom the last delivery is made, having taken the charge thereof, the preceding officers in charge thereof will be credited, in the items of their accounts, with the receipts for the deliveries, as aforesaid.

Law III.

Don Philip II, at Toledo, August 11th, 1596.

That the transmission of quicksilver from Guancavelica to Potosi, be made by the Royal officers under the direction of the Viceroy.

The carriage and transmission of quicksilver which shall have to be taken from Guancavelica to Potosi, must be on our account, by us and through our Royal officers, these, and the Viceroy of Peru, taking great care, that the officers of Guancavelica send the quicksilver to those of Chincha, at a time most favorable, and with the proper security and care, and that those of Chincha send it to those of Arica, and these latter to those of Potosi, causing all to do this, as a thing of the greatest importance: and the same we command to those of Guancavelica and Potosi, and the courts of Chincha and Arica, and that the Viceroy do not overlook any negligence or omission of the above, and that he punish with severity and example, the faults which may be proved.

Law IV.

Don Philip III, at Barcelona, June 13th, 1599.

That quicksilver be delivered pure, in good condition, and to responsible persons.

The quicksilver, which shall be received on our account at the Mines thereof, must be pure and in good order, and that which shall be for transportation to the Indies, and from one province to another, shall be delivered to responsible persons, who will act without fraud, and with perfect fidelity.

Law V.

Don Philip IV, at Madrid, August 22d, 1629.

That the officers of Vizcaya have the Administration of quicksilver.

The Viceroys of New Spain will leave the administration and distribution of the quicksilver, which is taken to the province of Nueva Vizcaya for distribution among the

Miners, to the officers of our Royal Treasury, who will administer and distribute it.

Law VI.

Don Philip II, at Aranjuez, May 31st, 1579.

That quicksilver be put up and sent in flasks, containing a quintal each and no more.

We command that the quicksilver, which is sent from these kingdoms to the Indies, and from one province to another, be put up in such a manner that each vessel shall weigh only one quintal, and with these shall be sent the directions (*badanas*) necessary for working it.

Law VII.

Don Philip IV, at Madrid, April 29th, 1639.

That the Royal officers immediately dispatch the quicksilver, and send it to the places to which it is consigned.

The flasks of quicksilver reach the Indies greatly diminished, on account of its bad storage and the injuries received from dampness. And to remedy this, we command our officers, into whose charge it may come, to immediately, and without any delay, send it to the place of its consignment, and during the time it may be necessarily detained, it must be kept in a dry place and without exposure to moisture, its care being preferred to that of any other kind of cargo or merchandise: and as some may arrive requiring repairs, the Presidents and official Judges of the *casa de contratacion* of Seville will send, with each package of quicksilver, the directions (*badanas*) for prevention, which have resulted from experience.

Law VIII.

Don Philip III, at Ventosilla, October 17th, 1617; Don Philip IV, at Madrid, July 13th, 1627; Vide also note at the end of this Title.

The price at which quicksilver is to be delivered in New Spain and New Granada.

To the Miners of New Spain there will be charged, for quicksilver placed in the city of Mexico, at the rate of sixty ducats per quintal, a price now regarded as moderate, considering the very great cost of freight, leakage, risks and other expenses incurred in placing it in that city: and to the Miners of the New Kingdom of Granada there will be charged, per quintal, eighty ducats, exclusive of the three dollars fees of the Alcaldes of Mines for opening (*de laxas*), which is the price which it costs to place it in said Mines.

Law IX.

Don Philip III, at Madrid, July 12th, 1616; Don Philip IV, at Madrid, June 15th, 1622; at Seville, March 10th, 1624; at Madrid, June 20th, 1626, and March 7th, 1630.

That quicksilver be delivered in Honduras at the same price as in New Spain.

For the quicksilver which may be delivered, by the officers of our Royal Treasury of the province of Honduras, to the miners thereof, for the reduction of their metals; it is our will that, at present there be charged sixty ducats per quintal, which is the same price, at which it is delivered to the Miners of New Spain.

Law X.

Don Philip II, at Aranjuez, May 18th, 1572; at Madrid, March 26th, 1577.

That the quicksilver which is distributed among the Miners, be one-half for cash, and one-half on credit.

All the quicksilver which, on our account, is taken to New Spain shall be deposited in our store-houses, and, a list being made of all the Miners of that Government and of New Galicia, it shall be delivered to them, one-half for cash, and the other half on credit, so that one-half of its proceeds may be brought to these kingdoms in the first fleet in which it may be carried out; and the other half in the second fleet; with good bonds and security: and that which shall be distributed in Peru shall be delivered in the same manner, the half for cash and the other on credit, with the shortest times, which shall not exceed, nor be limited by, the time that the workings of the Mines may cease.

Law XI.

Don Philip III, at Aranda, August 14th, 1610.

That great care be taken to collect the dues from quicksilver.

The Viceroys, Presidents and Governors will take great care in the distribution and advance of quicksilver, and that collection be made, with the greatest punctuality possible, of what the Miners may owe, as well for past advances as for those that may be made hereafter, of which a very particular report shall be made by the Council of the Indies, with an account of what shall be ordered, to give effect to the aforesaid.

Law XII.

Don Philip II, at Aranjuez, November 19th, 1589.

That reports be made of the quicksilver provided for the Mines, and of the silver which is produced.

The Viceroys and President-Governors will remit to us a very detailed account, drawn from the results of continuous years of all the fleets and galleons, of the quicksilver which has been provided for each *Asiento* of Mines and its products; and also of the silver which is generally taken out, and of that which belongs to our Royal fifths,—all with great clearness and in duplicate.

NOTE.—By cedula of September 7th, 1679, it is ordered that in New Spain quicksilver be delivered to the Miners at the price of sixty ducats per quintal, and that the distribution be made by the Viceroys, notwithstanding the *cedulas* of August 12th, 1675, and June 18th, 1678, which gave different directions, and which are revoked and annulled.

February 6th, 1705.

[From "Pandectas Hispano-Mexicanas," No. 586.]

Prohibiting clerigos and Religiosos from acquiring interest in Mines in New Spain.

The King.——Whereas, by *cedula* of the 14th of March, 1703, I commanded the Duke of Albuquerque, my present Viceroy of the Provinces of New Spain, to report to me on the excesses committed by the Administrators of the Mines of Guanajuata, and on their having taken the Mine of Rayas from Don Lorenzo Cano and having given it to Doctor Don Juan Diaz de Bracamonte, Presbyter, in compliance with which, he gave an account of what had been done, and stated that in that kingdom they had not enforced the prohibition of the law which prescribes that *clerigos* cannot hold Mines, since in many *Reales* there are *clerigos* who are Miners, either because of their having acquired them by patrimony, or because necessity has caused the matter to be overlooked. And having seen, in my Royal Council of the Indies, all the papers on the subject, and heard on it my *Fiscal*, and having in view Law IV, Title XII, Book I of the Recopilacion de Indias, which prohibits *Religiosos y clerigos* from working Mines; and considering this to be as grave an offense as it is declared to be in the law, I have resolved to order and command, as by these presents I do order and command, my Viceroys of New Spain and Peru, and I require and charge, the Archbishops and Bishops of the two kingdoms, to observe what is contained in Law IV, Title XII, Book I, above cited, and to cause it to be observed in the form and precisely as it is expressed; the said Viceroys and ecclesiastic prelates, of both kingdoms and provinces, acting together, each in his own place, for the better and effectual remedy of these violations; and that, in compliance with the said law, neither the one nor the other permit, in their provinces and dioceses, such relaxation, but that they watch, with due punctuality and vigilance, over the precise and punctual observance of the provision of the said law, without permitting, consenting to, or allowing it, at any time, to be contravened in its contents, on account of the grave injury which results therefrom to my Royal Treasury and to the better administration and preservation of the Mines. And of the receipt of this despatch and of the form in which each one shall fulfill it in their provinces, dioceses and jurisdictions, they will give me a very particular account on the first opportunities which may offer, for the reason that it pleases me to be fully informed of its effectual observance.

Given at Madrid, February 6th, 1705.

I, THE KING.

By command of the King our Lord.

DON JUAN DE APERREGUI.

July 5th, 1718.

Cedula, dated at San Lorenzo, July 5th, 1718.

[From the comentarios de Gamboa, cap. II § 40.]

Order to shut up and prevent the working of the quicksilver Mines of Quernavaca.

To my Viceroy, the President and Oidores of my *Audiencia* of the city of Mexico, in the Provinces of New Spain.

By a despatch of this date I have been pleased to charge and order Don Juan Joseph de Veitia to close and prevent the working of the Mines or veins of quicksilver discovered in the jurisdiction of the Villa of Quernavaca; in which matter and in all that concerns the business of the

15

administration of quicksilver in that kingdom, which I have placed in his charge, he is to proceed and act exclusively of you, my Viceroy, and of you, my Oidores of that *Audiencia*, without your interfering with, or taking cognizance of this matter or of any part thereof, under any pretext whatever: of which I have desired to warn you and to order you, as by these presents I do, that you so observe it, without interposing any hinderance or impediment; forewarning you, that if you act otherwise, it will be to my displeasure, and I will take the most severe measures against you; and that, on the contrary, you give the favor and assistance which he may ask of you, and which may be necessary for the execution and observance of the aforesaid; and of the receipt and execution of this order you will give me an account, through the superintendency general of quicksilver, on the first opportunity that may offer.

Dated at San Lorenzo, July 5th, 1718, and countersigned by Don Andres de Corobarrutia y Zupide.

April 3d, 1747.

Don Ferdinand VI, by Decree of April 3d, 1747.

[From Book IX, Title I, Law VII, Novisima Recopilacion.]

Annexation of matters of Mines to the Junta General of trade and coin.

Considering that the matters of Mines of different metals, which are found in these kingdoms, are very proper and suitable for the institution of the *Junta* of trade and coin, where more attention, than in the other tribunals, may be given to the quality of the metals, and to the qualifications of the assayers who have to report thereon, according to the laws; I have resolved, to commit to this *Junta*, cognizance of all matters respecting Mines and their incidents, to the exclusion of all other tribunals and Judges, and consequently I order, that the Council of Treasury and the *Junta* of Mines of Guadalcanal shall not, in future, take cognizance of these matters, and that they transfer to the said *Junta*, all *expedientes* and papers which they may have pertaining thereto.*

December 19th, 1754.

Don Ferdinand VI, at Buen-Retiro, December 19th, 1754.

[From Book IX, Title XVIII, Law V, Novisima Recopilacion.]

Exclusive jurisdiction of the Superintendent of the Mines of Almaden within the circuit of ten leagues.

The preservation of the Mines of Almaden being of such great importance, and desiring that the measures which

* The following is appended as a note to this law in the Novisima Recopilacion:

"By Royal order of October 15th 1788 communicated to the Council, because of the representation made to the King by the Junta General of trade, coin and Mines, that the Royal Audiencia of Aragon had refused to comply with two of its despatches for the transmission of the proceedings (*Autos*) in which it was engaged respecting Mines of Iron in the Jurisdiction of the Villa of Bielsa; His Majesty resolved that the Audiencia should transmit the said *autos* with the proceedings had; directing that in future it abstain from taking cognizance of matters relating to Mines and belonging to the Junta by its institution."

have been taken to render them more productive, may have due effect, without those embarrassments caused by the repeated conflicts (*competencias*) which, on different occasions, have arisen between the Superintendent General of Mines, and the Sub-delegates of the Royal *Cabaña* and the wagoners, the Courts of the Pueblos included in the pasture grounds assigned for their use, the *Comendadores* and owners of these; I have resolved to establish as a general point and rule, that, within the ten leagues around, counted from the four which are considered as the mouth of Mines, enclosures and earth-heaps, the Superintendent shall have exclusive jurisdiction, with respect to pastures for the oxen employed in their works, and also for the cutting of timber and wood necessary for working them; and that with respect to the said jurisdiction there cannot be claimed any right of cognizance (*competencia*) by the said Sub-delegates and other persons aforesaid.

May 30th, 1778.

[From Note to Preamble of Mexican "Ordenanzas de Mineria."]

Royal approval of the formal erection of the Tribunal and Body of Mining.

With your official letter of yesterday, I have received the Ordinances formed for the regulation and government ot Mining of this kingdom, in consequence of what was ordered by His Majesty and directed in those orders which, for its execution, I have issued to you.

By Royal order of the 29th of December last, His Majesty was pleased to approve the formal erection of the Tribunal and Body of Mining, published by *bando* of the 11th of August last; and by another, of the 20th of January of this year, his Royal will is pleased to repeat strictly the direction therein contained, for the conclusion and formation of the ordinances; and inasmuch as you in this part have complied with said direction, I pass to you a certified copy of said Royal orders, in order that you may know what His Majesty determined, and that the Tribunal may have the evidence which corresponds.

God preserve you many years.

Mexico May 30th 1778.

R. B°. Fr. Don Antonio Bucareli y Ursua.

May 20th, 1780.

Don Carlos III, by res. in council, May 20th, and cedula of the Junta de Comercio, August 15th, 1780.

[From Book IX, Title XX, Law I, Novisima Recopilacion.]

On the working of Mines of stone-coal; and the grant of privileges and powers, for 20 years, to encourage it.

Considering the abundance of Mines of mineral coal (*carbon de piedra*) in these dominions, and the great advantages which may result to my subjects from working them, on account of the scarcity of forests, and the increased consumption of firewood, which daily takes place in manufactories and Pueblos which are continually increasing, coal, for this reason, being reckoned among articles of prime necessity; and desiring the encouragement and extension of these useful establishments, I have granted, generally (*por punto general,*) and for the term of twenty years, as well to those interested in the mineral coal of Villanueva del Rio, as to all others of my subjects, who desire to work other Mines of this kind, in other Pueblos and provinces of these kingdoms, the following favors and franchises:

Art. 1. That no person shall impede or prevent the said parties interested in the coal Mines of Villanueva del Rio, or any others of my subjects who may desire to devote themselves to the discovery of this class of Mines, from making the examinations, works, and pits, which they may find expedient for the use and enjoyment of those which they may meet with, agreeably to the laws and ordinances of Mines; without any other difference, than that they shall not be subject to the duty of one-fifth, tenth, thirtieth, or any other, which is usually required by the Royal Treasury, in metalic Mines.

Art. 2. If the said parties interested in the coal Mine of Villanueva del Rio, or any others of my subjects, shall, by that means, discover any Mines of metal, they may immediately denounce them; and, this proceeding being carried on, the corresponding *cedula* shall be issued to them for their use and enjoyment, agreeably to the ordinances of Mines.

Art. 3. It being indispensable for the service of these Mines, in their works and in their transportation by land and water, to keep a large number of oxen, beasts of burthen, and vessels; it is my will, that said cattle be permitted to pasture, without any restriction, on the two leagues around the Mines, as working cattle; respecting the pastures (*dehesas*), inclosures, (*cotos*), and sowing lands (*sembrados*), if there be such, like the other citizens of the Pueblos; as is granted and ordered in Article 50 of Law IV, Title XVIII; the said cattle and vessels being also exempt from all executions (*embargo*) and encumbrances (*gravamenes*), as is prescribed in Article 7, Law IV, Title XIII, Book VI, of the Recopilacion, and Article 83 of said Law IV, Title XVIII, on account of the damage which, from the contrary, might result by the suspension of the workings of said Mines.

Arts. 4 and 5. (The original text of these two Articles is omitted from every copy of this Edict in the possession of the Translator. The substance of them is stated in a note of the Novisima Recopilacion as follows: "In the omitted Articles, 4 and 5, there is granted, for the period of twenty years, to stone-coal, from the Mines worked, exemption from all the duties of export and import, in the ports of the kingdom, except those of general revenue (*Rentas Generales,*) on what is carried out of the kingdom; and exemption from *Alcabala* and municipal and special duties on the first sales made by persons interested in their works (*fabricas*); and it is ordered that the Judges of Revenue (*Jueces de Rentas*) do not impede the transportation and utensils for the use of said Mines, nor charge them new duty.")

Art. 6. That the amount of gunpowder and sulphur which is required for the operations of said Mines, and the salt necessary for all the operatives, shall be delivered to the parties interested, for the said term of twenty years, in the respective capitals of their district, on the certificate of the Sub-delegate who may be nearest thereto, or of the person who is appointed by my general *Junta* of commerce, at its cost to the Royal Treasury; the said parties interested being permitted to use, in the Mines and in the store-houses which they may establish in the cities, *villas* and places, the Shield of the Royal Arms, with the inscription *Reales Minas*, and the name of the Pueblo of the territory where they may be established: and also, the parties interested may have, for the custody of their Mines and territories, the number of retired soldiers which may be necessary, these being placed at the orders of the director of the Mines.

Art. 7. It being indispensable, for the working and preservation of these Mines, to have the timbers necessary for their operations, and for the establishment of dwellings and store-houses; the parties interested shall have authority to mark in the forests on Royal, Señorial, and vacant lands, whatever trees they may require, excepting those intended for the use of the Royal Navy: it being understood that this authority is to be limited by that which the contractors have of this kind.

Art. 8. No private porter, who leaves the Mines or goes to them with a sealed letter, shall be molested by the officers who watch over this kind of duty, provided that they have a seal with the inscription of the Mine, in the terms which have been mentioned in the sixth Article.

Art. 9. The interested parties may have, in any city, *villa* or place of these kingdoms, store-houses of coal, in order to supply the consumption, and so that those who require it may never be without it.

Art. 10. When the first discoverers and workers of these Mines find it expedient to admit into their partnership any other persons interested, they may do so; any one of them having also the right to cede, sell, or alienate the share which he may have in the partnership, to any person or persons he may please, either during his life or at his death.

Art. 11. For the maintenance and preservation of these privileges and franchises, which I grant in favor of those who dedicate themselves to the discovery and working of the said Mines of mineral coal, those persons shall be appointed for Conservator Judges (*Jueces Conservadores*) in first instance, who shall be nominated to the Minister or proposed to my said *Junta*, the corresponding commission being issued to them by the *Junta*; with the provision, that they shall only take cognizance of matters touching the preservation of the establishment, and other matters of good order, without meddling, in private matters or contracts, of persons in charge of the working of the Mines, or of the other persons employed in them, with the other prohibitions which shall be included in said commission; and in second instance, no Tribunal, *Audiencia* or *Chancilleria*, shall have jurisdiction, except my said *Junta*, which shall have, of these matters, the exclusive jurisdiction (in second instance); without its being necessary for this reason, that he shall be a titled Notary who draws up the proceedings and decrees of the said office of Judge-Conservator, it being my will that the parties interested may avail themselves of the Notary whom they may deem most satisfactory and suitable; the Judge charging, as fees, the same as is, by Royal tariff, fixed for those of general jurisdiction (*los ordinarios;*) observing the same rule, when he is obliged to go to the Mines from the Pueblo of his residence.

Art. 12. The Judges-Conservators who shall be appointed, may delegate to the Director, or to any other intelligent person who may be in the respective Mines under his charge, the authority to arrest and send to a secure prison, or to the one nearest, any person who may excite quarrels, commit theft, homicide, or mutiny, or be found with prohibited arms, taking this prompt measure, with the obligation to report to the Judge-Conservator within twenty-four hours, in order that the preliminary examination (*sumaria*) may be had, which, upon the information of said person, shall be made in writing; which same form shall be observed for the imprisonment of any other person outside of the limits of the Mines, who shall be suspected, either of having committed an offense within such limits, or of having damaged or injured any of the works, storehouses, or other buildings, which are constructed, within their vicinity, for the better use of the Mines: and if, by the preliminary examination, the Judges-Conservators should find that the offenses committed do not constitute a case for the exercise of their conservatorship, or that they are not offenses against the substance, progress and privileges of the Mines, they will transfer the culprits and proceedings to the Judge or Courts to whom the cognizance of them belongs.

Art. 13. If the works, structures, excavations, and operations undertaken in the Mines, should meet with any difficulties, which cannot now be foreseen, in the ultimate perfection of their establishment, the parties interested shall have my Royal protection and aid, in proportion to the merit they may manifest by their industry and by the capital expended, to enable them to overcome any difficulties or impediments which may occur; they, in such case, proposing the means which experience shall have proved to be necessary and proper.

Art. 14. Finally, it is my will, that, if the said parties interested in the Mines of mineral coal of Villanueva del Rio, and any others who may work other Mines of this class which may be discovered, shall cease working them for the term of six months, this not resulting from any extraordinary accident of caving in or of inundation, the right which they may have acquired in them shall be considered as terminated, and that it shall pass to others who may wish to work them.

PART III.

MINING LAWS OF SPAIN AND MEXICO;

ORDINANCES OF NEW SPAIN, 1783.

PART III.

MINING LAWS OF SPAIN AND MEXICO;

ORDINANCES OF NEW SPAIN, 1783.

May 22d, 1783.

ROYAL ORDINANCES,

FOR THE

DIRECTION, REGULATION AND GOVERNMENT

OF THE

IMPORTANT BODY OF MINING OF NEW SPAIN,

AND OF ITS

ROYAL TRIBUNAL GENERAL.

(NOTE.—This translation is made from an official copy printed at Madrid in 1783, and authenticated by the Rubric of Josef de Galvez.)

THE KING.

In a letter of the 24th of December, 1771, my Viceroy of New Spain represented to me, among other things: That, in order to improve the falling condition of Mining in that kingdom, to correct radically and suitably the mischievous abuses introduced among Miners and operatives, and consequently to prevent the mutual complaints resulting therefrom, he considered as a matter of great and urgent importance, the formation of new General Ordinances for that Body, so that they might be uniform and embrace a better method of government in all their parts; proposing, at the same time, the means which he deemed most likely to secure the object, in the execution of so important a work. On his information, and from what my Supreme Council of the Indies laid before me on the subject in a *consulta* of the 12th of June, 1773, I thought fit to determine and to order my said Viceroy, by *Cedula* of the 20th of July next following, among other things, that there be formed the new ordinances which he proposed, explaining, declaring, or adding what might be necessary, in view of the actual state of affairs, and after consultation with the Miners and delegation of skillful persons (*Peritos*,) having present all the papers relating thereto and referred to in his said Letter, and also the Laws of the *Recopilacion* of my said Dominions, and especially those which are referred to by the said *Cedula*. Afterward, conformably to a *consulta* of the 7th of August of the said year 1773, which was laid before me by a *Junta*, which I ordered to be formed of four Ministers, with my entire approbation, it was commanded to my said Viceroy, by a Royal Order of the 12th of the next November, that in the ordinances which, in consequence of the said *Cedula*, were to be formed for said Mining (*Mineria*,) there should be regulated and established a Body (*Cuerpo*) formal and united like the *Consulados* of Commerce, in order, by this mode, to secure to its members the permanence, encouragement, and support which they required. Afterward, in a letter of the 26th of September, 1774, my said Viceroy represented to me: That the Miners of my said Dominions had petitioned, in a printed Representation, which he transmitted therewith, dated the 25th of February of the same year, not only to be formed into a Body like the *Consulado*, as had already been ordered, but that there should be established a Bank of Supplies (*Banco de Avios*) for the encouragement of the Mines; that a College of Metallurgy be created, for instructing those who might construct machinery and execute other professional operations; and that there be formed a new Code of Mining Ordinances; proposing as a fund for the support of said establishments, the proceeds of the double duty of seniorage payable on the metals, and from which they hoped to be relieved, by reason of what they had also stated in their said Representation; my said Viceroy stating to me, on all and each ot these points, what he thought expedient. In view whereof, and considering the *consulta* of my Supreme Council of the Indies, dated the 23d of April, 1776, I was pleased to determine, among other things, and to order by my Royal *Cedula* of the 1st of July of the same year, that the important Guild (*Gremio*) of Mining of New Spain should be erected, and should erect itself, into a formal Body like the *Consulados* of Commerce of my Dominions, giving to it, for that purpose, my Royal consent and necessary permission, and granting to it the power of levying, upon its silver, the one-half or two-thirds of the double duty of seniorage which was contributed to my Royal Treasury, and from which I relieved it by the same *Cedula*: In consequence of all which, by an *Acta* which the Representative Deputies of said Guild passed on the 4th of May, 1777, it proceeded to erect itself into a formal Body, to determine the offices of which the corresponding Tribunal should be composed, and to appoint the persons who were to fill them; and what they had done, they reported to the Viceroy, who, in my Royal name, and by his Decree of the 21st of June of the same year, approved it, permitting to the Tribunal so erected, until I should make known my sovereign pleasure concerning it, the exercise of all the power and authority of administration, direction, and management which the *Consulados* of the Monarchy enjoy, according to its laws, so far as they may be conformable to my Royal will, prohibiting it however only from exercising the contentious and exclusive jurisdiction (*jurisdiccion contenciosa y privativa*) permitted to the same *Consulados* of Commerce, and until the new ordinances of Mining are formed, as was ordered, and approved by me. And the Viceroy having reported to me all this, by Letter, dated the 27th of August, of the said year 1777, I thereon thought fit to confirm it, by my Royal Order of the 29th of December following, directed to the proper Viceroy, commanding him, moreover, by that and by another of the 20th of January, 1778, that if the new Tribunal of Mining had not as yet formed and presented to him their ordinances, he should cause it to be done with the utmost dispatch: which was done on the 21st of May of the said year, and the Viceroy transmitted them to my Royal hands, with a Letter of the 26th of August, 1779, in order that, after considering them and what might be represented concerning them, by the Fiscal of that Royal *Audiencia* and the Assessor General of the Viceroyalty, I should determine what might be my Royal will respecting their approval. Considering all this, and having heard, on this grave and important subject, Ministers of approved zeal and probity, and having considered the mode of reconciling most justly the true interest of the State and the particular welfare of the said important Body of Mining, I have ordered to be issued, for the direction, regulation, and administration of that Body and of its Tribunal, the following Ordinances:

Title I.

Of the Tribunal General of Mining in New Spain.

Art. 1. This shall be styled *The Royal Tribunal General of the important Body of the Mining of New Spain*, and shall be held and esteemed by all others with that respect, which is so conducive and suitable to the very useful purposes for which my Sovereign Pleasure has created it.

Art. 2. The Tribunal shall be preserved and maintained perpetually, in conformity with the Act (*Acta*) of its said creation, which I have approved ; and consequently it shall always be composed of an Administrator-General, who is to be its President, of a Director-General, and of three Deputies-General, whose number may be reduced to two in case of convenience, but cannot be increased.

Art. 3. The said offices must be filled only by practical Miners, intelligent, and qualified by proper experience acquired in that business for more than ten years, without, in any case, dispensing with this quality, which must be united, in all of them, with that of being good American or European Spaniards, free from all mixed blood, sons or grandsons of ancient Christians, and born in lawful matrimony, preferring, under these circumstances, those who have been Judges and territorial Deputies of Mining, or in other honorable occupations in that profession, and well experienced in it.

Art. 4. The Administrator-General and Director-General of this new and first creation,—in consideration of their well known extraordinary merit in having planned and promoted the reform of Mining, and the establishment and preservation of its Body, applying and promoting, for many years before, the measures and means most efficacious and conducive to that end ; and in consideration also of the special instruction and application which they have, and have shown, in these matters; of their long service (*antigüedad*) in the profession of Mining, their families having followed no other since they first settled in New Spain; and finally, that, in order to carry into complete effect and to perfect such projects, a considerable time is required, and that no persons can be more fit to carry them out than those who planned and commenced them,—shall hold their said offices for life; but the Deputies-General who now serve, shall only continue in office the time which corresponds to them, over and above that which has already elapsed since their appointment, according to what shall be laid down with respect to their succession.

Art. 5. For the elections of Administrator-General and Director-General, in case of vacancy, and of Deputies-General hereafter, there shall meet in Mexico every three years, counting from the present time, and in the beginning of the month of December, one Deputy for each *Real* of Mines with sufficient power from its Miners; and if a Deputy should not be able to go from any place, on account of its remoteness, or inability to pay the expense of his journey to Mexico, and of his sojourn there, it will suffice to send full power and instruction to a person resident in said Capital, provided he be not a Deputy or Agent (*Apoderado*) of any other *Real* of Mines; but he must have the quality of being an owner or supplier (*Dueño ó Aviador*) of Mines.

Art. 6. In order that Places (*Lugares*) ot Mines may have a vote in the election, it must be proved that they have a Town (*Poblacion*) laid out, a Church, a Curate or substitute, a Royal Judge (*Juez Real*) and Deputies of Mining, six Mines in actual working, and four reducing Establishments (*Haciendas de beneficio*).

Art. 7. The city of Guanajuato shall have six votes in said election; that of Zacatecas four; that of San Luis Potosi three; that of Pachuca y Real del Monte three; and generally, the *Reales* of Mines which have the title of City (*Ciudad*) shall always have three votes each, and those which have the title of Village (*Villa*,) or which have in them Royal Treasuries (*Caxas Reales*,) shall have two votes each.

Art. 8. Before proceeding to the election, there shall be held three scrutinies, on three several days, for determining the qualification of persons eligible to said offices, with this proviso, that the Administrator-General must always be one of those who have been Deputies-General in one of the preceding triennial periods, except in case of reëlection, when it will be necessary to observe what is prescribed in Article 10 of this Title: it also being understood, that in each triennial period, there is to be appointed and installed anew, only one of the three Deputies-General, in order to supply the place of the one whose term expires, who must be, in the first triennial period, that one who, in the Act (*Acta*) of creation, was elected by the smallest number of votes with respect to the other two, the same rule being followed with respect to these in the second triennial period, and the term of the last of the three Deputies elected in said Act (*Acta*) expiring in the third triennial period, so that, in each one of the successive triennial periods, it shall be the senior who by rule and preference is to be replaced by the new Deputy; consequently, in conformity with this regulation, each one will have and hold, in future, said office for nine years, except in case of the death of one of them before the expiration of that time, in which case, at the first triennial *Junta*, there will be named, besides the Deputy who is to replace him whose term of nine years is to expire, another to fill the vacancy, which latter, in point of seniority, shall stand in place of his predecessor, so as not to depart from the order which is established as the rule of succession by seniority.

Art. 9. The *Junta* of electors will be presided over by the Administrator-General, the Director-General, and the Deputies-General, who shall also be entitled to votes, and the election shall take place on the 31st day of December, by secret ballots, (*cedulas secretas,*) and those shall be elected who receive the greater number of votes; and in case of any dispute, the person shall be considered elected for whom the Administrator-General shall declare his vote.

Art. 10. In order that any person may be re-elected to any of the said offices of the Royal Tribunal, an interval of three years must have elapsed since he held the office, and he must have received more than one-half of all the votes.

Art. 11. No person elected to any of said offices can excuse himself from serving in it, and besides, he must accept it on the same day before sunset, under a penalty of $2,000, and of being obliged, after paying this penalty, to serve in it.

Art. 12. In case of the death of the Administrator-General, the Director-General, or of any of the Deputies-General, or in case of their resignation (which can be accepted only for most just and indispensable causes), the other members of the Tribunal shall elect a person *ad interim*, who shall hold the office till the expiration of that triennial period, and till the respective General *Junta* takes place, by which the regular person shall be elected, as provided by Article 8th of this Title.

Art. 13. Those who may be elected at, the proper time, as Administrator-General and Director-General, after those now in office, and so on successively, shall hold these offices, the first for six years, and the second for nine years, with attention to the requisite that, in addition to the circumstances already explained and applicable to other individuals of the Tribunal, the Director ought to be best acquainted with all the interests, concerns, and resources of his Body with respect to the working and management of Mining, and with the theory and practice of the sciences relating thereto; which cannot be acquired in a short time.

Art. 14. The Factor, *Assessor*, and Notary (*Escribano*) of the Royal Tribunal, may be appointed by it, and be removed, with or without cause, at its pleasure.

Art. 15. In the first General *Junta* which is to take place in Mexico in order to carry into effect these ordinances, there shall be elected twelve Advisers, experienced Miners or Suppliers of Mines, skilled, distinguished, and of the highest reputation, of whom four shall be persons who ordinarily reside in Mexico; and the Royal Tribunal may consult all, or any of them, in difficult cases, whenever it shall be necessary, or shall appear advantageous. And in order to make these offices temporary, and also to avoid the inconveniences which might arise from their being all changed at the end of each triennial period, there shall be appointed, in the successive General *Juntas*, six Advisers, to replace, in the second triennial period, the six who, in the said first General *Junta*, may be elected by the smallest number of votes, and, in the third and each successive triennial period, to replace the six longest in office, so that all shall respectively cease to hold office on being replaced by those newly elected, and the number of twelve remain complete: it being declared, as I declare it, that, in the said General *Juntas*, the re-election of such Advisers shall be permitted, without the necessity of observing those intervals of time and other formalities prescribed, in Article 10 of this Title, with respect to the offices therein mentioned; provided, that those re-elected reckon their seniority from the time of their re-election. And I grant to the said Advisers seats in the public sessions of the said Royal Tribunal next after the Deputies-General. And if any Territorial (Deputy) of any of the *Reales* of Mines shall be in Mexico, I grant to him also the honor, distinction and office of Adviser of the said Royal Tribunal, so long as he shall remain there.

Art. 16. On the days of scrutiny, and before proceeding to the election, there shall be presented to the General *Junta* of Mining a correct and clear statement of the endowment Fund, its receipts and expenses for the preceding triennial period, and also of the Bank of Supplies (*Banco de Avios,*) its receipts and losses, showing the condition, at that time, of the common interests of the Body, the amounts of metals, money, and goods, its claims, affairs, and dues.

Art. 17. Before proceeding to the scrutinies, they must obtain the permission of the Viceroy, and after the elections are concluded they will render an account to him, following in this respect the practice of the *Consulado* of Commerce of that capital.

Art. 18. To the Director-General shall belong the offices of *Fiscal* and *Promotor* of the important Body of Mining, and, consequently, he shall represent, advise, and propose, to the Royal Tribunal, whatever may appear to him conducive to the progress, security and welfare of said Body, giving timely advice and caution, so as to remove whatever may be considered adverse and prejudicial to said objects.

Art. 19. The Royal Tribunal will report to me annually, through the Viceroy, respecting the working of the Mines, and the condition of the things pertaining to the Body of Miners, and moreover, it may also do so extraordinarily, through the same channel, on all grave matters when it may appear necessary.

Art. 20. The Royal Tribunal may have an Agent (*Apoderado*) in the city and court of Madrid, for the protection of its interests and affairs. And in case it be necessary to send a confidential person to the same court, for any grave matters or important claims, it cannot do so without first satisfying the Viceroy of the gravity of the matter which requires such expense, and his approval of it must be reported to me, and must precede my Royal License.

Art. 21. The Notary (*Escribano*) of the Royal Tribunal will keep a Book of Resolutions, among others which may be necessary, in which there shall be entered whatever is discussed or determined, with respect to matters of administration or management, whether it be a temporary provision, or an absolute and perpetual resolution.

Art. 22. In the Royal Tribunal there shall be preserved the originals of the Royal *cedulas*, orders and directions, which have been, or may be, sent to it directly by me, and also the official documents from the Viceroys, and the copies of orders which may be received through them, and finally, all the papers and fundamental documents of its creation, and which relate to its administration: all which shall be kept and preserved in the Archives, and a Book shall be kept, in which all shall be authentically attested, so as to establish their validity whensoever or howsoever that may be required: prohibiting, as I prohibit, the originals from being, in any case, exhibited or taken out, but permitting only copies or authenticated copies (*Testimonios*), when they are required to be given, duly attested, corrected, and compared in legal form and according to law.

Art. 23. Before proceeding to the triennial elections, an Inventory shall be made, and the papers of the Archives and of the Notary's office (*Escribania*) shall be examined by two of the Deputies, comparing the existing papers with the Inventory of the preceding triennial period, and there shall be added to it those which have been received within the said last three years.

Art. 24. The secretary of the Royal Tribunal shall be one of the Royal Notaries (*Escribanos*), who is well instructed and prompt in his office, and who has all the other qualifications prescribed by the laws, as requsites for obtaining and holding that office; and, moreover, he must be a man of good birth and character, and corresponding education, of proper conduct, and well known good habits; so that, by these circumstances, the office may held to be *honorable*, and he who may hold it shall be respected and esteemed in the Royal Tribunal and elsewhere, and he shall always be addressed by the title of *Don.*

Art. 25. The Secretary shall propose to the Royal Tribunal three persons, one of whom is to be appointed chief clerk (*Official Mayor*), and another, second clerk, if at any time one shall be required; but he shall have full authority to appoint and remove, at his own pleasure, the writer (*Escribiente*) or writers whom he may have to employ.

Art. 26. The Royal Tribunal will appoint two Porters who are also to be executive officers (*Ministros Executòres,*) provided that they are persons of good character and Spaniards.

Art. 27. The Royal Tribunal shall draw up a Tariff (*Arancel*) of fees which may be justly received by the officers in Mexico, and in the *Reales* of Mines; but it is forbidden to carry it into effect until it has been laid before the Royal *Audiencia* of the respective district, which shall approve and mark those which ought to be exacted, reporting it to me for my sovereign approbation.

Art. 28. The Administrator-General, Director-General and Deputies-General of Mexico, and all other officers, at the time of entering upon their respective offices, shall take an oath, that they will perform their duties truly, faithfully, and honestly, and that they will observe, and cause to be observed, these ordinances, and that they will maintain secrecy in the causes and affairs which may come to their knowledge; and also that they will defend the mystery of the immaculate conception of our Lady.

Title II.

Of the Judges and Deputies of the Reales of Mines.

Art. 1. The respective Royal Judges (*Justicias Reales*) shall be Judges of Mines (*Jueces de Minas,*) conformably to the laws of the Recopilacion de Indias, in all cases not referred by these ordinances to the Deputations of the Body of Mining (*Cuerpo de Mineria.*)

Art. 2. All those who, for more than one year, shall have worked one or more Mines, expending on them, as owners thereof in whole or in part, their capital, their labor, or their personal attention and care, shall be enrolled (*matriculados*) as Miners of that Place (*Lugar*), and their names shall be entered in the Book of enrolled Miners which shall be kept by the Judge and Notary of that Mining place (*Mineria*).

Art. 3. The Miners so enrolled, and the Suppliers (*Avia-*

dores), being Miners; the millers (*Maquileros*,) and the owners of *Haciendas* for grinding and smelting ores in each Place (*Lugar*), shall assemble in the beginning of January of each year, according to custom, in the house of the Judge of Mines, to elect persons who are to fill the office of Deputies of said Mining place (*Mineria*,) and who must be, or must have been, Miners, that is, owners of Mines, of the greatest experience and intelligence in them, men of good conduct, trustworthy, and possessing all the other qualifications requisite for such offices.

Art. 4. Each one of the enrolled Miners shall be entitled to a vote at such elections; but of the Suppliers (*Aviodores*), being Miners as aforesaid, the Millers (*Maquileros*) and the owners of *Hacienda* mentioned in the preceding Article, every two shall have one vote, and they shall not be voted for as Deputies of Mining, unless at the same time they are Miners and have the necessary qualifications.

Art. 5. Where there is a great number of members (*vocales*), as in Guanajuato, the practice, which has hitherto prevailed, will be continued, of choosing beforehand, in that *Real*, Electors who will proceed to the election of Deputies.

Art. 6. The Administrators of Mines may vote in place of their principals (*Amos*,) who are not residents in that neighborhood (*territorio*,)they having, for that purpose, sufficient authority; and they may also be elected as Deputies, if their occupations permit, and they have the necessary qualifications.

Art. 7. The Judge of Mines of each *Real* or *Asiento*, and the Deputies of the preceding year, shall preside over and regulate the election, and shall be entitled to vote; and, in case of a disagreement, the vote of the Judge of Mines shall decide it; it being understood that those persons who receive the greatest number of votes shall always be considered elected, if they have the qualifications and requisites prescribed.

Art. 8. In each *Real* or *Asiento* of Mines there shall be a Deputation composed of two Deputies; and in order that these offices may be biennial, and that one of them may always be held by a person properly instructed in the duties of his office, only in the first year of carrying this measure into effect, shall both Deputies be appointed; but in each of the succeeding years, only one, to replace the Deputy oldest in office: and, as this rule cannot be applied to the second year of said elections, one of the two appointed at first, who was elected by the greatest number of votes, will continue with the Deputy who comes in anew; consequently the other will serve in said office for one year only.

Art. 9. There shall also be elected in each *Real* or *Asiento* of Mines, and in the same manner, four Substitutes who are to supply the place and fill the office of the Deputies, in case of their refusal to serve, their death, sickness, necessary absence, or other just impediment, and to assist in the several Courts of Appeal (*Juzgados de Alzadas,*) in such cases and under such circumstances as will be mentioned in the proper place; but where electors are appointed in conformity with Article 5 of this Title, the four who shall have been elected by the greatest number of votes shall be Substitutes for the first year: it being understood that these said officers are also to be biennial, and that, in each successive year, only two new ones are to be chosen, observing, in this respect, what is prescribed in the preceding Article with respect to the Deputies. And for greater perspicuity, and to prevent all arbitrary acts, in cases of succession to the office, either of said Substitutes or of the Advisers, by any of the substitutions directed by the various Articles of these ordinances, it shall be a general rule, that the order of preference be given to those for whom the greater number of votes has been given at their respective elections, when they are of the same date; but, the elections not being of the same date, then, the one longest in office shall be preferred.

Art. 10. The said Substitutes shall be, at the same time, District Attorneys (*Sindicos Procuradores*) of their respective *Reales* of Mines, and shall represent, demand and procure, whatever may to them appear conducive to the common welfare of those Miners and residents, and their services shall be regarded and considered, with respect to electing them to the office of Deputies, and to other offices of Mining.

Art. 11. Those elected as Deputies must accept the office within the third day, under the penalty of one thousand dollars, to be paid into the fund of that *Real*, and of being compelled to accept after having paid the penalty; but, if it appear that they have good and legitimate cause for refusing, they must nevertheless accept the office and serve in it until the cause is admitted by the Royal Tribunal General, to which body they shall represent it.

Art. 12. I prohibit the re-election of the same person to any of the said offices, until the expiration of two years from his last serving in such office; and the person re-elected after such interval cannot refuse to accept, under penalty of five hundred dollars, to be paid into the fund of the same *Real*, and he shall be compelled to accept after having paid it, without prejudice, however, to any sufficient causes of exemption which he may have, and he may represent them to the Royal Tribunal General of Mexico, on condition that in the meantime he accept and serve in the office, as prescribed in the preceding Article.

Art. 13. All the Miners, Suppliers, Millers, (*Maquileros*,) and owners of *Hacienda* of the respective Places (*Lugares*,) will confer upon the Deputies newly elected, power to promote their interests and claims, and for all other customary purposes, and they shall give and swear obedience to them, in whatever relates to the exercise of their office; and the said Deputies elect shall accept and swear to perform the duties of their office conformably to law, and also, to observe these ordinances, (which shall be read at every election, at the time those chosen enter upon their offices,) and to maintain secrecy in the causes which come within their cognizance.

Art. 14. As soon as the election is over, a report and notice of it shall be immediately given to the Royal Tribunal General of Mining, in order that, there being no defect, or certain and proved irregularity in it, it may have the approval of the Supreme Government of New Spain; but it is declared that no fees whatever shall be charged for such approvals, nor for the acts and formalities which precede them.

Art. 15. The territorial Deputies, and the Inspectors and Experts (*Peritos*) of Mines, shall receive no pay from my Royal Treasury for their services, but they shall be paid from the profits of the respective Mines, agreeably to the laws on that subject; for which purpose, the Royal Tribunal General of Mexico shall impose duties, just, moderate, and suited to the condition and circumstances of each *Real* of Mines, in the manner and under the rule laid down in Article 36 of Title III of these ordinances.

Art. 16. In February of each year the territorial Deputations will report to the Royal Tribunal General of Mexico upon the condition of the Mines and Miners of their respective districts and their dependencies, proposing whatever may appear to them conducive to their re-establishment, preservation and greater progress; and also the produce of silver, and consumption of quicksilver of the preceding year; the number which may be in actual working, and those which have been abandoned, and the causes of abandonment, and those newly discovered and re-established: asking, for this purpose, of the courts, Royal Treasuries and other offices, the certificates, attestations (*Testimonios*,) and other documents which they shall require. And I order that said reports and documents be laid before the Viceroy, in order that, having made himself acquainted with their contents, he may inform me of all, with the proofs, for the measures which may be required, and which may be agreeable to my Sovereign will.

Title III.

Of the Jurisdiction of Causes relating to Mines and Miners, and of the mode of taking cognizance, of proceeding, and of passing judgment and sentence in the same, in First, Second, and Third Instances.

Art. 1. I grant to the Royal Tribunal General of Mining the power to take cognizance of, and to decide all matters pertaining to that Body which relate to its administration, direction and management; and I therefore declare that the Deputations of all the *Reales* and *Asientos* of Mines, shall observe a strict and entire subordination to it, in all the aforesaid matters which are purely administrative.

Art. 2. Moreover, the Royal Tribunal General shall have exclusive cognizance of causes involving questions of discoveries, denouncements, *pertenencias*, measurements, drainage, desertions and want of supports (*despilaramientos*) of Mines, and whatever may occur in them to the prejudice of working them, and in contravention of these ordinances; and also whatever relates to the Supplies (*Avios*) of Mines, the exchange of metalic ores, (*rescates de Metales en piedras*), or of silver and gold, copper, lead and other mineral substances, to mills (*maquilas*), and other things of that nature; but I declare that the said contentious jurisdiction shall only be exercised by said Royal Tribunal General within the district of twenty-five leagues around the Capital of Mexico.

Art. 3. Without prejudice to the exclusive administrative jurisdiction which, by Article 1st of this Title, I grant to the said Royal Tribunal General, the Deputations of the *Reales* of Mines may, provided the two Deputies always act jointly and together, also use and exercise it in their respective territories, in the cases and matters relating to the encouragement and progress of the working of Mines in their particular district; to the interest and advantage of the owners of Mines; to the preservation and increase of the population; to the due administration of justice; to the welfare of the inhabitants, and to the relief of the poor; it being understood that all is to be in immediate subordination to the Royal Tribunal General, as prescribed in the said Article, and provided that they are not to interfere by any formal acts of jurisdiction, except in the cases and matters which are expressly conceded to them by these ordinances.

Art. 4. The territorial Deputations, in their respective districts, shall have the exclusive contentious jurisdiction which I declare and grant, in Article 2d of this Title, to the Royal Tribunal General, and in the particular causes and matters which are there expressed, proceeding in and deciding them, with absolute independence of the said Royal Tribunal, since, in the exercise of such contentious jurisdiction, they are not to recognize, in any manner, any subordination, the said Royal Tribunal being, as I wish it to be, prohibited from interfering to take cognizance of, or engage in, said causes and proceedings which are instituted out of its ditrict.

Art. 5. As said classes of causes and suits ought to be determined between the parties briefly and summarily, according to manifest truth and good faith, as in commercial transactions, without allowing delays, declarations or writings of lawyers, it is my will that whenever any person appears in said Royal Tribunal, or before the territorial Deputation of any one of the *Reales* or *Asientos* of Mines, to institute any action, they shall not admit any complaint or petition in writing, until after they have cited before them, if it be possible, the parties, so that, hearing orally their complaints and answers, they may settle, with the greatest dispatch, the suits or dispute between them; aud not being able to succeed in this, and the matter in question exceeding the value of two hundred dollars, (for if less than that amount the cases must be determined orally even though the parties refuse their consent,) petitions in writing will be admitted, provided they are not drawn up, arranged, or signed by lawyers. And if the suit must be tried, because it cannot be compromised or settled orally between the parties, the demand or petition of the Plaintiff is to be attended to before any other of the Defendant.

Art. 6. In consideration of the object, above expressed, of dispensing justice in suits and differences, briefly and summarily, and according to manifest truth and good faith, I order and command, for the better attainment of that end, that in the proceedings of the court of first instance, both in said Royal Tribunal and in the territorial Deputations, and in the proceedings on appeal, and in the judgments which may be pronounced, no consideration shall be paid to any default in observing the minute formalities of the law, to inaccuracies or other defects; but, in whatever stage of the proceedings the truth may be ascertained, the causes shall be decided and adjudged, and for that purpose they shall officially examine the necessary witnesses, provided they do not exceed the number of ten, and shall take the depositions of the parties where they may appear to said judges calculated to elicit the truth, and they may then proceed to pronounce their decision and judgment.

Art. 7. To avoid malicious appeals, and those which are interposed for the sole purpose of delaying judicial proceedings, by perverting their course and summary character, I order, that no one shall be allowed to appeal from the Judges of the said Royal Tribunal, and from the territorial Deputations, except it be from a definitive judgment, or from an interlocutory decree which involves irreparable injury; and that any appeal which, in contravention of this Article, may be interposed, shall not be valid, and neither the Judges of the said Royal Tribunal, nor the territorial Deputations, shall or can be prohibited from taking cognizance of the cause, but they may proceed in it to the pronouncing of final judgment.

Art. 8. The interlocutory decrees and judgments which are given, must be signed by the Administrator-General and the two Deputies-General of said Royal Tribunal, even though the vote of one of them should not conform to that of the other two; for the Administrator-General and one Deputy-General, or the two Deputies-General, must give the decision and judgment, which the other one cannot refuse to sign.

Art. 9. The territorial Deputies may each separately proceed in causes, so as not to interfere with that dispatch in them which is of so much interest to the Body of Mining; but in pronouncing definitive judgments, and in giving interlocutory decrees which cause or may cause irreparable injury, they must act in conjunction; and if they should not agree, the Substitute, whose duty it shall be according to the rule already laid down, will act with them, in order that, the disagreement being obviated, it may be decided by the majority of votes, and the decision be signed by all three, as prescribed in the preceding Article.

Art. 10. In points of law, where they are not clearly laid down in these ordinances, the Royal Tribunal General will avail itself of the assistance of some learned and consciencious lawyer of its own selection, and the territorial Deputations will avail themselves of the assistance of any one who may be in the Place or Pueblo where they reside; and if there should be no such person, or in case of exception, they shall avail themselves of the assistance of the professional Judge (*Juez Letrado*) of the respective province appointed by me, who cannot be excepted to (*ser recusado,*) unless an associate can be appointed: declaring, as I declare, both with regard to this and the preceding Article, that he who has given his opinion in first instance (*instancia*) cannot give it in the second.

Art. 11. When the suits are closed and in condition to be decided, or whenever the Judges of said Royal Tribunal or territorial Deputations may deem proper, they shall be submitted to the court by the Notaries (*Escribanos,*) before whom they have passed, and who will report on them in the usual form, and with all the brevity possible, and which is so desirable and suitable to Miners.

Art. 12. The Decrees and Judgments which are given in the said Royal Tribunal General, and by the territorial Deputations, not being appealed from and having become *res judicata*, shall be carried into execution briefly and summarily: those of the Royal Tribunal, by means of the two Porters (*Porteros,*) whom it is to have, and who are to perform the functions of executive Bailiffs (*Algu-*

aciles Executores); and those of the territorial Deputations, by means of the ordinary Bailiffs (*Alguaciles*) of the Pueblos where they reside, both of them, for this purpose, issuing the necessary orders and mandates to the other Judges and courts, whom it may concern, to afford them the aid and assistance which may be required.

Art. 13. If, from such Judgments or definitive Decrees, an appeal is taken by any of the parties, where the amount in dispute exceeds four hundred dollars, (for, if the amount be less, no appeal shall be admissible, and the final action of the Judges of the Royal Tribunal, or territorial Deputations, shall be carried into execution,) those from the Royal Tribunal General shall be received by the Court of Appeals (*Juzgado de Alzadas*,) which is to be established in Mexico, and to be composed of an *Oidor* of that Royal *Audiencia*, by nomination of the Viceroy, in the same form, and for the same term, as the one destined for the Royal Tribunal of that *Consulado* of Commerce, of the Director-General of Mining, and of another Miner who, for this end, ought also to be elected, every third year, by the *Junta* General of Mining, from those who have been Administrators-General, Directors-General, or Deputies General, or one of the four Advisers (*Consultores*,) (out of the twelve,) who are to reside in Mexico, as ordered in its place. And the appeals from the territorial Deputations, within the district of twenty leagues in all directions from the city of Guadalajara, shall, in the same manner, be received by the Court of Appeals which I have ordered to be created there, and which is to be composed of one of the *Oidores* of its Royal *Audiencia*, to be appointed by the Regent President of the said Tribunal, for the same term and in the same form, as the one for the *Consulado* of Commerce of Mexico, and of two Miners of probity and possessing the other requisite qualifications, who shall be appointed Associate Judges of Appeals in the same city of Guadalajara, from those who reside there, by the said *Junta* General of Mining, which, every three years, is to meet in Mexico, as is before prescribed. But if in said city there be no resident Miners with the nececessary qualifications for Associate Judges, then, at said triennial election others who reside elsewhere may be elected, provided that, under the same circumstances of aptitude and qualification, preference be given to those who are at the least distance, even though they be Substitutes of Deputies of any *Real* or *Asiento* of Mines: it being understood that Appeals from all the other territorial Deputations shall be admitted in said form by the respective Court of Appeal to be erected in each Province, and to be composed of the Judge best qualified, and appointed by me, who may be in it, and of the two Substitute Miners upon whom the duty may devolve according to the rule already laid down, out of the four of the *Real* or *Asiento* of Mines nearest to the residence of said Judge: provided, that, if in the same place or in another equally distant, there shall reside one or more of the said twelve Advisers, they shall be preferred as Assaciate Judges of Appeal. And whenever said Judge shall not be of the profession (*Letrado*,) the court shall, in all points and matters where it may be requisite, avail itself of the assistance of a learned and conscientious lawyer.

Art. 14. In said causes of appeal, the proceedings shall be brief and summary, as in commercial causes, without allowing new terms for delays or proofs, or admitting pleadings and writings of lawyers, or anything but the statement of the grievances of the Appellant, and the answer of the other party or parties, only having regard to manifest truth and good faith, as in commercial transactions; and in this form they shall decide the cause.

Art. 15. Such appeals shall be commenced on or before

the third day after notification of the decree or judgment, and in no other manner; and I grant that they may be commenced by a letter from the Appellant, stating that he will transmit full power for the formality of the judicial proceeding, or that he will appear in person.

Art. 16. If the judgments of the Royal Tribunal General of Mining, or of the territorial Deputations in their respective causes appealed from, are confirmed by the Courts of Appeal, no further appeal, revision, or recourse, (*apelacion, agravio ni recurso,*) shall be admitted, and they shall be ordered to be really and effectually executed, and for that purpose, the processes shall be returned to the respective Judges.

Art. 17. But if they should reverse them, wholly or in part, and any one of the litigants should appeal or apply for a new trial, the Judges of Appeal will appoint, each one in his cause, two other Associate Judges, who are to be selected in Mexico, from the four Advisers residing in that capital; in Guadalajara, from the other Miners who reside there, with preference to those who are Advisers, if there should be any in that city; and, in default of both, the election may fall upon Miners who reside elsewhere, under the considerations which are expressed in relation thereto in Article 13 of this Title; and in all the other Courts of Appeal, the Judge may appoint any of the four respective Substitutes; it being understood, in both cases, that there be no objection or legal disqualification; and if there be such established against all, then the said nomination may fall on other Miners having the requisite qualifications: provided, however, that where any one or more of the twelve Advisers of the Royal Tribunal General reside, they shall be preferred to the Substitutes.

Art. 18. From the judgment which may be given in this third instance (whether confirming, reversing, or altering the judgment appealed from, in whole or in part,) no further appeal, new-trial, revision, or recourse shall be admitted, and the cause shall be sent back to the respective court for its fulfillment and execution, in which the proceeding shall be brief and summary as is already prescribed. But I declare that there shall remain open to the parties the legal remedy of the second petition (*segunda suplicacion*) to my Royal Person, in my Supreme Council of the Indies, provided it shall be shown that the matter in litigation amounts to or exceeds the sum of twenty thousand dollars; but they must give the security which the law requires, and without prejudice to the execution of what was ordered in the judgment from which this proceeding is taken, and first giving other security to abide by the last judgment which may be pronounced.

Art. 19. In the decisions of the said causes, on appeal, the judgment shall be determined by two of the three members (*Vocales*,) whether they be the Judge and one of the Associate Judges of the respective Court of Appeals, or the two Associate Judges, without the concurrence of the presiding Judge, and in either case the judgment shall be signed by all three.

Art. 20. Causes of possession and ownership must be determined together; but, first of all, restitution shall be made to any person who has been forcibly dispossessed, unless he has been deprived of his possession by a decree or judgment of a Judge, even though the same may be charged with injustice.

Art. 21. For no cause or motive, shall any Mine in litigation be shut up, or its working be suspended, even at the request of any of the parties, but an *Interventor* may be appointed for the satisfaction of the party who may ask one; but without removing from the Mine the party who may be in possession, and provided that, if the latter offer sufficient security to the satisfaction of the opposite party, the *Interventor* may be dispensed with. And I declare that the working of the Mine can be suspended only when the charge is made of its being in a ruinous condition, without supports or the necessary timbering, and it shall be so proved in the opinion of Experts (*Peritos*), who must immediately, and without the loss of a moment, proceed to make it secure, so that, being placed in good condition, its working may be resumed without danger.

Art. 22. All executive demands shall be proceeded in conformably to general jurisprudence and Royal Laws (*derecho y Leyes Reales*,) with respect to the order of process, always according to good faith and truth, without admitting the delays and technicalities which interrupt and retard the despatch of causes of this kind.

Art. 23. Whenever justice requires the issuing of an execution against any Mine or Reducing Establishment (*Hacienda de Beneficio*,) the same shall not be enjoined (*se embargará*), nor sold at auction, nor the machines, iron-work, tools, slaves, animals, structures, materials, or any of the necessary provisions; but the execution shall be levied upon the metals of silver and gold, and other products, deducting what may be necessary to cover and pay the expenses and working of said metals, which must by no means cease; for which purpose an *Interventor* approved by the Plaintiff, if the latter should not desire to carry on the Mine himself, and by the Defendant, if the Plaintiff will take it on his account, shall be appointed, and the intervention shall immediately cease when the debt is paid; and in either case the said *Interventor* shall make a weekly statement of the expenses and produce of the Mine, to be, at the proper time, laid before the Judges in the cause, with the respective vouchers, and the corresponding oath to such items as cannot be otherwise verified, to be applied to whichever party may be declared the real owner by the definitive judgment which may become final.

Art. 24. When the defendant shall surrender up his property, and it shall comprise any Mine or Mines, the creditor or creditors shall be notified to undertake the working thereof on their own account, and such working shall not be suspended, under the penalty, that, at the expiration of the time which shall be fixed by these ordinances, the Mines shall be considered as deserted and abandoned, and they shall belong to the first person who may denounce them, notwithstanding the fact of their being in litigation or under attachment for debt (*concursadas*).

Art. 25. The cost of working the Mines or *Haciendas* under such attachment, and the salary of the *Interventor* shall, in no manner, form a part of the execution debt, but shall be paid promptly and from the first proceeds, even though they should not exceed that amount.

Art. 26. In case of a want of supplies (*habilitacion*), and any one of the creditors shall offer to undertake it on his own capital, because the others refuse to contribute their *pro rata*, such creditor shall be preferred to the other restorers (*refaccionarios*), not only in satisfaction of his new advances, but also with respect to his old credits, even though they may not have resulted from any restoration (*refaccion*) or supplies of the Mine or *Hacienda*.

Art. 27. Whenever in other courts, by reason of causes of inventories, hereditary successions, general partnerships, proceedings of creditors, or surrender of property, there may be comprised Mines, their *Haciendas*, or other appurtenances or dependencies, with other property pertaining to said cause, I order that the Judge, who has jurisdiction of it, shall remit a letter of justice, official order, or note (*Carta de justicia, Oficio, ó Villete,*) to the respective Court of Mines, in order that, merely taking cognizance of the working of that Mine or *Hacienda*, it may continue and be kept up, without prejudice to the right and actions of the party or parties interested; it being the duty of the said Court of Mining to reserve their products for the disposal of the principal Judge of said causes; and also, in case widows, minors, or absentees are interested in such causes, to effectually protect and support their actions, in order thereby to maintain that true and reciprocal union which contributes to the preservation, welfare and prosperity of the whole Body.

Art. 28. In causes and suits respecting Mines restitution must be conceded for the full term (*conceder la restitucion del termino cumplido*); but on condition that the restitution for the full term allowed by law has no interruption (*no tenga hueco la restitucion por todo el termino del derecho*), except where it is for the relief of privileged persons (*los privilegiados*), when an extension is allowed to them for the half of the term.

Art. 29. In criminal causes, in thefts of ores, silver or gold, lead, tools and other things pertaining to Mines and the reduction of their ores; in offenses committed in the Mines themselves or in *Haciendas* of reduction, whether by one operative against another, or a breach of subordination on their part to the officers placed over them, or misconduct of either of these classes toward their masters and the owners of the Mines; and lastly, in causes of insult, injury, or want of respect toward said Courts of Mines, cognizance will be taken by the Royal Tribunal of Mexico with respect to its district, and by the territorial Deputations with respect to what pertains to the district of each one, proceeding in and deciding the causes of least consequence and gravity summarily, according to law and to the nature of these suits, and in strict regard to truth and good faith, according to the order already established for civil causes. But in those which, on account of their gravity and malice (*malicia*), require by law the infliction of ordinary punishment, the mutilation of limb, or other corporeal punishment (*corporis afflictiva*), the Courts of Mining are allowed only the limited jurisdiction of apprehending the criminals, of drawing up the process (*sumaria*), and transmitting it, with the criminals, to the Royal Judges of the respective Provinces, in order that the latter may, at the proper time, make report to the Royal Chamber (*Sala*) of crime of the *Audiencia* of the district, for its final decision.

Art. 30. In that class of criminal causes of minor importance which are treated of in the foregoing Article, and of which jurisdiction is granted to the Courts of Mining for their cognizance and decision, provided they do substantial justice and decide in the manner directed, if any of the parties appeal, these legal remedies shall be allowed, and they shall be decided by the Courts of Appeal, in the manner and form which is already prescribed for civil causes, observing the order which is required by the nature of these other cases.

Art. 31. Whenever conflicts arise between the Tribunal General of Mining or the territorial Courts of Mining, and other courts or tribunals, upon a plea to the jurisdiction, I order and command, that the Viceroy of New Spain shall decide the same, and that they shall observe and comply with his decision without appeal or revision; and that the Viceroys shall, in such cases, take the opinion of officers or professional lawyers (*Ministros ó Letrados,*) who have no connection with those tribunals between which the conflict arises.

Art. 32. I absolutely prohibit the arbitrary disposition of pecuniary fines which may be imposed in the exercise of both the civil and criminal jurisdiction which I grant to said Courts of Mining, and I order that they be strictly applied, by third parts, to my Royal exchequer, to the expenses of Courts, and to other objects which the law directs.

Art. 33. The Administrator-General and the Deputies-General will assemble to form a Tribunal, every day (except on holidays and those on which it is required to hear Mass,) from eight o'clock to eleven, and also, on extraordinary occasions, in the evening, and on any other day whatever, when the urgency and importance of any affair may require it.

Art. 34. The Director-General shall have a vote in all matters of direction, administration, and management, of which cognizance is granted to the Royal Tribunal General of Mexico, and, in order that he may attend when they are to be considered, due and special notice shall be sent to him; but I declare that he shall have no vote in the proceedings and decisions of suits and causes, except in cases of appeal to the Court of Appeals, in which he is already made one of the members of the court to be established in the capital of Mexico.

Art. 35. Matters of supplies (*Abastos*), public works and roads, and other objects of that nature, are to be under the exclusive cognizance and jurisd'ction of the Royal Judges and public Magistrates of each district. But the Royal Tribunal General of Mexico and the territorial Deputations, shall inform the said Judges and Magistrates of what they consider expedient, in order that said branches and works may be apportioned in the most equitable and fair manner, both proceeding therein with a mutual understanding, and in perfect harmony.

Art. 36. The taxes, or other duties and imposts, as well

public as private between individuals of the Body of Mining, which necessarily affect the advancement and working of the Mines and *Haciendas* of reduction, or the remuneration of the work of the territorial Courts of Mining, or of the employees in the new authorities (*Facultades*), offices, and other situations treated of in these ordinances, will be proposed, regulated, and assessed by the Royal Tribunal General of Mexico, so far as concerns its district, and by the territorial Deputations, in what relates to their districts respectively, although the latter shall be obliged to lay them, with the proper attestations, before the Royal court (*Justicia Real*) of the territory for its sanction. But none of these taxes, duties, and imposts can be established or carried into effect, without having been first submitted to the Viceroy of New Spain, in order that, the proceedings (*Expediente*) being examined into by his superior Government, as the nature of the subject may require, it may be determined upon and receive my Sovereign resolution, for which purpose a report will be made to me by the said Viceroy.

Art. 37. The Royal Tribunal of Mexico shall also immediately present to the Viceroy an exact estimate of the endowments and salaries allowed to the principal individuals who compose it, and to the subalterns who are appointed, or who may be elected, in pursuance of these ordinances, in order that the said Viceroy may transmit it to me, with his report, and that it may receive my Royal approval, which is necessary and proper for the security of the said Tribunal.

Title IV.

Of the order of proceeding in the substantiation and decision of contentious suits, in cases of the incompetency or vacancy of any of the Judges of Mining, and of exceptions (recusaciones) in first, second and third Instances.

Art. 1. The Royal Tribunal General of Mining shall not proceed to act in any contentious matter, without the indispensible presence of three of its Members; and if, on account of illness, necessary absence, or any other just and legal impediment, as that of having an interest in the matter in question, or of being related to the litigants, that number of Judges cannot, on any occasion, be assembled, the vacancies shall be supplied by those Advisers upon whom, of the four who must reside in the said capitol of Mexico, the duty devolves, according to the order already laid down; and the same thing shall be done to substitute and complete, in like cases, the number of Judges of Appeals, which can never be less than three, as is already prescribed in these ordinances. And whenever, in consequence of any of the impediments indicated, any of the territorial Deputies cannot, or ought not, to act as Judge in the matter in controversy, the Substitute, to whom it corresponds, shall supply his place.

Art. 2. I prohibit that peremptory exception be taken to all the Judges of the said Royal Tribunal General and to those of Appeals; but one or two particular members may be excepted to, reasons and security being given, provided, that those excepted to shall never be heard, nor any appeal be admitted from what is decided on that point.

Art. 3. Nor can, in any case, both of the territorial Deputies who, as is already provided, must be Judges of Mining, be excepted to; but either one of them in particular may be.

Art. 4. In cases where the corresponding exception shall be legal and admitted, as well in cases in first instance as in those on appeal and the respective proceedings thereon in the Courts of Appeals (*Alzadas*), the person excepted to shall be substituted, in the first case, according to the provisions of Article 1 of this Title, and in the second, the respective Judge of Appeals shall appoint, conformably to the provisions of Article 17 of Title III, those who are to supply the places of the persons excepted to.

Title V.

Of the fundamental ownership (dominio radical) of Mines; of their concession to individuals; and of the duties which for this are to be paid.

Art. 1. Mines are the property of my Royal Crown, as well by their nature and origin, as by their reunion declared in Law IV, Title XIII, Book VI, of the Nueva Recopilacion.

Art. 2. Without separating them from my Royal Patrimony, I grant them to my subjects in property (*en propiedad*) and possession, in such manner that they may sell them, exchange them, rent them, donate them, pass them by will, either in the way of inheritance or legacy, or in any other manner alienate the right which in the Mines belongs to them on the same terms on which they themselves possess it, and to persons capable of acquiring it.

Art. 3. Let this grant be understood to be upon two conditions: first, that they shall contribute to my Royal Treasury the prescribed portion of metals; and second, that they shall work and enjoy the Mines complying with what is prescribed in these ordinances, so that they shall be considered forfeited (*perdidas*) whenever a failure shall occur in complying with those (ordinances) in which it is so provided, and that they may be granted to any person who for that cause may denounce them.

Title VI.

Of the modes of acquiring Mines: of new discoveries, registries of Veins, and denouncements of Mines abandoned or forfeited.

Art. 1. As it is very just and proper to reward with particularity and distinction those who dedicate themselves to the discovery of new mineral places (*Minerales*), and metalic veins which exist (*se crian*) in them, in proportion to the merit, importance, and utility of such discovery; I order and command that the discoverers of one or more mineral hills (*cerros*) absolutely new in which there is no Mine nor trial-pit (*cata*) open, may acquire, in the principal vein which they may select, as much as three *pertenencias* continuous or interrupted, according to the measurements which hereinafter shall be prescribed; and that, if they may have discovered more veins, they may have one *pertenencia* in each vein, said *pertenencias* being determined and marked out within the term of ten days.

Art. 2. The discoverer of a new vein in a hill (*cerro*) known and worked in other parts, may have in it two *pertenencias* continuous, or separated by other Mines, provided that he specifies them also within ten days, as is mentioned in the preceding Article.

Art. 3. He who asks for a new Mine in a vein known,

and worked in other places, shall not be considered a discoverer.

Art. 4. Those mentioned in the preceding articles must appear with a written statement before the Deputation of Mining of that territory (*territorio*), or the one nearest if there should be none there, stating in it their names, and those of their partners (*compañeros*) if they have any, the place of their birth, their residence, profession and employment, and the most particular and distinguishing features of the place (*sitio*,) hill (*cerro*), or vein of which they ask the grant (*adjudicacion*): all of which circumstances, and the hour in which the discoverer presents himself, shall be noted in a Book of registry which the Deputation and Notary (*Escribano*) of Mines, if there be one, shall keep; and, this being done, his written statement shall be returned to the discoverer for his due security, and notices (*carteles*) shall be affixed to the doors of the Church, Government-houses (*Casas Reales*), and other public places of the Town for due information. And I order that, within ninety days, he shall have made in the vein or veins of his registry, a pit (*pozo*) of a *vara* and a half wide or in diameter at the mouth, and ten *varas* down or in depth; and that, as soon as this is done, one of the Deputies shall personally go, accompanied by the Notary (*Escribano*) if there be one, and if there be none, by two assisting witnesses, and a professional Mining Expert (*Perito facultativo de Mineria*), of that territory, to inspect the course and direction of the vein, its width, its inclination to the horizon, which is called *echado* or *recuesto*, its hardness or softness, the greater or less firmness of its sides, and the species or principal indications (*pintas*) of the mineral, taking an exact account of all this in order that it may be added to the corresponding part of its registry, with the evidence (*fé*) of possession which shall immediately be given in my Royal name, measuring to him his *pertenencias*, and causing him to fix stakes (*Estacas*) in his boundaries, as will hereafter be mentioned; which being done, there will be delivered to him an attested copy of the proceedings, as a corresponding Title.

Art. 5. If, during the said ninety days, any one shall appear, pretending to have a right to said discovery, he shall have a brief hearing in court (*en justicia*), and it shall be adjudicated to the one who best proves his claim; but if he appear after that time, he shall not be heard.

Art. 6. The restorers of old mining places (*Minerales*), which are decayed and abandoned, shall have the same privilege as discoverers, selecting and enjoying three *pertenencias* in the principal vein, and one in each of the others; and both shall be especially rewarded and receive the preference in parity of circumstances, and whenever there may be an opportunity.

Art. 7. If the question shall arise as to who has been the first discoverer of a vein, he shall be held as such, who first found metal in it, although others may have made experimental openings (*cateado*) previously; and in case of doubt, he shall be considered the discoverer, who may first have it registered.

Art. 8. He who denounces a Mine as deserted and abandoned (*despoblada*) in the terms which will be hereafter mentioned, shall have his denouncement received, provided that he expresses in it the circumstances prescribed in Article 4 of this Title, the particular locality of the Mine, its last possessor, if there be any information respecting him, and the possessors of the neighboring Mines, if they are occupied, all of whom shall be lawfully summoned; and if, within ten days, they do not appear, the denouncement shall be publicly declared on the three following Sundays, and, there being no opposition, the denouncer shall be notified, that, within sixty days, he must have cleaned and put in order some working (*labor*) of considerable depth, or at least ten *varas* perpendicular and within the sides (*respaldos*) of the vein, in order that the professional Expert of Mines (*Perito facultativo de Minas*) may inspect and examine its course, inclination (*echado*), and other circumstances, as declared in said Article 4: it, moreover, being the duty of said professional Expert (*Perito facultativo*) to examine, if possible, the pits (*pozos*) and different workings of the Mine; to see whether any of them are decayed, destroyed, or inundated; whether it has a draft-pit (*tiro*), or adit (*socabon*), or can have such; whether it has an ore-shed (*galera*), a whim (*malacate*), or other machines, habitable rooms, and stables; and of all these circumstances an account will be taken and noted in the corresponding Book of denouncements, which should be kept separately. And the said examination being made, and the *pertenencias* being measured and marked out with stakes (*Estacas*) as will hereafter be mentioned, possession shall be given to the denouncer, notwithstanding any opposition, which shall not be heard, if not made within all the periods before prescribed; but if, during those periods, any opposition may have been brought forward, the parties shall be heard in court (*en justicia*) briefly, and as is prescribed in the proper place.

Art. 9. If the former owner (*dueño*) of the Mine shall appear and oppose the denouncement, after the term of the proclamations (*pregones*) has expired, while the denouncer is already enjoying the sixty days allowed for putting in order the pit (*pozo*) of ten *varas*, he shall not be heard as to the possession, but only as to the right of property (*propiedad*); and, if he succeed in establishing that, he must repay the denouncer the expenses which he may have incurred, unless it be proved that the latter acted in bad faith, in which case he must lose them.

Art. 10. If the denouncer does not put in order the pit (*pozo*) or working (*labor*), as already prescribed, nor take possession, within the sixty days, he shall lose his right, and another person may denounce the Mine. But if, on account of the Mine being entirely in ruins, or of some other difficult and insurmountable circumstance, or on account of some real and serious obstacle, he shall not be able to establish the pit or working within the said sixty days, he must apply to the respective Deputation, which, the cause being examined into and proved, may extend the term as much as may be necessary, and no more; it being understood, that no opposition to the denouncement is to be admitted on this account any more than within the sixty days of the ordinary term.

Art. 11. If any one denounce a Mine as forfeited (*perdida*) on account of the non-observance of any of the ordinances which carry this penalty imposed, it shall be granted to him, provided that any of the prescribed causes shall be lawfully established and proved.

Art. 12. If the former possessor of the Mine, or any one claiming under him, shall reclaim, as having left in it, any exterior or moveable works made at his expense, as ore-sheds (*cubiertas de galera*), machines or other things of this class, of which the denouncer may avail himself, he shall pay their owners for them, according to the valuation of the Experts (*Peritos*).

Art. 13. If any one denounce surplus spaces (*demasias*) in the limits of occupied Mines, they can be granted to him, only in case the owners of the neighboring Mines, or one of them, shall not desire them for themselves; but if these owners have not occupied them, and do not occupy them, with their workings, within the time which, taking into consideration all the circumstances of the case, the Deputation of that territory may prescribe to them, they may be adjudged to the denouncer,

Art. 14. Any one may discover and denounce a vein or Mine, not only in common land, but also in the private lands of any individual, provided he pays for the land of which he occupies the surface, and the damage which immediately ensues therefrom, according to the valuation of the Experts (*Peritos*) appointed by both parties, and a third in case of disagreement: the same being understood with respect to him who denounces a place (*sitio*) or waters for establishing works and moving the machines necessary for the reduction of ores, which are called *Haciendas*, provided they do not include more land nor use more water than may be necessary.

Art 15. But if any one denounce a Mine or *Hacienda* within a Town (*Poblacion*), so that injury may result to its principal edifices, or other similar inconvenience, the denouncement shall not be granted without previous application to the Royal Tribunal General of Mexico, in order that, consulting with the Superior Government, this may decide the case with due consideration and circumspection.

Art. 16. Any one may denounce an old site (*sitio*) of *Hacienda* without paying anything for the same, even though there still be found in it the walls of aquaducts (*Targeas*), drains (*cauces*), yard (*Patio*), washing-places (*Lavadero*), furnaces, chimneys, dwelling-house, &c., provided they are entirely destitute of roofs, machines, tools and serviceable timbers; but if there be any, the former owner shall be notified to re-establish, sell, or rent them, within the term of four months, and if he fail to do this, they shall be granted to the denouncer, the latter being under obligation to pay the owner (*Dueño*) for what may be removable and useful, according to the valuation and appraisement of Experts (*Peritos*).

Art. 17. I prohibit any one who is not the discoverer from denouncing two contiguous Mines on the same vein; but I permit any person to acquire and possess one by denouncement, and another, or more, by purchase, gift, inheritance, or other just title whatsoever. And I further declare, that if any person desires to re-establish (*habilitacion*) many mines inundated or decayed, or any other considerable enterprise (*empresa*) of this kind, and, for this purpose, desires that there be granted to him by denouncement many *pertenencias*, even though they be contiguous and on the same vein, he must apply and represent the case to the Royal Tribunal General of Mexico, in order that, the merit and circumstances of the undertaking (*empresa*) being proved, it may report upon the matter to the Viceroy, who, finding therein nothing prejudicial to the Body of Mining, to the Public, or to my Royal Treasury, but rather advantageous, will grant to him this and other privileges, exemptions and assistance which may be given, provided, that my Royal approval be previously obtained for all such gifts (*gracias*) as do not come within the ordinary authority of the Viceroy.

Art. 18. *Placeres*, and any kind of deposits (*criaderos*) of gold and silver, shall be subject to discovery, registry, and denouncement, in the same form as Mines in veins, the same being understood of all kinds of metals.

Art. 19. Inasmuch as the refuse ores (*Desechaderos*) and rubbish-heaps (*Terreros*) of abandoned Mines are generally the means of support for the widows and orphans of the Mining operatives, for old men and invalids, and for other poor people of that employment, and even for all the inhabitants of the Place (*Lugar*), when the Mines are not in operation, I prohibit any person from denouncing them in order to make exclusive use of them, unless he, at the same time, denounce the Mines to which they pertain.

Art. 20. The same prohibition must be understood with respect to the slag (*escoriales*), refuse (*escombros*), and slime (*lameros*) of the smelting-houses (*Fundiciones*) and *Haciendas* of which nothing is left but the walls; but I order that, in case they have an owner, he shall be notified, and a certain time be allowed him, in which if he does not avail himself of the scoria (*Graseros*), residium (*Resocas*), and other refuse matter, or if the community does not use them, they may be granted to any one who shall denounce them.

Art. 21. Although in regular veins, or in *Placeres*, deposits (*criaderos*), or extraordinary pockets (*rebosaderos*), great natural masses of gold and virgin silver are found, I declare that the owners of the Mines shall acquire and enjoy them, paying the proper duties. And I also declare that there shall be considered as Treasures, only ancient deposits of coin or jewels, of bars or ingots, and other pieces smelted by man or buried by thieves, or in some other manner, from time immemorial, so that the owner thereof is unknown.

Art. 22. I also grant, that any one may discover, petition for, register, and denounce in the prescribed form, not only Mines of gold and silver, but also those of precious stones, copper, lead, tin, quicksilver, antimony, zinc, bismuth, rock salt, and any other fossils, whether perfect metals or half-minerals, bitumen or mineral tar (*jugos de la tierra*), they being given to him for his enjoyment, benefit and working, according to the particular circumstances of the cases. But I declare that, although the free discovery and denouncement of Mines of quicksilver is permitted, it must be on the express condition of giving an account of them to the Viceroy and to the Superintendent-Subdelegate of quicksilver in Mexico, in order that it may be considered and determined, whether such Mine or Mines shall be worked and carried on, at the expense of the particular subject who discovered and denounced them, he punctually delivering the quicksilver extracted from them into the Royal store-houses on the terms and prices stipulated; or whether it shall be done at the expense of my Royal Treasury, indemnifying the party for it by some equitable reward, according to the circumstances of the said discovery and denouncement, the whole of this important subject being regulated according to my Sovereign intentions recently declared on that subject.

Title VII.

Of the Persons who may, or may not, discover, denounce, and work Mines.

Art. 1. To all the subjects of my Dominions of Spain and the Indies, of whatever rank or condition they may be, I grant the Minés of all kinds of metals, with the conditions already stated, and which shall be expressed hereafter; but I prohibit foreigners from acquiring or working Mines as their own property (*propias*) in my said Dominions, unless they be naturalized, or tolerated therein by my express Royal License.

Art. 2. I also prohibit persons belonging to religious orders (*Regulares*) of both sexes, from denouncing, or in any manner acquiring, for themselves, or for their convents or communities, any Mines whatever: it being understood also that the working of Mines cannot devolve on the secular ecclesiastics, as being contrary to the laws, to the orders of the Mexican Council (*Concilio Mexicano*), and to the sanctity and practice of their profession; and therefore, in consequence of this prohibition, such secular ecclesiastics shall be absolutely obliged to sell and transfer to the hands of lay-subjects the Mines, or *Haciendas* for grinding and reducing metals, which, by title of inheritance or any other cause, may devolve on them, this being done within the term of six months, or such term as may be considered necessary in order to insure a useful result, and as may be fixed by the Viceroy with a previous report of the Royal Tribunal General of Mining, provided that, if it be proved that the provisions of this Article are evaded by artifice or fraud, to the prejudice of the working of such Mines and *Haciendas*, in which the State is so much interested, they may be denounced and disposed of in the same form as is prescribed for other cases.

Art. 3. Nor shall Governors, Intendents, Municipal Magistrates (*Corregidores*), chief Alcaldes (*Alcaldes Mayores*), or any other Judicial officers (*Justicias*) of *Reales* or *Asientos* of Mines, nor even their Notaries (*Escribanos*), hold any Mines therein; but I permit them to hold them in any territory out of their own jurisdiction.

Art. 4. Nor shall Administrators, Stewards *Mayordomos*), Watchmen (*Veladores*), Tallymen (*Rayadores*), Miners, or Minekeepers, and generally any servant or workmen of owners of Mines, whether regularly or temporarily employed, register, denounce, or in other manner acquire Mines within a thousand varas around those of their Masters (*Amos*); but I permit them to denounce any Mine for their said Masters, although they have no authority from them, provided the latter shall ratify the denouncement within the terms prescribed in Article 8, Title VI of these ordinances, without prejudice to the course of procedure.

Art. 5. No one shall denounce a Mine for another clandestinely or fraudulently, nor even publicly, if he has no power or written order, as is the usage.

Art. 6. Nor shall any one denounce a Mine for himself alone, having entered into a contract of partnership before the denouncement; and I order that the denouncer shall declare his partners in the said denouncement which he may make, under penalty of forfeiting (*de perder*) his share if he fail to do so.

Title VIII.

Of pertenencias and intermediate spaces, and of the dimensions which in future Mines are to have.

Art. 1. Experience having proved that the equality of the dimensions of Mines established on the surface cannot be observed under-ground (*en la profundidad*), where they are really of value, it being certain that the greater or less inclination of the vein to the plane of the horizon, makes the *pertenencias* of the Mines larger or smaller, so that there does not result that true and effective equality, which it has been desired to establish, between subjects of equal merit, as it may well happen that when a Miner, after much expense and labor, reaches the boundaries where the metal begins to be abundant and rich, another person compels him to turn back, as being already at the boundaries of his pertenencia, because of his having denounced the next Mine, and having placed himself at that point with more cunning than labor; so that from this arises one of the greatest and most frequent causes of litigation and dissention among Miners: Wherefore, and considering also that the limits established in the Mines of these Kingdoms, and by which those of New Spain have, until now, been regulated, are very confined in proportion to the multitude, abundance, and richness of the metalic veins which, of His great bounty, the Creator has been pleased to bestow upon those Regions, I order and command, that in Mines which shall hereafter be discovered in new vein, or without neighbors (*sin vecinos*), the following dimensions shall be observed:

Art. 2. On the thread, direction or course of the vein, whether of gold, silver or any other metal, I grant to every Miner, without distinction as to discoverers, (who already have their reward specified), two hundred Castillian *varas*, which are called *de medir* (long or running measure), taken on a level, and as has hitherto been understood.

Art. 3. To make what is called a square (*quadra*), that is, to make a rectangle with the preceding measurement, supposing the dip or inclination (*echado ó recuesto*) of the vein to be sufficiently shown in the pit of ten *varas*, the *pertenencia* will be measured by the following rule:

Art. 4. The vein being perpendicular to the horizon, (which rarely happens), one hundred *varas* shall be measured on a level, either side of the vein, or divided on both sides, as the Miner may desire them.

Art. 5. But the vein beng inclined, which is usual, attention shall be given to its greater or less dip, in the following manner:

Art. 6. If to one *vara* in depth the inclination be from three fingers (*dedos*) to two hands (*palmos*), the square given shall be of the same one hundred *varas*.

Art. 7. But if to the said one *vara* in depth there be an inclination of,

2 hands	and	3 fingers,	the square (*quadra*)			shall	be	$112\frac{1}{2}$	*varas.*
2	"	"	6	"	"	"	"	"	"	125	"
2	"	"	9	"	"	"	"	"	"	$137\frac{1}{2}$	"
3	"	"	0	"	"	"	"	"	"	150	"
3	"	"	3	"	"	"	"	"	"	$162\frac{1}{2}$	"
3	"	"	6	"	"	"	"	"	"	175	"
3	"	"	9	"	"	"	"	"	"	$187\frac{1}{2}$	"
4	"	"	0	"	"	"	"	"	"	200	"

So that, if to one *vara* in depth there be an inclination of four hands, which is one *vara*, there shall be granted to the Miner two hundred *varas* square (*por la quadra*) and on the dip of the vein; and so on with the remainder.

Art. 8. And considering that, in the manner prescribed, any Miner may reach the perpendicular depth of two hundred *varas* without exceeding the limits of his *pertenencia*, by which time, usually, he will have considerably exhausted the vien; and considering that those which have a greater inclination than that of vara for vara, that is, 45°, are either barren or of little duration, it is my Sovereign will that, although the dip or inclination of the vein may be greater than the dimensions indicated before, the square (*quadra*) shall never exceed two hnndred *varas* on the level, and that these dimensions shall always be the width of the said beds or lodes (*Mantas ó Vetas*), along the whole length of the other two hundred *varas*, which have been determined above.

Art. 9. But if any Miner, suspecting any other vein of a contrary inclination or variation from his own, which rarely occurs,) shall desire to have given to him any part of his square in a direction opposite to the inclination of the principal vein which he has denounced, it may be granted to him, provided that no fraud be shown, or prejudice result to a third person, and not otherwise.

Art. 10. In the *Placeres*, pockets (*Rebosaderos*), or other irregular deposits (*Criaderos*) of silver and gold, I order that the *pertenencias* and measurements shall be regulated by the respective territorial Deputations of Mining, attention being given to the extent and richness of the place (*sitio*), and to the number of applicants, preferring and distinguishing only the discoverers; but with the obligation that said Deputations give an exact account thereof to the Royal Tribunal General of Mexico, in order that it may determine thereupon, as it may think and deem most efficacious for avoiding all collusion.

Art. 11. The *pertenencias* being regulated in the manner prescribed, there shall be measured to the denouncer his *pertenencia* at the time of taking possession of the Mine, he being required to set up, in his boundaries, stakes (*Estacas*), or land-marks (*Mojones*), secure and easily distin guished, with the obligation of having them kept and preserved perpetually, without the power to change them, even though he allege that his vein has varied in course or inclination, (which are unusual things), but he must be content with the lot which Providence has assigned to him, enjoying it without disturbing his neighbors; if, however, he shall have no neighbors, or if he can, without injury to them, improve the location of his stakes (*mejora de Estacas*), or change his boundaries, he may be permitted to do so for such causes, with the previous intervention, knowledge and authority of the Deputation of the district, which shall summon and hear the parties as to whether they may be and are legitimate.

Art. 12. In Mines already opened and worked, the old dimensions shall be observed in their *pertenencias;* but they may be enlarged to the dimensions prescribed in these ordinances, whenever it can be done without prejudice to a third party.

Art. 13. The immutability of the stakes (*Estacas*) defined

in Article 11 of this Title, will also be observed henceforth even in Mines which are now worked, or which may be denounced as deserted or forfeited, fixing the measurements of those which have none, and preferring in their order the oldest Mines to those which are more recent; and if there result vacant spaces (*demasias*), what is prescribed in Article 13 of Title VI will be observed.

Art. 14. Whereas experience has shown, that the license or permission to enter into the *pertenencia* of another, working at a greater depth and within the vein, following its metal, and enjoying it until its owner can dig down to it, has been, and is, the most fruitful cause of the most bitter law-suits, dissentions and disturbances among Miners; and, moreover, that this intrusion more often occurs from fraud or accident than from the merit or industry of the intruder, resulting generally in nothing but the serious injury or total ruin of the two Mines and the two neighboring Miners, to the great prejudice of the public and of my Royal Treasury, I order and command, that no Miner shall enter upon the *pertenencia* of another, even though it be by going to a greater depth and by working his own vein, but that each one shall keep and observe the boundaries of his own, unless he amicably agrees and stipulates with his neighbor for the right to work in his *pertenencia.*

Art. 15. But if any Miner, pursuing his working fairly, reaches the *pertenencia* of another while following the metal which it bears, or discovering it at the time without the owner of the *pertenencia* having on his part discovered it, he shall be obliged to give the latter immediate notice of it, and thenceforward the metal and the expenses shall be divided between the two neighbors in equal parts: the one, for the merit of the discovery, and the other, as the owner of the *pertenencia*: all which shall be observed until an opening or communication is made into the *pertenencia*, either by the vein, or by a cross-lode (*crucero*), or in any manner that may be most easy and convenient; in which case, a boundary (*Guarda-raya*) being established, each one shall confine himself within his own *pertenencia*. But if he who discovers or follows the metal within the *pertenencia* of another shall not give prompt notice to his neighbor, he shall not only forfeit the right to one-half of all that he may take out, but he shall also pay double the value for what he may have taken out; it being understood, that for the imposition of this penalty there must be previously proved, in the plainest manner possible, and according to the order prescribed in Title III, the fraudulent intent (*mala fé*) in the taking out of said metal.

Art. 16. And in case any Miner shall have advanced so far in his subterranean workings as to have passed the boundaries of his *pertenencia*, either in its length or width, I declare that he shall not, on this account, be obliged to turn back or stop the work, provided, that he be in ground where no Mine has been taken up (*terreno virgen*), or in the *pertenencia* of a deserted Mine; but he shall be obliged to denounce the new *pertenencia*, which shall be granted to him so as not to exceed in each grant the dimensions which were previously granted to him, and with the obligation of removing his stakes (*Estacas*) to the new boundaries, in order that they may be thereafter known.

Art. 17. The Miner shall be the owner, not only of the space of the particular vein which he denounced, but also of all those which in any form, figure or situation, are found within his *pertenencia;* so that, if a vein begins (*sacare la cabesa*) in one *pertenencia*, and passing on, terminates (*llevare la cola*) in another, each owner shall enjoy the portion which passes through his respective boundaries, and neither the first, nor any other, on account of having discovered it in his boundaries, or of its beginning there, can claim it in its whole extent, or wherever it may happen to be.

Title IX.

Of the manner in which Mines ought to be worked, secured, and protected.

Art. 1. It being of the greatest importance that the lives of the workmen, and other persons who are obliged continually to enter and go out of the subterranean works of Mines, should not be endangered, and that the Mines should be kept in a state of security and convenience necessary to the progress of their works, even those which their first owners abandon as useless, or because they are unable to carry them on (*habilitarlas*); and it being impossible to establish any general and absolute rule on the subject, because the variety of circumstances of each Mine, in the greater or less firmness, tenacity and cohesiveness of the sides (*respaldos*) and of the substance itself of the vein, its greater or less inclination (*echado*), the width and depth of its workings (*labores*), lead to much diversity, in the dimensions and number of the pillars, arches, beams (*testeras*), cross-beams (*intermedios*), and other supports, which must be left or constructed in order to sustain the sides; and also in the arrangement of the works required for proper ventilation, and for the convenient removal of the substances which are to be taken out of the Mines; all of which cannot be effected without true practical skill and science in the working of Mines; I order and command as follows:

Art. 2. It shall not be permitted to any one to work Mines, without the direction and continual assistance of one of the intelligent and practical Experts (*Peritos*) who, in New Spain, are called *Mineros* or *Guarda-minas*, who must be examined, licensed and approved by one of the Professors of Mining (*Facultativos de Mineria*) which each *Real* or *Asiento* must have, as will be mentioned hereafter. But in Places very poor or remote, where, on that account, there is, as yet, no Professor of Mines, nor other Expert (*Perito*), licensed or examined, it is permitted to proceed under the direction of one of the most intelligent and reputable persons who may be found there, until these, or others, can be examined and licensed; the same being understood in all cases which require the direction or intervention of an Expert (*Perito*), it being thus directed in judicial proceedings, in order to give them the faith and credit which they ought to have.

Art. 3. In order to mark out and determine the shafts (*tiros*), communications (*contra-minas,*) or adits (*socabones*), and other great and difficult works, which, if found to be wrong after they are made, render useless the great expenses which they have caused, the direction of one or more *Mineros* or *Guarda-minas* shall not be sufficient, but the inspection or intervention of one of the said Professors of Mining (*Facultativos de Mineria*) shall be necessary, with the obligation, on his part, of visiting the work once every month, or every two months, as the progress of it may require, in order that, if he should observe any error in the execution, he may correct it in time, and before it occasions greater expenses.

Art. 4. In Mines opened on veins, of which the sides and interior substance are soft, or of so little tenacity or cohesiveness that they crumble and crack, and open into fissures or crevices, by the effect of air or time, or which, for any other cause, are known to be insufficient, of themselves, to maintain the security and firmness of the Mine, I order and command, that the workings (*labores*) be lined and secured with strong and solid timbers of known durability, or little liable to decay under ground, framed and strengthened according to the rules of art; or with good masonry of stone and mortar, if the richness and other circumstances of the Mine require or permit it; for which purpose, in all Places, *Asientos* or *Reales* of Mines, there ought to be a supply of those artificers, carpenters and masons, who are called liners (*Ademadores*), and these ought to have journeymen and apprentices, in order to preserve and continue so important a trade, and one which ought to be encouraged and well-paid.

Art. 5. In order to prevent the introduction of artificers who have not the proper intelligence and experience in subterranean architecture, none shall be admitted who have not been examined and approved by the Professor of Mines (*Facultativo de Minas*) appointed for that Place, or for some other.

Art. 6. If any Miner, in consequence of the great richness of the metalic substance of his vein, is desirous of substituting, in place of the pillars, arches or other supports of the vein itself which are sufficiently firm and tenacious, others of stone-masonry, he will be permitted to do so, after the inspection of one of the Deputies of the district, assisted by the Notary (*Escribano*), and the approval of the regular Professor (*Facultativo*) of the district.

Art. 7. I strictly prohibit any one from removing entirely, or even weakening and diminishing, the pillars, arches and necessary supports of the Mines, under the penalty of ten years imprisonment, which, according to and in the form declared in Title III of these ordinances, shall be imposed by the respective Judge upon any workman (*operario*), searcher (*Buscon*), or tester (*Cateador*) who shall commit the offense, and the same upon any Miner or Mine-Keeper (*Minero ó Guarda Minas*) who shall permit it; and upon the owner of the Mine, the forfeiture of it (*de perderla*) ,and also the half of his property, he being forever excluded from any employment in Mining.

Art. 8. I order and command, that the Mines shall be kept clean and unobstructed, and that the workings which are useful or necessary for the circulation of air, for the carriage and extraction of the metal, or for other purposes, although they contain no other ore than that in the pillars or division walls, shall not be occupied with rubbish and refuse (*atierres y tepetates*), but that these shall be carried out and thrown into the rubbish-heap (*Terrero*) of its own proper *pertenencia*, and on no account into the *pertenencia* of another, without the permission and consent of its owner.

Art. 9. There must be in the Mines sufficient and secure ladders (*escaleras*), of the kind and number considered necessary by the Mining Expert (*Perito Minero*), for the purpose of ascending and descending conveniently to the furthest workings, so that the lives of those who use them may never be endangered, in any manner, by their being weak, insecure, rotten, or much worn.

Art. 10. In order to prevent the violation of any of the Articles contained in this Title, it is my sovereign will, that the Deputies of Mining, accompanied by the Professor of Mines of that district, and by the Notary (*Escribano*), if there be one, and in default of him, by two assisting witnesses (*testigos de asistencia*), shall, once in every six months, or once a year in Places where they cannot do it oftener, visit all the Mines in their jurisdiction which are in actual working; and if they find that there is any failure in the points defined in the said Articles, or in others which have regard to the security and preservation of the Mines, and to their better working, they shall immediately adopt measures to have the defect remedied and amended within a proper time, and the means of assuring themselves that the remedies are properly carried into effect. And if there be a failure to do this, or if the same offense occur again, they will impose upon the offenders the corresponding penalties, multiplying and increasing them even to the forfeiture (*hasta la perdida*) of the Mine, leaving it for the first who may denounce it, provided that the Deputies proceed according to the form prescribed in Title III of these ordinances.

Art. 11. I most rigorously prohibit that permission be given to any one to dig through adits (*socabones*) cross-cuts (*cruceros*), or any other passages (*cañones*,) from other workings above and full of water, or to leave between them such slight supports as to allow the water to burst through, but they shall be obliged to drain with machines the inundated workings before connecting them with new ones, unless in the judgment of the Professor of Mines the piercing through may be executed without danger to the workmen engaged in it.

Art. 12. I also prohibit any one from venturing to introduce operatives into workings filled with noxious vapors, until they have ventilated them according to the rules of art.

Art. 13. As Mines require to be worked continually and incessantly in order to procure their metals, and as they require in them works and operations which can be executed only in a long time, and as their re-establishment, if their working be suspended and interrupted, will cost as much as in their original undertaking: Therefore, to obviate this inconvenience, and also to prevent any owners of Mines, who cannot or will not work them, from keeping them without use and for a long time, impeding, by a pretended working, the real and effective labor which others might bestow upon them, I order and command, that any one who shall, for four consecutive months, fail to work a Mine with four operatives, regularly employed (*rayados*), and occupied in some interior or exterior work of real utility and advantage, shall thereby forfeit (*por el mismo hecho pierda*) the right which he may have to the Mine, and it shall belong to the denouncer who proves its desertion according to the provisions of Title VI.

Art. 14. Experience having shown that the provision of the foregoing Article has been eluded by many owners of Mines by the artful and fraudulent means of having them worked some days of every four months, keeping them in this manner many years in possession, I order also that whosoever shall fail to work his Mine, in the manner prescribed in said Article, for eight months in a year, reckoning from the day of his possession, even though the said eight months should be interrupted by some days or weeks of labor, shall thereby forfeit (*pierda por el mismo hecho*) such Mine, and it may be adjudicated to the first person who denounces it, and proves this second species of desertion, unless, for this and for the one mentioned in the preceding Article, there have occurred causes of pestilence, famine, or war, in the Place (*Lugar*) itself of Mines, or within twenty leagues thereof.

Art. 15. Considering that many Miners, who formerly worked their Mines with care, spending large fortunes in shafts (*tiros*,) adits (*socabones*,) and other very expensive works, may happen to suspend working them for some time, soliciting supplies (*avios*), or for want of operatives, or of the necessary provisions, and for other just causes, which, combined with their former merit, render them worthy of some equitable consideration, I declare, that if any of the said Miners may have abandoned their Mine for the times and in the manner above described, they shall not thereby forfeit (*pierdan por el mismo hecho*) them, like the others; but their Mines must, nevertheless, be subject to denouncement before the respective new Courts of Mining, in order that, the parties being heard, and the alleged merits and causes being proved, justice may be done to him to whom it is due.

Art. 16. Whereas, many Miners abandon their Mines, either because they want the capital to carry on the work, or because they are unwilling to consume what they have derived from the same Mines, or because they have not the spirit to venture upon pursuing unfruitful workings where they had conceived great hopes, or for other causes, while persons are not wanting who might desire to take them, if they had notice of their abandonment, it being much easier to maintain the actual working of a Mine, than to reestablish it after it has suffered the injuries of time, it is my will, that no person shall abandon the working of his Mine or Mines, without first informing the Deputation of the district, in order that it may publish the same, by fixing notices on the doors of the churches and other accustomed places, so that all may have notice of it.

Art. 17. To avoid false and equivocal reports by which some abandoned Mines are often recommended, and the evil results of which increase the distrust which ordinarily attaches to this profession, deterring from it some persons who do not otherwise want an inclination to engage in it, I order the following:

Art. 18. That no one shall abandon the working of his Mine without giving notice to the respective Deputation, in order that an inspection of it may be immediately made by the Deputies, accompanied by the Notary (*Escribano*) and Experts (*Peritos*), who must examine and measure the Mine, particularizing all its circumstances, and making Maps which represent its plans and profiles; which, with all the particular information indicated, shall be preserved in the Archives, in order that they may there be freely examined by any one who may wish to see them, or to take copies of them.

Title X.

Of Mines which require draining.

Art. 1. Whereas, in the greater part of Mines, springs and veins of water are encountered, from which it often runs perennially and in such abundance as, in a short time, to fill and inundate all their workings, impeding their progress and the extraction of their metals;—I will and command, that the owners of such Mines shall maintain in them, continually, the drainage or clearing of the workings, so that they shall always be in a condition to be worked, and the metals which they may have, to be extracted.

Art. 2. As it is much the most convenient and least expensive method of draining veins, to countermine them by means of adits (*socabones*), I order, that in all Mines which require draining, and whose situation will admit of it, and where, in the judgment of the Professor (*Facultativo*) of the district, advantage must result from it, the owners shall be obliged to make an adit (*socabon*) sufficient for the draining and clearing of the workings; provided, that the richness and abundance of the metals justify it, and are likely to repay the expenses.

Art. 3. If, by such adit (*socabon*), several Mines can be cleared and kept in working order, I declare that, although each singly cannot bear the expense of the construction of the adit, it shall be made and paid for by all together, the expense being apportioned among them, according to the benefit each shall derive from it; and, if this cannot be ascertained at the time, then, until it is ascertained, they shall bear the expense equally, it being arranged according to what the poorest Mine can pay; and if this latter should improve its condition, then arranging the expense according to what the poorest of the others can pay; so that the work on the adit may not be suspended, and all shall be assessed, approved, and regulated by the Deputation of the district, and according to the judgment of its respective Professor (*Facultativo*) of Mines.

Art. 4. If any individual shall offer to construct an adit (*socabon*) by which one or more veins, or the Mines opened in them may be put in working order, even though he may not be an owner, in whole or in part, of any one of them, nevertheless his denouncement, if made in due form, shall be admitted, and notice shall be immediately given to the owners of the said Mines, who, if they will bind themselves to execute the said work, shall always be preferred; but if they refuse. it shall be assigned to the adventurer (*Aventurero*), on the following conditions:

Art. 5. That the adit (*socabon*) shall be truly useful and practicable, in the judgment of the Professor of Mines (*Facultativo de Minas*), who must be charged with laying out and planning the work, and with directing its execution as was ordered.

Art. 6. That the countermine shall be made, as nearly

as possible, in a straight line, and at the shortest distance from the vein or veins which are to be cleared, or in the line or direction of some one of them.

Art. 7. That there shall be constructed corresponding light-pits (*Lumbreros*), or a communication (*Contracañon*), or some other means, sufficient always to maintain, in the work, free ventilation and a supply of air for the operatives.

Art. 8. That its size shall be determined by the Professor, according to the circumstances; but it shall never exceed two varas in width and three in hight, being always well secured and lined (*ademado*).

Art. 9. That if the adventurer shall meet, in the progress of his work, one or more new veins, he shall enjoy in them the right of a discoverer, and the reward which, in these ordinances, is assigned to such. But if they are veins already known and opened in other places, I grant to him the right to acquire one *pertenencia* in each of them, and, if there be not sufficient room for that purpose, then the vacant space (*demasia*), until he comes to the *pertenencia* of another person.

Art. 10. That if the work pass through deserted Mines, the adventurer shall thereby become the owner of them, and may denounce them as soon as he projects the work; it being understood, that these and the new *pertenencias* are to be maintained, while the work is carried on, as much as that will permit. But I declare, that as soon as the work is completed, he must maintain them separately, under the penalty of forfeiting them (*perderlas*), as is already provided.

Art. 11. And finally, that if the adit passes through occupied Mines, and is in the line of the vein, the adventurer shall be entitled to half of the metals which he takes out of it, and the other half shall belong to the owner of the *per-*

tenencia, but all the expenses are to be paid by the adventurer: nevertheless he must not exceed in his adit the prescribed dimensions, nor undertake other workings, without the consent of the owner, in which case the expenses shall be equally divided between them both. But if the adit passes across the vein, the adventurer may open workings along the vein, dividing the metals and the expenses equally between the two, until in some way an opening be made into the workings by the owner of the Mine; and if the adventurer shall not immediately notify such owner of his discovery of metal, he shall not only lose his right to the half, but shall restore all that he may have taken out, and double its value, the fraud and mal-intent being previously proved, according to the form established in Title III.

Art. 12. All the provisions of this Title, from Article 5 inclusive, with respect to adventurers, shall also be understood, so far as they may be applicable, of owners of Mines who are disposed to clear their own Mines and those belonging to others, by means of an adit or general countermine, whether the work be undertaken by all together, or by some without the others, or in partnership with adventurers, observing punctually, in all of these cases, the stipulations which may be entered into, provided, they are not repugnant to the provisions and objects of these ordinances.

Art. 13. The owners of Mines which require draining, and whose situation do not permit countermining by means of an adit, must make in them a general and continuous shaft, which in New Spain is called a *tiro*, and serves to draw off, by means of engines or machines, the water, metals and other substances of the Mines; which, consequently, must be made in such situation, and of such dimensions, and with such securities, as the Pro-

24

fessor of the district may order and direct. And the territorial Deputations are charged to give to this matter, very special care in their visits, imposing and increasing the corresponding penalties in proportion to the offense which is proved.

Art. 14. Whereas, experience has shown the general utility of said works, and also that by neglect and inattention, and to economise their cost, they have been left higher than the workings of the Mine; that such undertakings afterward become much more difficult and expensive; and if capital be wanting to complete them, the owners are forced to clear the lowest workings by interior drains, raising the water to the shaft (*tiro*) by means of machines moved by men with little effect and great expense, and sometimes with intolerable fatigue to human strength; I order and command, that all owners of Mines which require draining shall always sink the bottom or base of the *tiro* deeper than the lowest workings and pits, in such manner, that there may remain to them a sufficient support to enable them to be worked, and room enough in the *tiro* for the water: the observance of which shall be watched with particular care by the territorial Deputations in their visits, imposing penalties as provided in the preceding Article.

Art. 15. If any owner of Mines which require draining is unwilling to keep them drained, contenting himself with working the upper portions which are not reached by the inundation, and another person shall denounce the Mine or Mines, offering to drain and clear the deepest workings, notice shall immediately be given to the possessor of such Mine, in order that, if he be either unwilling or unable to establish the draining within the term of four months, it may be adjudicated to the denouncer, he giving security for the expenses of the draining according to the estimate of Experts (*Peritos*), and to the satisfaction of the Deputies of the district.

Art. 16. If the owner of any Mine of which the workings are lower than those of his neighbors, whether from its situation or its greater progress, shall be put to greater expense for his draining by his neighbors either not draining, or draining insufficiently, their higher Mines, and water shall communicate from one to the other, I order and command that the owners of the higher Mines shall keep up all the draining which they require, or, in default thereof, shall respectively pay to the owners of the lower Mines, in silver or current money, the damage which they may cause to the latter, it being assessed by Experts (*Peritos*), who shall previously examine into the case, and make experiments with the greatest possible accuracy.

Art. 17. To all persons who shall undertake to defray the expense of draining and clearing many Mines, by constructing general *Tiros* or other works, and causing to be constructed and maintaining costly machines, because adits are not practicable, I grant the ownership of all the deserted Mines and *pertenencias* which they shall effectually clear, even though they be contiguous upon the same vein; and I command that the Viceroy, on the recommendation of the Royal Tribunal General of Mexico, shall grant to them, all the privileges, exemptions and assistance which can be dispensed. But I declare, that the owners of occupied Mines, and those who in any manner derive benefits from such works, shall only be obliged to contribute to them, in proportion to the benefits which their Mines receive, according to the valuation of Experts (*Peritos*), with the intervention of the Deputies of the district.

Title XI.

Of Mines worked in Partnership.

Art. 1. Inasmuch as many Mines are worked by several Miners united, their partnership beginning from the time they denounce them, or being subsequently entered into in various ways, and as this is of great advantage and utility in working Mines, since it is is more easy to engage therein where many unite together, each contributing his part of the capital, or because, where the capital of one alone is not sufficient for great undertakings (*empresas*), that of all the partners may be ample; I will and command, that such special or general partnerships be encouraged, promoted, and protected by all convenient measures, my Viceroy granting to those who may form them, all the favors (*gracias*), assistance and exemptions which can be granted to them, according to the judgment and discretion of the Royal Tribunal of Mining, and without detriment to the interests of the public and of my Royal Treasury.

Art. 2. Although, by these ordinances, I prohibit any one individual Miner, who works in the ordinary manner, from denouncing two contiguous Mines on the same vein; nevertheless I grant to those who may work in partnership, although they be not discoverers, and without prejudice to the right which by reason thereof they may have in case they are such, the right to denounce four new *pertenencias*, or Mines which have been worked and abandoned, even when they are contiguous and on the same course.

Art. 3. The common usage in New Spain of dividing a Mine into twenty-four imaginary equal parts, called *Barras* (shares), also of subdividing each of these into suitable smaller parts, shall continue to be observed as heretofore, without any change.

Art. 4. Consequently no one of the partners can pretend to have a right to work the part A, or any determined part of the Mine, while another works the part B, nor each place therein a certain number ot workmen; but they must work in common, so far as the Mine will admit, and must make a division of the expenses, sharing the amount proportionally among all the partners, and the same of the produce in metals of every kind and quality, whether in the rough state (*en bruto*), or after they have been reduced in common, as may be agreed upon.

Art. 5. To avoid the disputes and differences which ordinarily occur in partnership Mines, about the direction of the works, applications for supplies, the administration, and other points connected with their working; I order and command, that all the measures necessary to be taken, shall be determined by a plurality of votes, with the intervention of one of the Deputies of the district, who shall always endeavor to preserve harmony.

Art. 6. The votes shall be valued and counted according to the number of *barras* which each partner may possess in the Mine; so that, if one or more be owners of one and the same *barra*, they shall have only one vote, and he who may have two *barras*, shall have two votes, and so on for the rest; but if one person alone is owner of twelve or more *barras*, his vote shall always be counted as one less than half of the whole.

Art. 7. In all cases where, by a tie of votes, or by any other cause, a dispute shall arise, it shall be decided by the Deputy of Mining who presides over the *Junta*, as is before declared, and I charge him to always attend to what is most just, and most to the common interest of all the partners.

Art. 8. If, while being worked, a Mine does not produce any profit, or fails, wholly or in part, to defray the current expenses, and any one of the partners is unwilling to contribute his portion of them, in that case, the others shall give notice to the respective Deputation, in order that the day may be noted on which such partner failed to contribute; and if he shall persist in refusing to pay for four months continuously, I declare, that by that act, and from the day on which he ceased to contribute, the share of the Mine, which he possessed, is to be considered abandoned, and it shall go to increase proportionally the shares of those who may have contributed, without the necessity of denouncing it; but if, before the expiration of the four months, he shall contribute his part of the expenses, it shall be received, provided he pays, to the satisfaction of the parties interested, all which he may owe as having accrued during the time he ceased to contribute.

Art. 9. If while the Mine is producing (*en frutos*), any one of the partners is unwilling to contribute to the expenses of dead works (*faenas muertas*), which have been decided on, according to the formalities before prescribed, on the ground that they would consume a part or all the produce of the Mine, the other partners may retain and devote to this purpose, a part or all of his share of the metals.

Art. 10. If one or more Mines be worked by two partners, and they desire to dissolve the partnership, on account of a disagreement or any other cause, they shall not, in such case, be reciprocally obliged to purchase or sell, the one to the other, their respective shares, but each one shall be at liberty to sell to any third party, the partner only having the right of preference at the same price.

Art. 11. A partnership in Mines is not to be considered as dissolved by the death of any one of the partners, for the heirs are obliged to continue in it; but they have the privilege of selling their share, in the manner prescribed in the preceding Article.

Art. 12. If a share of a Mine, or an entire Mine, should be sold, after having been appraised and valued by Experts (*Peritos*), according to its condition at the time, and it should afterward produce great riches, I declare that the sale shall not, on that account, be rescinded on the pleas of enormous or most enormous lesion (*lesion enorme ó enormisima*), or of restitution *in integrum* to a minor (*restitucion in integrum de Menor*), or of any other similar privilege.

Title XII.

Of the operatives in Mines, and in reducing establishments and works.

Art. 1. Inasmuch as it is both notorious and true, that the operatives in Mines are a class of people very poor, but useful to the state, and as they ought to be protected, and paid for their severe toil, according to justice and equity; I will and command, that no owner of Mines shall presume, for any pretext or cause, to alter the rates of wages (*Jornales*) established by the legitimate custom adopted in each *Real* of Mines, but that the same shall be inviolably observed, as well with respect to the operatives in the Mines, as those who work in the reducing *Haciendas* or reducing works (*Ingenios*), under the penalty of paying double, if at any time they shall diminish the said rates of wages (*jornales*); and the operatives shall be obliged to work at the rates which are so established.

Art. 2. The operatives in Mines must be registered by their own proper names, and be marked on the rolls (*rayarse*) every time they leave their work, by clear and distinct lines, so that they themselves may see and recognize the marks, even though they be unable to read: all this in the proper forms commonly used in New Spain.

Art. 3. The rolls of wages (*Memorias de los Jornales*) shall be paid weekly, to each operative according to his tallies (*Rayas*), and with the greatest punctuality, cash in hand, and in current coin, or in silver or gold bullion of good quality, if there be no coin, or with a part of the metal which they extract, if it has been so agreed upon. And I strictly prohibi ttheir being, in any manner, compelled to receive merchandise, goods, fruits or provisions.

Art. 4. At the time of paying them their *Rayas*, the operatives shall not be forced to satisfy their debts or charges, even though these be privileged (*privilegiadas*), without an order of court, except those which they may have contracted with the owner of the Mine to pay with their labor; and, even for these, he shall retain and take from them only one-fourth part of what their *Rayas* amount to.

Art. 5. I forbid that alms, charities, society-contributions (*cornadillos de cofradios*), or other similar things be sought from the operatives, until, having received their dues they may wish of their own accord to make such donations.

Art. 6. When the operatives are paid by weekly rations and monthly salaries, the rations supplied shall be good wholesome meat, wheat, maize, *pinole*, salt, red-pepper, and such other things as may be usual, by fixed (*señaladas*) and exact weights and measures: to all of which very particular attention shall be paid in the visits (of the Deputies).

Art. 7. Each operative or servant of Mines, referred to

in the foregoing Article, must have in his possession a pass-paper (*Papel*) in which shall be noted the parts of his monthly salary which are to be paid, and those which have been received in advance, all written by the Tallyman (*Rayador*) or Cashier (*Pagador*) of the Mine or *Hacienda*, and the dollars and *reales* designated by circles and lines, and half-circles and half-lines; so that each operative may understand and adjust his own account, and have in his possession evidence of it.

Art. 8. The duties or tasks (*Tequios ó Tareas*) of the operatives must be assigned by the captain of *Barras*, with reference to the hardness or softness, extent, scarcity, and other circumstances of the working, proceeding, with the greatest justice and equity, in the regulation of said tasks, in allowing good pay for the jobs (*destajos*), and in increasing it when the circumstances are changed; and in case any one of the two parties complain of grievance in this particular, the respective Deputation of Mining will proceed to redress whatsoever grievance, by a verbal order, or by a summary judicial proceeding, if that should not suffice to settle them: all of which must be done in the form prescribed in Title III of these ordinances.

Art. 9. It is also my royal will, that to the Indians of allotment (*repartimiento*) no advances shall be made, so that, as soon as they have served the time of their assignments (*Tandas*), they may return to their towns and habitations, and be replaced by others, as is prescribed by the laws; and that to the free Indians advances may be made only to the amount of five dollars each, according to a Decree (*Auto*) issued by my Royal *Audiencia* of Mexico: Although in case of any recognized urgency, as to celebrate their marriages, or to bury their wives or children, I permit that, on the same being proved to the owner of the Mine, to the administrator or the overseer (*Mandon*), by a certificate of the curate, they be furnished with what they require for that purpose.

Art. 10. Both the owners of Mines and the operatives shall be entirely at liberty to agree between themselves whether to work in them on shares (*á Partido*), or on salaries and shares. Although this liberty is reciprocal, when the Mine is not worked solely on shares (*á solo Partido*), its owner or Administrator shall pay the operatives, on account of their wages or salaries, such sum as shall be conformable to the provisions of Article 1 of this Title; and if any barman (*Barretero*) working on wages only, shall, after having completed his task (*Tarea ó Tequio*), voluntarily continue the whole, or a part, of the unexpired time of the *Tanda*, taking out metal, the owner of the Mine shall not be obliged to do more than to pay him, also in money, and at the rates of the *Tarea*, for all the additional metal he may extract from it. But if, in order to promote and stimulate the work of the operatives, the owner or Administrator of the Mine shall agree with them to pay them so much for every sack or bag (*Costal ó Tenate*) of metal which may be extracted in addition to the *Tequio*, or with a part of the same metal, they shall observe, in this case as well as in all others which are adjusted and agreed upon, the stipulations which they both have entered into, so long as, in the opinion of the respective Deputies of Mining, there is no material change of circumstances; and if these cannot agree, the Substitute shall decide according to the rule already established. Moreover, if, in regard to the agreement as to the terms upon which the operatives are to work in the Mine, there should arise, between them and its owner or steward, any dispute which is likely to do injury to its working and progress, and consequently, to the State, and in case either of the parties should make reclamation about it, the proper Deputation shall decide it, or in case of their disagreement, the said Substitute, in conformity with the practice which has been established in the Mine in question, or if it be a new one, in the *Real* to which it belongs.

Art. 11. The metal produced by task and share-work (*Tequios y Partidos*) must be received and attested by the Tallyman (*Rayador*,) or Watchman (*Velador*), or other officer which the owner of the Mine may appoint for that purpose; and if he find that the metal of any barman (*Barretero*) gotten out as share-work, is better and cleaner than that gotten out as task-work, both shall be mixed together, in the presence of the said interested operative, and be mingled to his satisfaction, in order that, from the part of the heap formed by said mixing which he may select and choose, there shall be filled up as many sacks, bags, or measures, as there may have been of his share; it being forbidden to the owner of the Mine, his steward, overseer, or other servants, to prevent, under any pretext, the said interested barmen (*Barreteros*) from being present during all of the said operation, or to cause the said sacks or bags to be filled with the mixed metals, from any other part of the pile than that which the barmen (*Barreteros*) may select.

Art. 12. The watchman (*Velador*) may examine all those who enter or come out of the Mines, observing with the greatest care whether any who enter are intoxicated, or whether they carry with them any intoxicating liquors; and he may also examine all that goes in or out of the Mine under the head of breakfasts, dinners, &c.; and if he shall find any stolen metal, tools, powder, or other like thing, he may preventively arrest the thief, iron him and secure him, and, this being done, report him to the territorial Deputation, in order that, conformably to the provisions of Title III of these ordinances, respecting criminal causes, due proceedings may be had.

Art. 13. Idlers and Vagabonds, of whatsoever class or condition, who may be found in the *Reales* of Mines and the neighboring places, may be seized and forced to work in them, as also the operatives who, from mere idleness, leave their employment, without engaging in any other business; for which purpose the owners of Mines may keep Police officers (*Recogedores*), with the license of the Courts and of the territorial Deputation of Mining, as is usual; but it is to be understood, that no Spaniard, nor Spanish half-breed (*Mestizo de Español*), for such are to be reputed as Spaniards, can be thus dealt with, as both are exempt by the laws, and when such persons by their idleness or offenses are to be corrected, other penalties are to be imposed by the proper Judge, according to the nature of their excesses.

Art. 14. In the distribution and allotment (*repartimiento*) of the Indians of the Pueblos near the *Reales* of Mines, who are called *of Quatequil* or *of Mita*, to the *Haciendas* for reducing ores, the orders and superior grants (*Despachos y Providencias superiores*) obtained at different times by the owners of said *Haciendas* which are in operation and have continued so without intermission, shall be observed; but in respect to those deserted and abandoned *Haciendas*, whose allotment has passed to others more recently established, the latter are to be maintained in the possession of the same, and the former, in case of their re-establishment, can only reclaim the *Quatequil* of the Pueblos which formerly was theirs and which may not have passed to some one recently occupied, observing the same with respect to the working bands (*Quadrillas*) of Mines and *Haciendas;* but in neither case shall such distribution and allotment of Indians of *Quatequil* or of *Mita* exceed four per cent., conformably to the practice followed in New Spain. And in order to soften the rigor of the *Mitas*, as much as possible, in favor of the Indians, I order and command, that in the execution and fulfillment of Law I, Title XV, of Book VI, and Law IV, Title XV, Book VII, Negroes and free Mulattoes who are vagabonds, and half-breeds (*Mestizos*) of the second order who have no employment, may be seized and forced to labor in the Mines; and that those who, for offenses, are condemned to any labor, not being excepted by the preceding Article, may be made to work in the Mines, provided, that the owners of the Mines are willing to receive them, as such owners are to be, on their part, entirely free to do so or not, according to the greater or less facility which they may have for guarding them during the intervals of the working.

Art. 15. The working bands (*Quadrillas*) of the abandoned *Haciendas* cannot easily erect themselves into Pueblos, even though they may construct a chapel and erect a belfry, because, by appropriating in this way the land and water of the *Hacienda*, for which purpose the place (*sitio*) was suitable, they impede, or even render impossible, its re-establishment; and in order to avoid this, I will and command, that while living there, they always remember that the *sitio* will be perpetually subject to denouncement, and that in case such *Hacienda* is re-established in it, they must become again members of the working band (*vecinos de Quadrilla*), and live dependent upon the owner of the *Hacienda*.

Art. 16. The operatives belonging to working bands (*Quadrillas*) of Mines or *Haciendas* shall be obliged to work, in preference, in the one to which they are attached (*aquadrillados*), and they can work in another place only with the consent of the owner of the band (*Quadrilla*), or when he has no occupation for them.

Art. 17. It being found by experience that Mines, which

are in dead works (i. e. unproductive), are generally deficient in operatives, because all flock to those which are yielding ores, especially if their owners allow them shares (*Partidos*), thereby interrupting, and even rendering impossible, the re-establishment (*habilitacion*) of the other Mines: in order to remedy this, I order and command, that the territorial Deputations shall cause such operatives as are vagrants and not organized into working bands (*aquadrillados*), to be divided in such a manner that, by distributing them alternately and successively to both these classes of Mines, they shall not fail to enjoy the advantage of those which are productive (*en bonanza*), nor deprive the others of labor. And with the same object, it is my Sovereign will, that no operative who shall leave one Mine to work in another, shall be admitted by the owner of the latter, without producing a certificate of good behavior from the Master (*Amo*) he has left or from his Administrator, under penalty, both to the Mine-owner who admits him, and to the operative, of being punished in proportion to the evil intent with which they have respectively acted: the observance of which shall be very strictly watched by the said territorial Deputations, as matters pertaining to their jurisdiction.

Art. 18. The operatives of Mines who, because they have contracted debts in any one of them, go to work and to be enrolled in another, shall be forced to return to the first, and to pay in it, by their labor, such debt, according to, and in the manner prescribed by, Article 4 of this Title, unless the creditor consent to the payment of the obligation by the owner of the other Mine.

Art. 19. Thefts committed by operatives of Mines or *Haciendas*, whether of metalic ores, tools, powder, or quicksilver, shall be punished according to the circumstances and gravity of the offenses, and to their frequency, if repeated, imposing the penalties established by law, and proportioning the punishments of the offenses which the Indians may commit, according to the damage done and the evil intent evinced; the respective Judges, in the cognizance of these cases, regulating themselves by what I have laid down and declared, in relation thereto, in Title III of these ordinances.

Art. 20. Operatives who, for slight offenses, or for debts or other causes, are accustomed to remain in prison a long time, thus injuring themselves and causing want to their families and even to the Mines, may be removed from prison and made to work in them, provided, that there be in the Mine or *Hacienda* where they are to be employed, means of confining and securing them, during the intervals of work, in order that, by this means, a portion of what they may earn being set apart for their own support and that of their families, the remainder may be applied for the payment of their debts, the celebration of their marriages, or for the pecuniary penalties imposed for the satisfaction of the party aggrieved, the owner or Administrator of the Mine or *Hacienda* keeping, of all this, a clear and separate account.

Art. 21. If any barman (*Barretero*), or other operative, or servant of Mines, carry on the working so as to leave the ore on the sides, or maliciously conceal it in any other manner, he shall be punished in the terms prescribed in Article 19 of this Title.

Title XIII.

Of the supply of water and provisions to Mining establishments.

Art. 1. The supply of drinking water in the *Reales* and *Asientos* of Mines being an object deserving the greatest attention, I ordain and command, that very particular care be given to its introduction, to the preservation of its source, the durability and cleanliness of the conduits, and that no use be made of water impregnated with mineral particles.

Art. 2. I most rigorously prohibit that the waters from the drainage of the Mines and from the washings of *Haciendas* and smelting houses be emptied into the streams or acqueducts which carry it to the Town; and I order that it be made to pass off by means of canals, or be disposed of in some other way.

Art. 3. I will and ordain, that in the immediate vicinity of the *Reales* of Mines there be sufficient commons and watering-places (*Egidos y Aguages*), for the pasturage of animals which propel the machinery required for the reduction of the ores, or which are employed in transporting the ore and other things required and used by the Miners, and which shall be for the use of all, without the possibility of being sold to any individual, church, or religious community. And I declare, that if any one of either of these classes of persons is now in the occupation of such lands, he shall be removed therefrom, paying him, if he be in lawful possession, according to the appraisement of Experts (*Peritos*) appointed by both parties, and a third, in case of their disagreement; but with the express condition, that the sales of the said lands shall be understood as applicable to those only which, according to law, may be granted, and to the extent which may be necessary for the said purpose, and no greater, unless the owners voluntarily consent to sell the surplus which may result.

Art. 4. All persons may also freely go with, and pass the said animals through all fields, meadows, and public and common lands, of other *Reales* of Mines, or of Places (Lugares) which have no Mines, without paying for this anything, even though their owners may not be residents of that territory (*territorio*,) enjoying the like exemption from paying, in those of private persons, if it be not customary for other muleteers (*Arrieros*) and passengers to pay; but, where it is the custom to do this, they are to pay only what is just and usual. And I declare, that those who go about to search and try for (*catar*) Mines, shall each be allowed a saddle-beast and one for packing, without paying for pasturage, either in common or private property, whether it be customary or not to pay for the same; but, in order that this exemption may not become burthensome, very particular care must be taken that there be no excesses, and if there should be any injury done to a third person, he may make reclamation before the respective Royal Court for the proper remedy.

Art. 5. In order to prevent any extraordinary rise in the price of provisions and clothing in the *Reales* of Mines, when they are in a flourishing condition (*en bonanza*), and that the same may be equitably regulated, according to the circumstances which ought to govern in them, the territorial Deputations will be careful to represent what they deem best to the courts of the district, as is directed in Article 35 of Title III of these ordinances, and also, for restraining and punishing monopolies, extortions, usuries, and all fraudulent, iniquitous, and unfair contracts, which may be discovered.

Art. 6. All persons shall be at liberty to carry to the Mines, maize, wheat, barley, and any other kinds of provisions and necessaries, as charcoal, wood, tallow, hides, &c., and especially, if the Miners themselves have sent to have them brought on their account: and for this purpose, I grant that they may get and bring such provisions and effects from all cities, villages (*villas*) and Places (*Lugares*), *Haciendas* and Ranches, even if situated in other territories, provinces, or governments, provided, that in any case there be no just and proper reason for preventing it: I therefore order, all Governors and Courts of such Places (*Lugares*) not to put any embarassment or impediment in the way, nor to permit, on this account, the prices of such things to be raised, but, on the contrary, to assist or favor them, in order that the Mines, and the persons employed in them, may always be provided and supplied with what is necessary.

Art. 7. Without prejudice to the jurisdiction and cognizance which I have granted to the Royal courts, in Article 35 of Title III of these ordinances, the territorial Deputations may frequently visit, inspect, and examine the fountains and perennial springs, which constitute the supply of water for propelling the machinery of Mining, in order that they may be able to make representation to the said courts, as occasion may require, and for their due information, so that they may prevent the cutting down, in these sources and their vicinity, of the forests which cover them, or the clearing away, for the purpose of cultivation, or the wasting of such forests, and may also prevent the making of excavations near to or below them, or the doing of anything else which may drain or diminish them, but, on the contrary, may cause them to be purified and cleaned with the precautions and means which art prescribes.

Art. 8. The said Deputations must also see that the rivers and streams keep up their supply, and preserve their ancient sources, giving notice to the Royal Courts, in time, and before the changes and obstructions which they themselves may form, shall become irremediable, whether by the currents forming islands or bars which have to be excavated, or principally by the floods which result from storms or other extraordinary causes, so that skill and diligence may, in many cases, prevent and remedy them. And, in order that the provisions of this and the preceding Articles may be carried into effect, the Deputies and Professional Experts (*Peritos Facultativos*) of each *Real* of Mines will visit the fountains and rivers within their district, twice every year, once a little before the rainy season, and once after, examining, on both occasions, with care, in order that, if they should be found to require any clearing, repairs, improvement, or change, for the preservation of their channels or direction, they may represent it to the Royal Courts, so that it may be ordered to be executed, with all possible dispatch, and with the intervention of the said Deputies and Professional Expert (*Perito Facultativo*), at the expense of the owners of *Haciendas* and of other persons interested in such waters; and, if there be none such, or their contributions be insufficient, the said Deputations will propose the means which they may consider proportionate and equitable, in order that, in the manner prescribed in Article 36 of Title III of these ordinances, it may be decided, whether or not, they are to be made at the public expense.

Art. 9. To the end that the Royal and common roads necessary for communication between Places (*Lugares*) of Mines, and those of the district of country upon which they depend for their supplies and provisions, may be kept in order, and made as safe as possible, considering that, generally, in all places near to *Reales* of Mines, they are rough, difficult and dangerous, especially in the rainy seasons, I order and command, that the territorial Deputations shall act, with the greatest zeal, before the Royal

Courts, respecting so important an object, whether it be done at the expense of the owners of Mines and *Haciendas*, and of the muleteers (*Arrieros*) and passengers, if this be just, according to the practice observed in this particular, or whether in some other proper way, provided, that, on this point, the Royal Court shall also be governed by the provisions of the said Article 36 of Title III.

Art. 10. For the repairs and security of private roads from some Place (*Lugar*) to Mines, and from one Mine to another, and from Mines to *Haciendas*, the provisions of the preceding Article shall be observed, although such works must be constructed by the owners of the respective Mines or *Haciendas;* but the territorial Deputations are charged to exercise the greatest zeal and care on this point, according to the results of the frequent visits which they shall make for that purpose, particularly as the said roads and paths, by nature narrow and broken, are made more perilous by travel, and by the carelessness and negligence of those who are obliged to pass over them.

Art. 11. Over the rivers, streams, and torrents, which must necessarily be crossed in going to and from *Reales* of Mines, there shall be built good bridges of masonry, or at least of wood on strong pillars of stone and cement, which usually are more easily made for this class of rivers, because, running between hills which are close together and high, they are deep and rapid, rather than wide and of great volume; and for the determination of their real necessity, the amount of their cost, and upon whom the tax shall be levied, proceedings will be had according to the provisions of the aforementioned Articles 35 and 36, of Title III, of these ordinances.

Art. 12. The forests and woods, in the neighborhood of Mines, shall serve to supply them with timber for their machinery, and with wood and charcoal for the reduction of their ores; the same being also understood with respect to those which are the private property of individuals, on condition that they be paid their proper value: Wherefore, the latter will be prohibited, as I hereby prohibit them, from carrying timber, wood, and charcoal, from their said properties to other towns which can be supplied from distinct localities.

Art. 13. The cutters and carriers of timbers shall not cut them at any other time, nor deliver them in any other manner, than shall be prescribed by the particular Regulations which the Royal Tribunal of Mining shall form, to which, punctually and strictly, they must conform, provided, that these Regulations are first approved by the Viceroy, and authorized by my Sovereign approbation.

Art. 14. I most rigorously prohibit dealers in firewood and charcoal, from cutting young shoots, for firewood and charcoal; and I order, that where there are none, plantations of young trees be formed, especially in localities and places which formerly had them, because, by their consumption and the neglect to re-produce them, the two articles most useful and necessary for the working of Mines and the reduction of their ores, have become scarce and dear; it being understood that, in order to attain so important a point, there will also be formed by the Royal Tribunal of Mining the proper instructions and a special ordinance, which must be punctually observed, under the penalties which may be provided therein, the formal approval and authorization being previously obtained, as provided in the preceding Article.

Art. 15. Wells of salt water and veins of rock salt, which are often found in some mineral provinces and territories of Mines, may be denounced, the greatest care and attention being used to have these discoveries made, and no Judge or private individual being permitted to impede them; but on the condition of giving an account of them, and of their denouncements, to the Superior Government, to the end that it may agree and determine on their working and reduction, and the distribution and price of the salt, so that no injury may result to my Royal Treasury, and as great advantage and benefit as possible may result to the Miners, and more particularly to the discoverer and denouncer; provided, that the Indians shall, in no manner, be deprived of the salt-works (*salinas*) which the law gives them, nor of the use of them for the purposes for which they are now permitted.

Art. 16. The Judge and Deputies of each *Real* of Mines will take particular care that, in the prices of timber, wood, charcoal, hides, tallow, cordage, salt, magistral, chalk (*greta*,) ashes (*cendrada*), barley, straw, and other things of indispensable necessity in Mining, the sellers shall not act with extortion; for which purpose, the said Royal Judge will, in concert with the said territorial Deputation, regulate the prices, with all due regard to justice and equity, so that the seller shall neither fail to enjoy that regular profit which he may justly expect from his trade, nor, on the other hand, charge those exorbitant prices which render fruitless the labor of the generality of Miners, who are not in flourishing circumstances (*en bonanza*).

Art. 17. The retail sale, or distribution of quicksilver, shall be immediately established, conformably to what is directed and approved by my Royal orders of November 12th, 1773, and October 5th, 1774.

Art. 18. Whosoever shall work Mines in a different Place (*Lugar*) from that of his residence, and shall derive much or considerable profit from those which he works, shall be obliged to build, or reconstruct, a house in the Place (*Lugar*) to which his Mines belong, or to construct some other work equivalent to it and useful to the public, according to the judgment of the respective Deputation of Mining, besides being subject to all the charges which are, or ought to be, borne by the residents and Miners of the same Place (*Lugar*.)

Art. 19. No trader or Miner shall, under any authority or pretext, go out on the roads to stop or intercept the vendors of grain, fruit, or any other commodity, even though they may pretend that they do not purchase to sell again, but for their own consumption; but I grant to Miners, who purchase such things in other Places (*Lugares*), the liberty of carrying them, on their own account, to the Mines, and to sellers, liberty to carry them thither voluntarily, without obstruction.

Title XIV.

Of the Millers (Maquileros) and Purchasers of Ores.

Art. 1. Considering the advantages which are afforded, not only to the greater progress of Mining, but also to the increase and support of its population, by the customs observed in New Spain, of making it lawful and free to any one to purchase and sell rough ores (*metales en piedra*), and to erect establishments in which to reduce them, although those who construct them may have no Mines of their own; it is my Sovereign will and I order, that both these customs be preserved and encouraged, provided, that, in their use, the provisions of the following eleven Articles be strictly and punctually observed.

Art. 2. I prohibit any one from purchasing ores in any other place than in the ore-sheds of the Mines, or in some public place adjoining them, and in the view, and with the knowledge and permission of the owner, Administrator, or Tallyman (*Rayador*) of the Mine, from whom the purchaser must obtain a ticket (*Boleta*), stating the day on which he purchased the ore, its weight, quality, and price, and whether it be the property of the Miner, or the portion of some servant or operative.

Art. 3. If any Miner make complaint, that ore in the possession of any purchaser has been stolen from his Mine, and the latter, admitting the appearance and qualities of the metal, should not promptly show by the ticket (*Boleta*), as prescribed in the foregoing Article, that he has purchased it, it shall be considered as stolen, without the necessity of any other proof, and shall be immediately restored to the Miner; but if the latter prove, by other means and clearly, that it has been stolen, and that there has been a repetition of the offense, in addition to the restoration to the Miner of the stolen property, there shall be inflicted upon the offender, by the proper Judge, the penalties directed in Article 29 of Title III of these ordinances, according to the circumstances, gravity, and malice which may be proved.

Art. 4. No person shall purchase of operatives or servants, quicksilver, either pure or impure, powder, ashes, chalk, or lead, in bars or sheets, under penalty, to the buyer, of paying double the amount on conviction, and to the seller, of being severely punished, according to the malice proved, even though there be no party to ask it.

Art. 5. In order that the owners of *Haciendas* which reduce ores by mills (*á Maquila*) may not injure the Miners, by raising excessively the price thereof, and, on the other hand, that such owners may not fail to realize the fair profit which may be due to them, I order and command, that the Judges of the respective *Reales* and *Asientos* of Mines, in necessary concert with the Deputation of the territory, shall regulate and establish, each year, the mill-due (*Maquila*) which, during the whole year, shall be charged per quintal of ore, regulating it according to the then price of timber, iron, labor, and to whatever else is to be considered, and establishing it by a Tariff of prices (*Arancel*) which the said Royal Judges of Mining shall draw up and authenticate, and cause to be fixed up and exhibited in public places, and to be kept in each *Hacienda* where ores of other persons are brought to be reduced, in order that they shall conform to it strictly.

Art. 6. The said Millers, under no pretense or pretext, shall charge the owners of the ores, a higher price for the quicksilver than that paid for it in the same *Real* of Mines, by the Miners who procure it on their own account, and for their own consumption.

Art. 7. With respect to salt, magistral, chalk, ashes, *temesquitate*, poor-lead (*plomo pobre*), charcoal, wood, and other ingredients, which are used in reductions by means of quicksilver and fire, the Millers shall not be allowed a profit of more than twelve per cent. on the actual and current prices of such articles in that Place (*Lugar*), paid by those who purchase them from first hands for their own use and consumption.

Art. 8. The tickets (*boletas*) which are usually given to the owners of the ores, and in which is stated the amount of the costs and products, shall not only express them generally, but must particularize the mill-work, the price charged for each ingredient, the cost of operatives, the consumption of quicksilver or alloy, and the produce of silver, gold, etc., which must be signed by the owner or

Administrator of the *Hacienda*, and by the amalgamater or smelter, who may be in it. And, in case of a violation or contravention of any of the preceding Articles, execution shall issue, on the mere verification of the ticket, against the owner or Administrator of the *Hacienda*, to indemnify the owner of the ores; and if it be proved that he acted with mal-intent and fraud, he shall pay three times the value.

Art. 9. No Miller shall oblige the owner of the ores to pay him the expenses of the reduction, in the silver or gold reduced, but only in current coin; but if they have voluntarily agreed that the payment shall be in bullion, it shall be taken at its full value, and not at a discount (*á precio de avios*), nor with any premium; the same being observed with respect to the silver-amalgam (*platas de azogue*), which is to remain at the *Hacienda*, as a pledge, during the continuance of such agreement.

Art. 10. To avoid the frauds and impositions which frequently arise from the uncertainty attending the reduction by quicksilver and by fire, which uncertainty often serves as a pretext for fraudulently depriving the owners of the ores, of a part of the silver and gold which they produce, at the same time injuring the Millers, when the ores are too poor to cover the expenses of the reduction; I order and command, that, until there be established in the *Reales* of Mines, as there should be, an office, public and authorized, to reduce, by way of assay, one or more quintals of ore, so as to determine its true *ley*, the owner of the ore or of the *Hacienda*, when there is any distrust or suspicion of the small yield from the general reduction, may select and deposit, at will, one or more quintals of the ore, in order to have it reduced afterward, if necessary, by Experts (*Peritos*) approved by the parties, and a third one in case these do not agree.

Art. 11. For the same object as that of the preceding Article, it is my sovereign will, that no owner of ore, who sends it to be reduced at the works of another's *Hacienda*, shall be prevented from attending, personally, or by a confidential agent, at all the operations of reduction, testing the amalgam, adding ingredients (*guias*), assaying the scoria or detached particles (*grasas ó plomos*), and doing everything that may appear necessary for the better direction of the reduction of his ore, and to satisfy himself of its exactness.

Art. 12. The amount of freight which is to be paid to the muleteers, who transport the ores from the Mines to the *Haciendas*, shall be settled, whenever any excessive charge is made, by the Royal Judge of each Mining establishment (*Mineria*), acting in concert with the territorial Deputies, with due regard to justice and equity, and making a difference between the dry and rainy seasons.

Art. 13. And if any of the said muleteers shall be proved to have stolen, or sold, any of the metal on the road, replacing it with rubbish (*tepetate*), or in any other way, they shall be proceeded against by the proper Judge, following the provisions of Article 29 of Title III of these ordinances, in the imposition of penalties, and also in case of a repetition of the offense, always having regard to the nature and gravity of the said offense, and deciding it conformably to law, and in the form and terms prescribed in the said Title III; it being understood, that if, in any of the cases comprised in the thirteen Articles of this Title there shall result the imposition of fines, or forfeiture of property, of animals, or of any thing else, the proceedings, in the application of such penalty, shall conform to Article 32 of Title III.

Title XV.

Of suppliers of Mines, and dealers in bullion.

Art. 1. It often happens, that Miners work their Mines with the capital of others, either because, at the outset, they have not sufficient funds of their own to put them in operation, or because their own means become exhausted, in constructions of dead works, before obtaining ore rich enough to yield a profit over and above the expenses of its extraction; and, in such cases, they usually contract with Mine suppliers (*Aviadores*) in one of the two following ways: that is, by letting the *Aviadores* have the gold and silver which they extract, at a price less than the legal and full value, the *Aviadores* deriving their profit from this difference; which method is called supplying by premium on bullion (*a premios de platas*); or, by giving the *Aviador* a part interest in the Mine itself, thus making him a perpetual owner in the Mine, or an owner, for a certain period, in its ores, by a species of partnership (*compañia*). And whereas, from the necessities of the Miners and the greediness of some *Aviadores*, it often happens that agreements are loosely entered into, which, from their iniquitous and usurious character, or from their not being well understood at the beginning, give rise to subsequent reclamations on both sides, from which result law-suits, and the stoppage of supplies, thus wasting the Mines and the capital invested in them;—it is my Sovereign will, that no Miner shall enter into any contract for Mine-supplies, (*Avios de Minas*), except by an agreement in writing (*contrata firmada*), it being optional with him to execute it or not, before a Notary, or witnesses, under the penalty, that, if made in any other manner, the special stipulations which may be alleged shall not be judicially enforced, but the agreement shall be determined only by the general rules of law.

Art. 2. In all agreements for premiums on bullion (*premios de platas*), mentioned in the foregoing Article, attention and consideration must be paid to the number of *marks* in each delivery, and how often the deliveries are to take place, so that, if, by the accidental circumstances of the Mines, the number of deliveries should increase or diminish considerably, either one of the contracting parties may be at liberty to increase or diminish the premium on the bullion, notwithstanding the contract, which was entered into at first with a different understanding; for which purpose, it shall always be specified, in the instrument originally entered into by the parties, what number of annual deliveries of bullion, and what number of *marks* in each one, may be equivalent to, or in satisfaction of, the stipulated premium of bullion; or, if the parties wish, they may renounce altogether any right to profit by this kind of accident; in which case the contract, entered into in said form, must have its full effect.

Art. 3. If the Miner shall secure the supplies (*Avios*), up to a certain amount, by means of mortgages or securities, to the satisfaction of the supplier, the latter shall not receive a higher premium than what is equivalent to five per cent. per annum, on the capital advanced.

Art. 4. The Mine suppliers (*Aviadores*) shall furnish the supplies, in ready money, or in bills payable at par; but if the Miner ask goods and effects, they shall send such as are of good quality and in good condition, and at the prices at which they are bought for cash, at the Place (*Lugar*) where the supplier (*Aviador*) resides, and they shall not furnish them in any other way.

Art. 5. The risks and accidents of the road, in the transportation of the supplies, and the freights and excise duties (*alcabalas*) paid thereon, shall be for account of the Miner, if the supplies be by premium on bullion (*á premio de *platas;*) but if they be a contract of partnership (*de compañia,*) these shall be for account of both parties, unless it has been otherwise expressly stipulated, by special provisions, in the instrument which they shall have executed.

Art. 6. If the capital so supplied (*caudal de Avios*) shall be entirely lost, or if it be in part uncovered, the Miner shall not be responsible for it, either in his person or in any other property which he may have, but only in the products of the Mine, and in the *Hacienda* of reduction, if that has been constructed with the capital so supplied; but the Mine, with its profits and products, after deducting expenses, shall go to the payment of the suppliers (*Aviadores*), one after another, beginning with the last or most recent; it being understood, however, that as this is a privilege which the law grants to credits given for repairs (*de refaccion*), its three requisites must all concur in order to enjoy it; but if the Miner, from necessity and without fraudulent intent, shall surrender the Mine, giving previous notice thereof to its creditors, it shall not remain liable for former debts, being now in the possession of another owner. And, moreover, I declare, that if the capital with which such Mine is supplied, and in regard to which such deficiency (*descubierto*) occurs, has not been furnished under contract of partnership between the supplier (*Aviador*) and the Miner, in which case there must be a common profit and loss, but as a loan, and the Miner, of his own will, or because the supplier asked it for greater security, has pledged his property, under such circumstances the said obligation shall be binding in all its parts, notwithstanding the general provisions of this Article.

Art. 7. If the mode of advancing and repaying the supplies has not been agreed upon at first, when they are on remium of bullion, the supplier (*Aviador*) cannot adopt such a mode or do it in such a manner, as to injure the Miner in working his Mine, by cutting off his supplies, nor shall he be obliged to receive back from the Miner, in small sums, those supplies which he may have advanced.

Art. 8. Although the Miner may not have observed, at any time, that his silver has a *ley* of gold, the separation of which would be expensive, or that there be silver in the ingots of gold of a small *ley*, and the supplier should discover it, by having them assayed, or in any other way, he must not for that reason, consider that this is his profit, but he must place it to the credit of the Miner or the owner of the ores, in the account which is to be kept with him.

Art. 9. When the supplies are made by way of partnership in the dominion and ownership (*dominio y propiedad*) of the Mine, it must be understood that the capital invested in it, prior to the time when the profits exceed the expenses, is not to be first deducted from the profits, but these are to be immediately divided, the said capital remaining so invested and active, during the continuance of the partnership.

Art. 10. The Merchants and Dealers in bullion, who receive it without having advanced supplies to the owners, or without any risk, must pay the full value thereof; and if they exchange for it goods from their stores, such goods must be at the current prices, and of good quality. But I order and strictly command, that said Merchants or Dealers in bullion must receive it from the owners of Mines, assayed and with the duty paid (*quintadas*,) as directed by law and repeatedly prescribed by my Royal decrees, in order to prevent its being secreted, or used in such a way as to defraud my Royal dues: declaring, as I declare, that in the *Reales* of Mines, in which it may not be easy to have such bullion assayed and the fifth deducted (*quinten*), on account of the distance of the Royal Treasuries or marking-offices (*Caxas Reales ó Caxas-Marcas*), the Merchants or Dealers shall bind themselves, before the Royal Court and territorial Deputation, to take it forthwith to the Treasury of the district, in order to comply with the obligation of paying my Royal dues, and to make good what is due on the quicksilver, according to the bond which is usually given for that purpose, in New Spain, the said court and Deputation prescribing to them a term within which all this is to be done, and, moreover, giving notice, to the respective Royal officers, of the aforesaid obligation, in order that, in case it is not complied with, said bullion shall be considered as confiscated, and they may proceed to enforce the confiscation, and to impose the other penalties, prescribed by the laws, on those who defraud my Royal dues.

Art. 11. All Merchants in the *Reales* of Mines must keep correct and light scales, in which shall only be weighed silver and gold, never weighing with steel-yards (*en Romana*), however large may be the masses or quantities of these metals; and they must also keep weights, marked and properly adjusted, according to those which they have lawfully received from the proper Royal authority. And I permit the respective Deputies of Mining to examine them, from time to time, (exclusive of the inspection which is incumbent on the Royal court and public magistracy), and to take care that the weighing be always justly and faithfully done, so that, if any fraud result and be proved, and also in case of a repetition of the offense, proceedings may be had, by the Royal court which is competent to take cognizance of these cases, for the imposition of the penalties proportioned to the nature and gravity of the offense, as provided by law, being under obligation to hear on the subject, by way of report, the Deputation of the district.

Art. 12. All Miners shall have their tools marked; and

any one who shall buy them of an operative, or receive them in pledge, shall pay for them double their value.

Art. 13. The said Merchants and suppliers may roast the cakes of silver-amalgam, to their satisfaction and to that of the owner, in a charcoal fire, but not in a flame, nor in any way so that it may be fused, unless in crucibles; and it is permitted to divide them, so as to examine them inside; provided, that this, as well as picking off the scales of smelted silver, be done on a table, or in such a manner that the owner may sweep up, and carry away, all the fragments, dust, or waste particles of his silver.

Art. 14. Every supplier may, at any time, appoint an inspector (*Interventor*) to every Miner whom he has contracted to supply, although it may not be so expressed in the contract of supplies; but it is to be understood, that such inspector is only to attend to the proper keeping of the accounts, and to have under his control the money and effects, without authority to interfere with the works of the Mine which the Miner may determine on, but he may defer their execution while he reports to the Deputies, asking for experts (*Peritos*), and this only in cases which will admit of such delay.

Art. 15. Inasmuch as, the current working of Mines cannot be suspended, without great injury, especially if they require draining, I order, that if the supplier, advancing supplies from time to time, shall fail to provide them in such a manner that, when the time of paying wages (*tiempo de la Raya*) arrives, there be no means for paying them, and if it shall have happened that the Miner, fearing and foreseeing this case, has called upon, and remonstrated with, such supplier, and made report to the Deputation, then, not only may he pay the wages, from what is best supplied to the Mine, even from the implements and tools, but the Miner may also demand executively from the supplier what may be due, and obtain money from another, or contract with a new supplier (*Aviador*); which debt shall be preferred to that of the preceding one, when the Mine begins to yield a profit.

Art. 16. Those who, under pretense of taking supplies (*Avios*) for Mines, divert, waste, or, in any way, apply to another purpose, the funds and effects which are advanced for working them, shall not only be bound to pay them, and all damages and interest, to the party, with their persons and their property of whatsoever kind, without being allowed the privileges of Miners, or any other exemptions, but they shall also be punished in a manner suitable to the gravity, character, and circumstances of the case, and especially if they have received the supplies in trust; the cognizance of these causes being governed by the provisions of Article 29 of Title III.

Art. 17. Triers of veins (*Cateadores*,) searchers (*Buscones*), or operatives, and any other persons whatsoever, who present ores and specimens, pretending that they are of a certain Mine, for which they solicit supplies (*Avios*), this being in fact false, and done merely with the object of defrauding and deceiving incautious persons, I order to be punished with the utmost rigor of the law, according to the circumstances, gravity, and mal-intent which may be proved in said offenses, by the court which may have cognizance thereof, according to the provisions of said Article 29 of Title III, of these ordinances.

Title XVI.

Of the Fund and Bank of Supplies of Mines.

Art. 1. Whereas, by my aforesaid Royal *Cedula* of the 1st of July, 1776, I saw fit to relieve the Body of Mining of New Spain from the double duty of one real in each *mark* of silver, which, by way of seigniorage, was paid to my Royal Treasury, granting to it, at the same time, the right to impose on its bullion the half, or two-thirds, of the same duty, in order to supply the convenient and necessary assistance to the new and commendable establishment which is the object of these ordinances; and also considering, that the destination of the same, most conformable to my beneficent intentions, is the formation, by that means, of an endowment fund for the supply of Mines, the present fluctuating and insecure condition of the general system of Mining being, for the most part, the result of a want of capital, with the aid of which it would no doubt be put in a more secure and flourishing state, with great advantage to my Royal Treasury and to the Public:—Therefore, keeping in view what has been proposed on this subject by the Royal Tribunal of the important Body of said Mining, I have been pleased to decree and command, that all the bullion which may enter my Royal Mint in Mexico, and in any other that may be established in the kingdom of New Spain, and all which may be remitted *en pasta*, to those in Spain on account of its individual owners, (which must always be assayed and the fifth deducted), shall henceforth pay two-thirds of a *real* [per mark], for the purpose of forming, preserving, and increasing the endowment fund of the said Mining; and that, from this contribution, no Miner shall be exempt, not even those to whom, for any just cause, there may have been, or may hereafter be, granted the remission or diminution of metalic duties which refer and appertain to my Royal Treasury.

Art. 2. The management, collection and custody of the funds which are, in this manner, to be raised, shall always be under the control and at the disposal of the said important Body of Mining to which they belong, through its Royal Tribunal General of Mexico, which represents it.

Art. 3. Deducting from these funds what may be necessary to maintain the said Royal Tribunal, and the college, and the instruction of young men intended for Mining, which will be treated of hereinafter, and to defray the extraordinary and necessary expenses incurred for the common interest and benefit of the said important Body of Mining, all the surplus, and the successive additions and products, must be used and invested solely in supplying the Mines of the kingdoms and provinces of New Spain, and in paying the expenses of working them, by establishing a Bank of bullion (*Banco de platas*), according to the rules laid down in the following Articles.

Art. 4. For the management and business of said Bank there shall be one Factor, or more if necessary, a man of intelligence and experience in the matter of supplying Mines, who shall be subject to and dependent upon the Royal Tribunal General of Mines, and be appointed by it, by a majority of votes, and removable in the same form and without the necessity of assigning the reason.

Art. 5. Such Factor may be paid a percentage on the profits of the Bank, or a fixed salary, or both, according as the said Royal Tribunal may direct, under different circumstances, on condition that he give sufficient securities and bonds, to be fixed and approved by the said Chiefs.

Art. 6. The gross sum of the capital of the Bank which may be in coin or in gold and silver bullion, shall be kept in chests, with four keys, which shall be held by four of the Chiefs who, at the time, belong to said Royal Tribunal;

but the goods and merchandise for the supply of Mines, and the part of the capital required for the current business and operations, shall be in the possession of said Factor, and under his charge and management, he and they being responsible, respectively, for what may be entrusted to them.

Art. 7. The Royal Tribunal General of Mines will cause to be made out in the Factory, annually, in the month of December, an account and inventory of the store-houses, and a cash account and balance, two of the Chiefs of the said Royal Tribunal assisting in these operations; and, moreover, the Tribunal will receive the accounts of the Factor, without prejudice to the right of asking for them on extraordinary occasions, when proper prudence and circumspection seem to require it.

Art. 8. The Royal Tribunal will keep the accounts and correspondence with the Miners, who are supplied by the Bank, receiving and answering their letters, and issuing, accordingly, the corresponding orders to the Factor.

Art. 9. For the business of the Factory, there shall be such number of clerks as the Factor may deem necessary, who are to be nominated by him; but the appointments are to be made, and their salaries fixed, by the Royal Tribunal, and they are to be paid by the Bank: the Factor having the power to dismiss these officers, giving a verbal notice thereof to the Royal Tribunal.

Art. 10. The Factor shall receive the bullion which the Miners who have been supplied (*Aviados*) may remit, and shall exchange it for coin, in the Mint of Mexico, paying beforehand, into the principal Treasury, the metalic dues on that which has not paid them in the provincial Treasuries (*Foraneas*); but on condition that, before it is sent to Mexico, the said Miners shall declare, at the Royal Treasuries or marking offices of the respective jurisdiction, the amount of bullion which they are about to remit without having paid the requisite metalic duties, procuring the proper permits for its free transportation, under a bond to return to the proper Treasuries a certificate of having paid said duties, in order thus to avoid all frauds, and to pay what is due for quicksilver, in each case, under penalty of confiscation of what is, in any other manner, carried away, and of incurring the other penalties which the laws impose upon the defrauders of my Royal dues; the Royal officers taking care to notify those in Mexico of this class of remittances, to the end that they may watch and see that the provisions of this Article are complied with.

Art. 11. The said Factor shall pay the dues on the capital which is received by the Bank on interest, the salaries of the employes, and all other sums, on warrants of the Royal Tribunal, with which, and the corresponding proper receipts, he shall justify this part of his accounts. But for the remittances to parties to whom supplies have been contracted (*Aviados*,) and with whom accounts current are to be kept, whether in money or effects, no particular warrants shall be required, but only the orders, which, by the same Tribunal, and in conformity with Article 8 of this Title, may be given with respect to the things which are under his charge and management, according to the provisions of Article 6.

Art. 12. It shall be the duty of the Factor to make purchases of the effects and merchandise required for the supplies of Mines, according to his judgment, and in conformity with the orders of the Royal Tribunal, entering them in a separate Book, and preserving the original invoices.

Art. 13. Goods which are delivered to Miners, on account of contracts of supply and which are received on account of the Bank, must be delivered and received in good condition, and be of good quality, and, in Mexico, at the prices in Mexico, and in the *Reales* of Mines, at the current prices in such *Reales*, if the Bank shall have storehouses there, or shall be the expense of the transportation.

Art. 14. In order to determine upon propositions or petitions for the supply of Mines, the Royal Tribunal will require their owners to produce the titles of property and possession, and certificates and reports, or any other proofs whatsoever sufficient to establish what they say of the Mine respecting its condition and circumstances, in order that, these papers being passed to the Assessor for his examination and approval, it may be determined, whether the proposition appears to be fair and in good faith; in which case, the Royal Tribunal must inquire officially and secretly, with the greatest prudence, sagacity and justice, taking, and causing to be taken, such measures, judicial and extra-judicial, as may appear proper, in order to proceed understandingly in deciding upon such supplies, preserving in its archives all these documents.

Art. 15. While the funds of the Bank are not sufficient to furnish all the Mines which propose for supplies with a fair and well-grounded claim, that Miner shall be first attended to and supplied who most needs it, without any distinction of persons, and without allowing any other preference than that of said necessity, and the utility of the working of the Mines, the Royal Tribunal acting in such cases with that justice and impartiality which it ought always to observe.

Art. 16. When the claim is determined to be a good one and admissible, the terms and stipulations, under which the supplies are to be furnished, shall be agreed upon with the owner of the Mine, and, before concluding the contract, the Royal Tribunal must approve them, in strict conformity with the provisions of Title XV of these ordinances, without pretending that the Bank of Mining has any privilege, to the prejudice of other Banks, or of private contractors for supplies (*Aviadores*): after which, the contract being thus approved, the instrument shall be executed before the Notary of Mining (*Escribano de Mineria*), and orders shall be issued for furnishing the supplies, conformably to its contents.

Art. 17. In Mines which are furnished (*habilitadas*) by the Bank, inspectors (*Interventores*) shall be placed, who shall be trust-worthy persons and of good reputation, in order that, jointly with the owner of the Mine, they may receive and keep the money and effects of the Bank, in vaults and chests, with two keys, issuing them according to the contract, and, being present at the payment of the wages, they will sign the Rolls (*Memorias*), watching and inspecting the operatives who enter the Mine, and the ores which are taken out, being present also at their reduction in the *Hacienda*, and, in fine, intervening, in the name of the Bank, in whatever is done, in strict conformity with the instructions which may be given to them, until the supplies advanced are covered and paid.

Art. 18. The inspectors (*interventores*) cannot oppose any arrangement of the owner or Administrator of the Mine, with respect to the management, direction or economy of working it, nor with respect to the works or constructions which, in the said Mine, may be determined on, it being understood, however, that if they are to be costly, they cannot be decided on or approved, without consulting the Royal Tribunal.

Art. 19. Nor shall they interfere in the election and appointment of the subaltern employes of the Mine; but they may watch their conduct, in order to inform the owners of anything they shall observe which requires to be remedied; and in case it should not be remedied, they will inform the Royal Tribunal, in order that the proper measures may be taken; and this body will, moreover, take care that the inspector and owner of the Mine are upon good terms, acting in concert, always uniting their efforts for the success and advancement of the operations.

Art. 20. The inspectors shall be paid weekly, the salary that is allowed them, on account of the supplies, and when these supplies are repaid, their merit shall be attended to, and they shall be rewarded, in proportion to the benefit which the Bank has received, and to the time, industry and good conduct of their service; but if, on the contrary, any fraud, usurpation, or improper proceeding, shall appear, whether to the prejudice of the Bank or of the owner of the Mine, they shall be severely punished, in proportion to their offense, by the proper court, according to the prvisions of Title III of these ordinances.

Art. 21. If any competition, with respect to furnishing a Mine, should arise between any private person and the said Bank, I declare that the individual supplier shall have the preference, under equal circumstances, in order that he may immediately commence supplying the Mine. And in order that the said Bank shall not monopolize the free power of supplying of Mines, I also declare that this kind of business shall continue as heretofore, the Bank having no other object than that of supplying the want or scarcity, and of making as constant and perpetual as possible the encouragement of Mining.

Title XVII.

Of Experts (Peritos) for the working of Mines and for the reduction of ores.

Art. 1. In order that Mines may be worked with effect and security, and their riches be made fully available, it is necessary that the operations be directed by men well instructed in the principles and rules of the natural and practical sciences, and the arts therewith connected, and be taught, by personal experience, their true and proper application. Wherefore, and in order that the owners of Mines may not be misled in the selection of the persons whom they employ, accounting as well informed those who have only a superficial knowledge, or an acquaintance with terms, or those who have no other qualification than that of having lived a length of time in the *Reales* of Mines, without possessing judgment or science, and having no other claim than the recommendation of their associates, their voluntary and malicious errors, being on the other hand very difficult to discover, which leads Miners into a blind and dangerous confidence on the most important matters, and has caused them great injuries:—to the end that these evils may be avoided and that the *Peritos* may be rendered worthy of public and judicial confidence in matters pertaining to their art, I ordain and command, that in each *Real* of Mines there shall be one or more well informed persons, instructed in, and practically acquainted with Geometry, Subterranean and Hydraulic Architecture, and also with Machinery, Carpentry, Smithery and Masonry, so far as these arts are necessary to the working of Mines, which persons are to be called *Peritos Facultativos de Minas;* and also other men skilled in the science of minerals, called Mineralogy, and in the proper mode of treating them in order to extract all their metals, and in the mode of reducing them, to a condition fit for use, as well in large as in small quantities, which is called Metallurgy, which persons are to be called *Peritos Benefciadores;* and both classes must be examined, approved, and appointed by the Royal Tribunal General of Mining, and in no other manner shall any faith or credit be given to them, in court or out of court, and they shall be regarded as intruders, and shall be excluded, and fined, whenever they interfere in what pertains to the art of Mining, even though they may allege that they are bachelors of art, surveyors, architects, or masters of works, or that they have been Administrators, officers, or operatives in Mines.

Art. 2. The said Professional Experts of Mines (*Peritos Facultativos de Minas*) shall have instruments necessary and sufficient for the cases which may occur in practically measuring Mines, whether under-ground or superficially, which must always be true, correct and according to rule, so that there shall be no want of due punctuality and regularity in the operations; for which purpose the instruments shall be inspected and examined, when they themselves are examined and receive their licenses, and afterward, on the extraordinary visits (of the Deputies.)

Art. 3. The Experts of reduction (*Peritos Beneficiadores*) shall have a suitable public laboratory, with furnaces and machines for grinding and washing ores, and also ingredients, utensils, correct scales, and proper weights, and whatever else may be requisite, not only for small assays, but also for reducing by fire or by quicksilver, one, two, or three quintals of ore.

Art. 4. The Professional Experts of Mines (*Peritos Facultativos de Minas*) must examine, at the proper time, and give certificates of examination, to all persons who devote themselves to that occupation, as Miners or Masters who direct and carry on under-ground works, and as Liners and Masons of Mines, and Carpenters and Smiths of machines. And I prohibit all persons from being employed in such occupations, and from exercising the office of Masters in any Place (*Lugar*) where this office may be already established, without having the aforesaid certificate of examination, under penalty of three months' imprisonment for the first offense, and, for the second, of being banished from the Place (*Lugar*): which penalty may be imposed by the respective territorial Deputies.

Art. 5. The reducing Experts (*Peritos Beneficiadores*) of each *Real* of Mines, shall examine, and give Letters of approval to those who apply and devote themselves to the business of Amalgamaters, Smelters and Refiners, without which requisite, no one can be employed in such capacities in any *Hacienda* or establishment for reducing ores, without incurring the same penalties contained in the preceding Article. And I declare that these examinations, and all others provided for in this Title, shall be made without any fees and entirely gratis.

Art. 6. If any person shall pass from one *Real* of Mines to another, after having been examined and approved in the one which he leaves, there shall be no necessity for examining him anew; but he shall produce his Letter (*Carta*) of examination, signed by the Expert by whom it is issued and attested by the Notary, or by the Deputation of that Mining place and two assisting witnesses in default of a Notary.

Art. 7. The said Mining and Reducing Experts (*Peritos Facultativos de Minas y Peritos Beneficiadores*), shall, at the time of receiving their certificates from the Royal Tribunal, take, before that body, a solemn oath, in due form but gratis, to discharge always, and in all cases, their respective offices, well and faithfully, and according to the best of their knowledge and understanding, without fraud, deceit, or any mal-intent; they not being required thereafter to take such oath in any proceedings, whether judicial or extra-judicial, because, having taken it once, as above directed, they are ever after bound by it.

Art. 8. To the said Professional Mining and Reducing Experts full faith and credit shall be given, in court and out of court, in all matters of their art; but they may be excepted to, when they have been appointed by the Judges, and when they are so excepted to by one of the parties to a suit, the other party shall also have the right to nominate a new Expert on his side, and the Judge to select the third, in case of their disagreeing, although they need not both be from the same district; avoiding successive exceptions and appointments of new Experts, when there is good reason to suspect that they are made through fraud or malice, or for the purpose of delaying the decision of the cause.

Art. 9. The Professional Mining and Reducing Experts shall be present at the visits made to the Mines and *Haciendas*, and shall do and observe whatever is prescribed in these ordinances, giving their assistance in all cases, properly belonging to their business, in which they are called by the Judges and the Deputation of Mining, charging therefor the proper fees which are marked and assessed in the Tariff of fees (*Arancel*) which the territorial Deputations shall propose to the Royal Tribunal General, in order that that body, after examining it, may consult the Viceroy, who, upon full information on the subject, will resolve and determine what fees may be exacted; and, without this requisite, the Tariff of fees shall not be carried into effect

Art. 10. Until the Seminary for the education and instruction of young men intended for the business of Metallurgy, Mineralogy, and the other sciences necessary for properly directing the operations of Mines (which establishment will be treated of in the next Title), shall supply a sufficient number of educated persons, such as are described in this Title and are necessary to perform the duties required by these ordinances;—I order, that all those who, at present, are employed in the operations of measuring Mines, projecting shafts and adits, and other important means for the working of Mines, whether they have the title of Surveyors and Measurers of Mines, or whether they have, without such title, been well received in Mining establishments on account of their practical knowledge, skill, and special study, shall go before the Royal Tribunal General and present themselves for examination, so that the corresponding commission (*titulo*) may be issued to them, without fees, as is provided in Article 5 of this Title, and they must exhibit the instruments which they use, in order that they may be inspected and examined, under penalty, in default thereof, of their not being entitled to faith or credit, in court or out of court; and further, if, in any work directed by such persons, damage shall result, the owner or Administrator who has employed them, shall not be exempt from the responsibilities and penalties imposed by these ordinances, and by the general laws, upon those who proceed without the direction of Experts in cases where they ought to follow such direction.

Art. 11. All persons who are appointed professional Mining Experts, and Reducing Experts, must be Spaniards, Spanish Mestizos, or noble Indians of known birth, parentage and education, and of good life and conduct; under which circumstances these employments and offices are always to be accounted as honorable, noble and meritorious; so that those who have served faithfully in them shall enjoy all the privileges of Miners, and shall be eligible to higher offices and occupations in Mining and elsewhere, having a seat in the public sessions next to the Judge and Deputies of the district, ranking among themselves according to the seniority of their appointments, and without distinction between professional Mining Experts and Reducing Experts, as both these are equally worthy of honors and distinctions.

Title XVIII.

Of the education and instruction of youth intended for the Mines, and of the encouragement of industry in them.

Art. 1. In order that there may never be wanting persons educated from their childhood in good morals, and instructed in all the learning necessary for properly working Mines, and that what has hitherto been acquired by the long and painful experience of many centuries and of different nations, and even by the particular and peculiar skill of American Miners, may be preserved, in a manner more exact and complete than, by mere tradition, which is generally meagre and unreliable, it is my sovereign will, and I command, that there be erected and established, (and if they are already established, that they be supported and encouraged with the greatest care and attention), the College and Schools which, for the said purposes, were proposed to me by the Deputies General of the said important Body of Mining, and in the form and manner which are ordered in the following articles.

Art. 2. There shall, for the present, be maintained and provided with board and suitable clothing, twenty-five children, Spaniards, or noble Indians of legitimate birth, preference being always given to the descendants or near relatives of Miners, and especially to those whose ancestors have resided in the *Reales* of Mines.

Art. 3. I grant free admission, and gratuitous instruction, in the Schools, to all children whose parents or guardians may desire to place them in this profession, they coming daily from their own homes, to attend the lessons; and I also command, that all children having the above mentioned requisites as to race and birth, who pay for their maintenance, shall be admitted to live in the College as pupils.

Art. 4. In said college there shall be appointed the necessary secular Professors, with good salaries, to teach the sciences, mathematics, and experimental physics, which may be conducive to the proper and good direction of all the operations of Mining.

Art. 5. There shall also be Masters of the mechanical arts necessary for preparing and working timbers, metals, stones, and other materials, for the construction of offices, machines, and instruments, used in the working of Mines and in the reduction of their ores, and also a Master of delineation and drawing.

Art. 6. The said college shall have the title of *Royal Seminary of Mining*, and two secular priests of proper age, one to be Chaplain-Rector and the other Vice-Rector, must reside in it, to take care of the education of the children in religion and government, to see that they study and duly improve their time, and to say Mass for them every day in the year.

Art. 7. The immediate direction and government of said Royal Seminary shall be intrusted to the Director-General of Mining, to whom I grant the power to propose to the Royal Tribunal the persons who are to be engaged as Master-Professors, or in other service, and to select the children who are to be admitted as collegians (*collegiales de ereccion*), or as boarders (*Pensionistas*) deciding upon their necessary qualifications; proposing also, after hearing the opinion of the respective Masters of the said college, the branches which are to be taught, and the method of teaching them, so that the Royal Tribunal may decide upon all these matters what it may deem best; it being also the duty of said Director to take particular care that all the employes properly perform the duties of their offices, and to draw up the particular Regulations for the government in detail of said college, which Regulations must be presented to the Royal Tribunal, in order that, when acted on by it, they may be passed to the Viceroy, who, after obtaining the necessary information on the subject, will report them to me for my Sovereign approval; and having received this, the said Regulations will be observed and carried into effect, with due punctuality and exactness.

Art. 8. The expenses of the erection, maintenance, and improvement (*fomento*) ot said Royal Seminary will be defrayed out of the endowment fund of Mining, according to the provisions of Article 3 of Title XVI.

Art. 9. The said Seminary shall be under my Royal protection, and be immediately subject to and dependent upon the Royal Tribunal General of Mining, in all its matters and affairs.

Art. 10. For the election and appointment of the Master-Professors of the sciences which are to be taught in the Schools of the College, there shall be issued, at a certain fixed time, edicts of convocation, and to those who present themselves, there shall be distributed, by lot, some problems in the respective sciences, the solutions of which must be returned within the third day; but with the condition, that before these problems are distributed and delivered, the Director shall present to the Royal Tribunal the solutions of all of them, folded in separate and sealed envelopes, which shall not be opened till each candidate has presented his solutions, in order that the proper comparison may be made of the two. And on the same day on which this takes place, the candidate shall deliver a public lecture, of two hours length, on points which the director shall at the moment propose, and in the presence of the Royal Tribunal and of the Notary, who shall certify to the act (*Acto*), and enter it in the respective register.

Art. 11. At the conclusion of the said public Act, the Director will propose three of the candidates for each professorship, from whom the Royal Tribunal shall elect one by secret ballot; and in case there be no choice, on account of a tie of votes, the one first proposed shall be considered elected.

Art. 12. The said Master-Professors of the college, besides teaching daily by theoretical and practical lessons, shall present, once every six months, a Memoir or Dissertation on some subject useful and advantageous to Mining, and pertaining to the sciences applicable thereto, which Memoirs shall be read to the Royal Tribunal, and perserved in its archives, in order that they may be printed and published whenever it shall appear expedient.

Art. 13. The Collegians and Students of the Seminary shall, every year, have public exercises (*Actos*), in the presence of the Royal Tribunal of Mining, in order that, showing thereby their respective progress, they may be rewarded and distinguished according to their merit.

Art. 14. When the said youths have completed their studies they shall go to assist in the *Reales* of Mines, for three years, and shall practice the operations with the professional Expert of Mines, or with the Reducing Expert of the district to which they are attached, in order that, having received a certificate signed by them and by the territorial Deputies, they may be examined in the Royal Tribunal, both on theoretical and practical knowledge, and, being approved, may receive their diploma (*Titulo*), without any fees being charged therefor; and they then may be appointed Professional Mining or Reducing Experts in the *Reales* of Mines, Inspectors (*Interventores*) in those supplied by the Bank, and to other suitable offices.

Art. 15. In order more effectually to advance instruction and education in the important objects of the said College with real advantage to Mining; I order and command, that the Owners or Suppliers of Mines, who take their bullion to Mexico, shall be obliged to deliver to said Mining College specimens of their ores, in quantities sufficient to permit their quality and properties to be there examined, and the most advantageous method of reduction to be determined, in order that, judging by the result of these operations, the Royal Tribunal may determine what will be most conducive to the improvements which are the object of these provisions.

Art. 16. Considering that industrial skill (*Industria*) makes useful to human life the ordinary productions, even those most common in nature, and that, on the contrary, without it, the advantages and profits to be expected from the most abundant riches of nature are ordinarily of little use or value;—I will and command, that industrial art (*Industria*) applicable to Mining, and which is of such vast importance to it, be excited, encouraged and promoted, with the greatest activity, intelligence and discretion, particular care and attention being given to the use and effect of machinery, to the operations and methods at present employed in its application, in order that whatever is really useful and perfect of its kind, may be preserved in full perfection, without any gradual loss or depreciation, as has happened and may happen; and that whatever, by comparison with the best and most certain methods, shall appear capable of amendment or improvement, may be actually brought to the greatest degree of perfection and practical efficacy: without suffering either ancient prejudices, bound to ignorance and caprice, to retard the progress of industrial skill, or ill founded innovations to replace what is worthy of preservation.

Art. 17. All persons who invent or discover any kind of machines, engines, expedients, operations or methods for improving the art of Mining, or who produce any advantage, even though at first it may appear small, shall be heard and attended to; and if, on account of their poverty, they shall be unable to make the requisite experiments of their inventions, the expenses thereof shall be defrayed from the fund of Mining, and also the expense of constructing the machines, provided, that having presented their project, they show and prove its effects, and that the Director General of Mining and the Masters of the College, approve it, and think it capable of being carried into practical effect. But ill-founded ideas, which result from ignorance of principles, or want of practical knowledge, and from which their deluded authors are easily led to expect great and imaginary advantages, shall be rejected as useless and unworthy of attention; and if the authors of such projects insist and renew their applications, they shall have no hearing, unless they make the experiments at their own expense, and thereby prove the utility of their inventions: retaining in the archives of the Royal Tribunal the documents relating to these and all other cases, for the purposes which they may serve.

Art. 18. Useful and approved inventions which, after being tried on a large scale, shall be established in general use for more than a year, shall be rewarded with a patent (*privilegio exclusivo*) for the lifetime of the author, in order that no one may use them, without his consent; and without allowing him a moderatc share of the profit and advantage, which are actually derived from the use of such invention.

Art. 19. He who by his own study, instruction or observation, or from having traveled in other countries, shall offer any machine, expedient, or process, used in other places or other times, and the same is approved by examination and experiment in the manner prescribed iu Article 17 of this Title, shall be attended to and rewarded, in the same manner as if he had been the inventor; since, although his ingenuity may have been less, his merit and labor may have been greater, and the public advantage will be the same, whether resulting from an invention absolutely new, or from the introduction or application of a practice not known in the place where it is established.

Title XIX.

Respecting the privileges of Miners.

Art. 1. Although the rules of administration, economy and industry, which are prescribed in these Ordinances, and are to be introduced in the Mining of New Spain, are calculated greatly to diminish the danger and difficulty, with which this most important business has been hitherto conducted; the riches of the Mines being rendered more accessible by these efficacious aids, and the proper mode of acquiring them, less hazardous; nevertheless, keeping in view, what should never be lost sight of, in these Ordinances, the hardships, difficulties and uncertainties peculiar to, and inherent in, this kind of work; and that, in its valuable products, consists principally, the special bounty which it has pleased Providence to bestow upon my dominions in Spanish America; and which is the main cause of the prosperity and happiness of my subjects, of the preservation and increase of my treasury, and of the commercial activity of these my kingdoms, and even, in great part, of that of the world; I have granted, and do hereby grant, to those persons who in New Spain devote themselves to the working of its Mines, all the favors (*Mercedes*) and privileges granted to the Miners of the kingdoms of Castile and Peru, so far as they may be adapted to the respective local circumstances, and not opposed to what is established by these ordinances.

Art. 2. Moreover, I declare in favor of the scientific Profession of Mining the privilege of Nobility, in order that all persons who devote themselves to this important study and occupation, may be considered and treated with all the distinction due to so noble a profession.

Art. 3. The owners of Mines cannot be imprisoned for debts, nor even their Administrators, overseers, tally-men, or other persons serving in Mines and *Haciendas*, provided, that any of these dependents, in their case, may be kept in confinement in the said Mine or *Hacienda* where they serve, with the obligation of their master (*Amo*) to pay one-third of their salaries and shares, toward extinguishing their debts, so long as they remain in his service; but if they leave that Mine or *Hacienda*, without engaging to serve in another, they may be taken to prison.

Art. 4. When, for the debts of Mine-owners, attachments are levied on their Mines or Mining *Haciendas*, there shall only be allowed to them from the products, while their debt is being paid with the bullion taken out, what is absolutely necessary for their support, according to the circumstances of their family and of the establishment levied on; but always with the proviso, that the creditor is not thereby to be placed in a worse condition than he occupied before the sequestration.

Art. 5. When an execution is levied upon their property of another kind, there shall always be reserved a horse, saddle and bridle, a pack-mule, arms, a bed and the wearing apparel of themselves, and of their wives and children, which may be absolutely indispensable for their decent appearance, leaving subject to seizure, all costly dresses, ornaments, jewelry, and trinkets of value.

Art. 6. The Royal Tribunal of Mining will inform me, through the Viceroy, of the names of meritorious persons in said profession, especially of those who have left it, on account of having lost in it their capital, or on account of being too old or infirm to pursue it, pointing out to me those whom it may consider persons most fit to be appointed by my Royal favor, if such should be my Sovereign pleasure, to the office of Judges of *Reales* and *Asientos* of Mines, in order, not only that their merit may be rewarded, but that these offices may be filled with practical and intelligent persons, as the laws require.

Art. 7. The children and descendents of Miners, and of those who have been suppliers of Mines to a considerable extent, also deserve distinguished consideration, and on that account, the Royal Tribunal will also inform me, through the Viceroy, of the merit of their ancestors, in order that my Sovereign clemency may give to them such political, military, and ecclesiastical appointments in America, as may be deemed proper.

Art. 8. I declare that Miners and Administrators of Mines shall not be prevented or impeded from holding and exercising, if they possess the necessary qualifications, the office of Judge and of Alderman (*Regidor*) in the cities, *Villas*, and *Pueblos* of Mines, or any others; but they shall not be obliged to accept them, nor shall they be fined for refusing, provided they are employed in their profession, and excuse themselves for the purpose of attending to it.

Art. 9. In the distribution of ground lots (*solares*) for building houses, and in renting those which are already built, and in supplying the *Plazas* and market-places of the Places (*Lugares*), *Reales* and *Asientos* of Mines, not only with things required for the Mines and Haciendas, but also with supplies and provisions for the consumption of their houses and families, Miners shall be attended to, in respect to other persons, as their useful profession deserves. And I grant, that they may hunt and fish in the Forests, Woods and Rivers, cut wood and make charcoal, and pasture their animals in the commons and watering places (*Egidos y Aguages*,) the same as any other inhabitant, provided that such Forests, Woods, *Egidos*, and *Aguages* are public and common, and, if they are private property, they must pay the just value, as is already prescribed; and lastly they may enjoy all the privileges and advantages which the inhabitants of the Place (*Lugar*) enjoy, although not residents thereof, provided only, that, in order to enjoy these favors, their Mines or reducing *Haciendas* must be in the same territory as the said Town (*Pueblo*).

Art. 10. It being both notorious and pernicious, that Miners are accustomed to spend their money with immoderate liberality, wasting it with the greatest imprudence and disorder, till they and their families are speedily reduced to want, and their means have passed into the hands of other persons, who do not invest them in working Mines, it is my Sovereign will and I command, that the Judges and Deputies of *Reales* and *Asientos* of Mines, advise, and, in case of necessity, admonish the Miners, and especially those who are in affluence, not to consume their capital in extravagant and idle expenses, or in vain liberality; and when this is not sufficient to correct them, they will report to the Royal Tribunal General of Mining, in order that, having satisfied itself of the reprehensible conduct of the Miner concerning whom the report is made, it may appoint a *Curator*, or in some other manner provide for the preservation of his property, as in the case ot a real spendthrift.

Art. 11. In order to avoid the disorders and dangers, spiritual as well as temporal, which are produced by games of stake and chance (*envite y azar*); and even by lawful games, when they are carried to excess, and also by other public diversions and festivities, I very strictly prohibit, in all *Reales* and *Asientos* of Mines, among their owners as well as the operatives, the practice of any games of cards which are prohibited by my repeated Royal Edicts and Decrees (*Pragmaticas y Cedulas*), and also all playing at those games which are permitted, for a larger stake than is usual for a fair relaxation and prudent diversion. And, with the same rigor, I prohibit the game of dice (*Dados*) sheep's shanks (*Tabas*), and cock-fighting, and all other scandalous diversions, as they not only occasion a loss of time which should be devoted to labor, but also the ruin of property, and sometimes many homicides and disorders.

Wherefore, I very strictly charge the Judges and Deputies of all *Reales* and *Asientos* of Mines to take care and watch, with the most vigilant application, the enforcement of this Article, under penalty of being themselves made wholly responsible for its non-observance, and of being liable to the punishments which the said Royal Edicts and Decrees (*Pragmaticas y Cedulas*) impose on those who violate them.

Art. 12. The Royal Tribunal General of Mining will observe and comply with what is contained in these present ordinances, and will cause them to be observed and complied with by all the subalterns, inferiors and dependents of its Body, each one in what respectively concerns him, without any forced constructions, which might alter and corrupt their true spirit and real intent, observing them themselves, and causing others to maintain them always in all their force and vigor. And the territorial Deputations of Mining will also, on their part, observe and comply with whatever in these said ordinances concerns them, and will cause them to be observed and executed with the greatest punctuality and exactness, without the right, any more than the Royal Tribunal General, to act themselves, or to permit others to act, in contravention, in any manner, of their tenor and meaning; and I only permit that, if any points or cases shall occur which are not provided for in them, nor in the Royal orders which I have issued on this subject, both courts shall be governed in their decision by the practice and proceedings of the *Consulados* of Commerce of my European and American Dominions, so far as they may be applicable. But the doubts which, in any manner, may arise, as to the true meaning of any one or more of these Articles, shall be proposed by the Royal Tribunal General to the Viceroy, in order that he, after obtaining the necessary information, may report to me, for my Soverign determination.

Art. 13. Finally, I order and command, the Governor and members of my Supreme council and Chamber of the Indies, the Royal *Audiencias* and Tribunals of New Spain, its Viceroy, Captains or Commandants-General, Governors, Intendants, Ministers, Judges, and other persons, whom the provisions and enactments of these ordinances do, or may, concern, strictly to conform thereto, executing and fulfilling them, each in what concerns him, with the greatest exactness, regarding all their contents as Law and Statute, positive and perpetual, observing them and causing them to be inviolably observed, notwithstanding any other Laws, Ordinances, Establishments, Customs or Practices, to the contrary, since, so far as there may be any, I revoke them expressly, and will that they have no effect; prohibiting, as I prohibit, that they be explained or annotated in any manner, for it is my will that they remain precisely as written and intended. And, in like manner, I very strictly command all Tribunals, Magistrates, and Courts, mentioned in this and the preceding Article, to aid and assist efficaciously in the punctual fulfillment of what is ordered and directed in these my Royal Ordinances, avoiding, by all possible means, all kinds of contentions and obstructions, which will always incur my Royal disapprobation, as being prejudicial to the administration of justice, and to the good government, tranquility, and happiness of the Important Body of Mining of those my Dominions: to which ends I have commanded the dispatch of this present *Cedula*, signed by my Royal hand, sealed with my private seal, and countersigned by my underwritten Secretary of State and of the General Department of the Indies (*Despacho universal de Indias*), which shall be entered in the Auditor-General's Office of the Indies, and in the corresponding offices of New Spain.

Given at Aranjuez, the twenty-second of May, seventeen hundred and eighty-three.

I, THE KING.

JOSEF DE GALVEZ

Entered in the Auditor General's office of the Indies, Madrid, the twenty-fifth of May, seventeen hundred and eighty-three.

D. FRANCISCO MACHADO.

A true copy of the original.

JOSEF DE GALVEZ.

PART IV.

MINING LAWS OF SPAIN AND MEXICO,

FROM 1783 TO 1821.

PART IV.

MINING LAWS OF SPAIN AND MEXICO;

From 1783 to 1821.

January 15th, 1784.

[From Note to Article 13, Title XIX, Mexican "Ordenanzas de Mineria."]

Proclamation of the Viceroy of New Spain, putting in force the Mining ordinances of May 22d, 1783.

Don Matias de Galvez, Lieutenant-General of the Royal Armies of His Majesty, Governor and Captain-General of New Spain, President of its Royal *Audiencia,* Superintendent-General of Royal Treasury and *Ramo* of Tobacco, Judge-Conservator thereof, President of its *Junta,* and Subdelegate General of Mails of the said Kingdom, &c.

With the most important object, premeditated many years ago, of regulating, encouraging, and attending to, with due speciality, the working and improvement of the very rich and innumerable Mines of New Spain, the indefatigable care and solicitude of the King, our Lord, (whom God protect,) have been pleased to remit to me, by the last mail, the new and appropriate Royal Ordinances for the Direction, Regulation and Government of the Important Body of Mining of these Kingdoms, and its Royal Tribunal General, embraced in Royal *Cedula* dated at Aranjuez, May 22d, 1783, in which, His Majesty, first referring to the Reports, Royal Orders, and Proceedings which preceded, treating of this most grave subject, and all else which occurred therein, includes all the ordinances embraced in XIX Titles, and concludes as follows:

"Finally, I order and command the Governor and members of my Supreme Council and Chamber of the Indies, "the Royal *Audiencias* and Tribunals of New Spain, its "Viceroy, Captains or Commandants General, Governors, "Intendents, Ministers, Judges, and other persons whom "the provisions and enactments of these Ordinances may "concern, to strictly conform thereto, executing and fulfilling them, each in what concerns him, with the greatest exactness, regarding all their contents as Law and "Statute, positive and perpetual, observing them and "causing them to be inviolably observed, notwithstanding "any other Laws, Ordinances, Establishments, Customs or "Practices to the contrary, since, so far as there may be "any, I revoke them expressly, and will that they have no "effect; prohibiting, as I prohibit, that they be explained "or annotated in any manner, for it is my will that they "remain precisely as written and intended. And, in like "manner, I very strictly command all Tribunals, Magistrates and Courts, mentioned in this and the preceding "Article, to aid and assist efficaciously in the punctual "fulfillment of what is ordered and directed in these my "Royal Ordinances, avoiding, by all possible means, all "kinds of contentions and disputes, which will always "incur my disapprobation, as being prejudicial to the "administration of justice, and to the good government, "tranquility and happiness of the Important Body of Min- "ing of those my dominions, to which ends I have com- "manded the despatch of this present *Cedula* signed by my "Royal hand and sealed with my private seal, and coun- "tersigned by my underwritten Secretary of State and of "the General Department of the Indies, which shall be "entered in the Auditor-General's office of the Indies, and "in the corresponding offices of New Spain.

"Given at Aranjuez, May 22d, 1783.

"I, THE KING.

"JOSEF DE GALVEZ.

"A true copy of the original.

"JOSEF DE GALVEZ."

And having transmitted the printed copies which came to the said Royal Tribunal General, with an official letter of December 19th, 1783, it asked me, the other day, to be so good as to direct that due obedience and observance be given to said Royal *Cedula* and ordinances, according to their contents: which I thus immediately decreed, in conformity with the corresponding previous report and request of the *Fiscal*, who said, among other things, as follows: "Lastly, it is proper that your Excellency order a Procla- "mation (*Bando*) to be drawn up, which shall be sent with- "out delay over the mountains to all the Governors, Cor- "regidores, Alcaldes Mayores, and other Courts of these "kingdoms, by which means will be made known to the "inhabitants thereof, the general purport of the Royal

"*Cedula* of May 22d of this year, and Your Excellency "will order them very strictly to contribute to, and efficaciously aid in, the fulfillment of what is ordered and "directed therein, and in all and each one of the Ordi-"nances contained in its XIX Titles; it being understood, "that, although they are not now sent to them, because "of there not being a sufficient number of copies, they "must not, on that account, neglect to comply with their "provisions, which will be opportunely communicated, "and in the mean time they can, in such cases, be advised "of them by the respective territorial Deputations.

"Mexico, December 23d, 1783.

"POSADA."

And, in consequence thereof, and in order that all may know, I command that it be thus promulgated and published, by Proclamation, in this capital and in the other cities, *Villas* and Places of these Dominions and in all the *Reales* of Mines, in order that the said Royal Ordinances may be observed, kept, and fulfilled, with due punctuality, as Laws and Statutes, positive, proper, and special, of Mining: of which printed copies have already been distributed to the Tribunals and Officers in this city, and sent to the principal Mining Places, as will be done to all others, as soon as a sufficient number of copies is received. And let the Royal Tribunal General of this Important Body be held and esteemed by all others, with that respect which is as conducive as appropriate to the very useful purposes for which the Royal Pleasure created it; and let it also enjoy and exercise all the jurisdiction, *fueros*, and powers, which are granted to it in said Royal Ordinances, and let it be respected and obeyed by all those whom, in any manner, it concerns or may concern, under the grave penalties which those incur who disobey Judges and transgress the Laws and Sovereign Orders of the King. And, for the same purposes, let there be sent copies of this Proclamation, as has been done of the Ordinances, to the Presidency and Regency of the Royal *Audiencia* of Guadalajara, and to the Commandancy General of the Interior Provinces, for their information and government, and, by their order, let the corresponding decrees be published.

Given at Mexico, January 15th, 1784.

MATIAS DE GALVEZ.

March 4th, 1785.

[From Note to Article 10, Title XVI, Mexican "Ordenanzas de Mineria."]

Decree prohibiting the exportation of silver from the Mines in the form of toys, &c., in fraud of the Royal revenue.

We, the President Regent, and Oidores of the Royal *Audiencia and Chancilleria* of this New Spain, in whom is now vested the Superior Government thereof: In Junta of Royal Treasury, held on the 29th of January last, it was enacted, and, by Decree of the 4th of February of this year, ordered to be executed, that, to prevent the abuse which, in all the *Reales* of Mines, is experienced from the exportation, without permits (*Guias*), of bullion in the form of images (*Muñecos*), pine-cones (*Piñas*), and toys (*Juguetes*), they shall be considered the same as any other smelted bullion, and that they must be manifested to the respective courts, who will issue their permit (*Guia*) to the corresponding Treasury (*caja*) in order that they may be there presented and the duties thereon paid, certificates being given to the parties interested for their security, it being understood, that anything which shall be seized without having this requisite, or the prescribed certificate, will be held as contraband, and be subject to the penalties prescribed by the laws.

Given in the city of Mexico, March 4th, 1785.

VICENTE DE HERRERA,
ANTONIO DE VILLA URRUTIA,
MIGUEL CALIXTO DE ASEDO,
RUPERTO VICENTE DE LUYANDO,
BALTAZAR LADRON DE GUEVARA,
JOAQUIN GALDIANO,
JOSE ANTONIO MINAFUERTES,
EUSEBIO VENTURA BELENA.

September 2d, 1785.

[From the "Pandectas Hispano-Mexicanas," No. 2314.]

Circular, exempting effects used in Mines from excise duties.

Art. 1. By decree of the 29th of January last, the Superior Government declared, and it was resolved in Junta of Royal Treasury, on the 15th of March of this year, and ordered to be executed, on the 11th of the following April, that upon chalk (*greta*), lead, ashes, and other alloys which result from the smeltings of metals, and also upon salt, *saltierra* and magistral, with which the ores are worked with quicksilver, there be imposed no excise duties (*alcabala*) even when not introduced on account of the Miners, provided, that those who purchase these things do so for the purpose of using them in the working and reduction of metals, and not to trade in them.

Art. 2. That all the tools, utensils and supplies, which are used directly in the working of Mines and the reduction of their ores, or for drainage, as are the eleven species which compose the framework (*cuartones*) of *Arrastre*, the different things which are called *de fondo*, charcoal, fire wood, oak timber, stones, flags, double and single tiles, iron, brass, animals, hair hides (*cueros al pelo*), grease, cordage, and others, are exempt from excise duty (*alcabala*) in the *Reales* of Mines, the said Miners introducing them for the precise purpose of working their Mines, and using and consuming them in the Mines; but they shall pay such duty, whenever the introduction is for the purpose of sale or trade.

Art. 3. That the personal privileges of the Indians and

poor people, to be exempt from this duty, in those things and fruits to which they are entitled, be observed and attended to, even where such things are not brought in on account of the Miners.

Art. 4. That excise duty (*alcabala*) be not imposed on corn, barley and other kinds of grain, which the Miners introduce on their account for feeding the animals which they use in the Mines and *Haciendas* for the reduction of ores: which is the only point in which any change is made with respect to what is directed in the circular orders of the 3d of January and the 3d of August, 1782.

Art. 5. I transmit this determination to you for your information, and I charge you to read it very attentively, for its exact fulfilment; repeating to you that, in the free entry of the effects, which individuals of Mining may make, for the purpose of working their Mines, and in the transmission of them from one custom-house to another, you do not omit the formalities of permits (*Guias*) and bonds (*responsivas*), and the precaution of sworn paper, and others which are prescribed by the said circular of January 3d, 1752.

Art. 6. I also notify you, that if, by having understood the declaration of the 24th of April, 1781, without subordination to what I have referred to in this order, you have failed to collect excise duty (*alcabala*) on the *cuartones de Arrastre*, on the different articles called *de fondo*, charcoal, fire wood, oak timber, stones, flags, *camones de Arrastre*, lime, sand, double and single tiles, iron, brass, animals, hair hides, grease, and cordage, and there be, in consequence, any thing due to the Royal Treasury, you will immediately direct its collection.

Art. 7. If upon the effects which have been mentioned, chalk, lead, refinings, and other alloys which result from the smelting of ores, or on salt, *saltierra*, and magistral, you have collected the duty of *alcabala*, contrary to what is provided in Articles 1, 2 and 3 of this Order, you will report to me, with all possible brevity and all possible clearness, the persons from whom you have made this exaction or these exactions, specifying the amounts and dates thereof, in order that the proper measures may be taken for their return, as the said superior decree of January 29th directs.

God preserve you many years.

Mexico, September 2d, 1785.

JUAN NAVARRO.

To the Administration of *alcabalas* of ——.

February 7th, 1786.

[From Note to Article 6, Title XIII, Mexican "Ordenanzas de Mineria."]

Regulations for supplying Salt to Miners.

In consequence of what was enacted by the *Junta* of Royal Treasury on the 27th of January last, I instruct your Honor (V. S.) to so order and direct, that, in all the *Reales* of Mines, the corresponding measures (*providencias*) be taken by the Deputies and Electors to make, every year and in due time, the necessary provisions of salt, purchasing it, for that purpose, of the Administrators and private individuals in first hands, at the store-houses and salines, and at cost, with the object of conferring greater benefit and assistance upon community, and of giving relief to poor Miners, with the understanding that it shall be issued and given to them, in preference to all others, even anticipating their wants, since by this Superior Government it will be so directed to the Royal officers and distributors (*expendedores*), without prejudice to the general liberty, to any persons of whatsoever class, to trade in salt, and to carry it for sale to the *Reales*, or wherever they may please, and to the right which the Miners and residents ought to enjoy, if they wish, of carrying it on their own account, and of supplying themselves, independently of the provision which is made in the name of the said Mining (*Mineria*).

Also, it is declared, that the authority asked for, to absolutely compel the *Arrieros* to carry salt to the Mines, cannot be granted, because it is commanded by his Majesty in the last Royal order in favor of this worthy and useful class, that they shall not be molested nor compelled to carry loads (*cargas*), without even excepting those of the Royal Treasury, to any destination, they not being disposed to do so of their own free will, and being paid fair freights, according to the times, seasons and conditions of the roads; of which resolutions I notify your Honor (V. S.) for your information, and in order that you may make dispositions for their most strict observance.

God preserve your Honor (V. S.) many years.

Mexico, February 7th, 1786.

EL CONDE DE GALVEZ.

May 9th, 1786.

[From Montemayor y Beleña's "Recopilacion Sumaria de Autos Acordados," Tomo II, No. 52.]

Respecting Excise duty (alcabalas) in matters of Mining.

In consequence of the various appeals of the Miners of that kingdom, asking exemption from excise duty (*Alcabalas*), and hearing the *Fiscal* of the Royal Treasury, Don Ramon de Posada, the Viceroy, Don Martin de Mayorga, declared on the 24th of April 1781, that all effects, instruments, utensils, and ingredients, which, directly or indirectly, conduce to the working of the Mines and the reduction of their ores, should be exempt from said duty of *acabala*; but not the sale and exchange of Mines, works, and *Haciendas*, which contracts do not immediately effect the abundance of gold and silver.

That metals, in ore or extracted, chalk, lead, ashes, and any matters which result therefrom, should be equally free, as also the gold and silver for coin, bars, or precious ornaments, according to what is provided in the Laws of Castile and of the Indies.

He also declared exempt from excise duty (*alcabala*) the eleven species included in the *Bando* of October 20th, 1780, approved by the King, besides the instruments of Mines, and the supplies of Iron, brass, animals, hides, grease and cordage, magistral and *saltierra;* but not the other provisions and supplies which ought to pay it, except maize and wheat, with respect to which, he ordered to be observed what was prescribed by me, when I was Visitor-General of that Kingdom, on the 18th of December, 1769, which provisions His Majesty approved, by Royal order of June 10th, 1770.

This determination being communicated by said Viceroy to the Tribunal of Mining and to the Direction of *Alcabalas*, and being put in force, he reported it to His Majesty, with *Testimonio* thereof, and of the *expedientes*, which gave reasons for His Royal approval.

The King, before deciding upon this subject, wished to hear the opinion of the Council, and for that purpose the said *Testimonios* were transmitted to it by His Royal order.

In this state of the case, two Representations were received, one from the Tribunal of Mining, dated January 28th, 1781, setting forth the appeals which it had made to the Viceroy, complaining of the injuries which the new Plan of the Administration of *Alcabalas* would produce and those which it already experienced; and the other of the Direction of *Alcabalas* of March 6th, 1782, in which it represented, that, in order to cut off the injuries which resulted from the equivocal understanding which was given to the said declaration of the Viceroy, Mayorga, of April 24th, 1781, it had issued a Circular Letter respecting what the Tribunal of Mining reclaimed without reason: It offered to show, at the proper time, that the effect of the said provision (*providencia*) of the Viceroy, was neither just nor expedient, inasmuch as it facilitated frauds, and proposed various means to prevent them.

In view of these Representations, the King resolved, and it was communicated to that Government, in Royal Order of September 30th, 1782, that, the Direction having made a representation to the Viceroy of what it had to offer on the matter, the *expediente* be passed to the *Fiscal* of the Royal Treasury; and, with his Opinion, it be taken without delay, to the *Junta*, and its resolution being carried into effect, the whole matter be transmitted for the Royal approval.

The Council, having complied with the directions of the Royal Order by which the said *Testimonios* were remitted to it, delivered its opinion, and, in accordance therewith, His Majesty approved, in all its parts, what was directed by the Viceroy, Mayorga, on the 24th of April, 1781, ordering, besides, that, in what had already occurred, there be observed and executed what was directed by the anterior order of September 30th, 1782, which resolution was communicated to the Viceroy, on the 13th of January, 1783.

The *Audiencia* Governing (*Gobernadora*), in Letter of April 25th, 1785, No. 379, reported, with *Testimonio* of what, on the 11th of February, 1783, the Tribunal of Mining moved, on the continuous appeals of different Miners, to be exempted from excise duty (*Alcabalas*) on the utensils, tools, and supplies of the Mines, in consequence of the Direction having declared that the Decree (*Providencia*) of the Viceroy, Mayorga, must be understood as applicable to the case where the Miner is himself the introducer, and the effects are for the working of the Mines: it applied to the Viceroy to declare, that all the inhabitants of the *Reales* should enjoy the exemption; that there should also be exempted from said duty provisions, and provender for animals, as the cost thereof was very considerable, and that lead, alloys, *saltierra,* and *magistral* ought to pay excise duty (*Acabala*) only in case of trade; and tools and utensils, when not intended for the working of the Mines.

That the Viceroy, Don Matias de Galvez, having received, at this time, the two anterior orders of September 30th, 1782, and January 13th, 1783, the direction manifested the motives which it had had, in issuing the said Circular Letter on the manner in which the Administrations of *Alcabalas* ought to understand the exemption declared on the 24th of April, 1781; and having heard the *Fiscal* of the Royal Treasury, was of opinion that in its main provisions, the Resolution, which was already approved by His Majesty, ought not in any way to be changed.

That, in order to enjoy exemption from excise duty (*Alcabala*), the utensils, tools, and supplies, must be introduced by the Miners themselves for the strict purpose of working the Mines.

And that, in the complaints of the contributors because the exemption was not conceded to them, and in the representations of the Administrators that they opposed the payment of what was considered due from them, it was decided, in first instance, in the Direction, allowing appeals to the Superintendency General.

That said *Audiencia*, on the 29th of January of the same year, 1785, acceded to the proposition of the *Fiscal*.

And lastly, that the *expediente*, thus drawn up, was passed to the *Junta* of Royal Treasury, and in session of March 15th it resolved:

That on the chalk (*Greta*), lead, ashes (*cendrada*), and other alloys (*ligas*) which result from the smelting of ores, and the *saltierra* and *magistral* used in reductions with quicksilver, no excise duty (*Alcabala*) is to be collected, although not introduced on account of Miners, provided, that the purchasers buy them for the purposes of consumption, and not for trade in these several kinds of articles.

That all the tools, utensils, and supplies which immediately serve for the working of Mines, reduction of their ores, or for draining them, meaning those of iron, brass, animals, hair-hides, grease, cordage and others, are equally exempt from excise duty (*Alcabala*) in the *Reales* of Mines, they being introduced for the precise object of working the Mines and for consumption therein; but that they must pay said duty, if introduced for trade.

That said exemption is also to be understood as applicable to the eleven species, contained in the Declaration of April 24th, 1781, of *Cuartones de Arrastre*, charcoal, wood, &c., observing, with respect to them, the personal privileges of the Indians and poor, relating to exemption of excise duty (*Alcabala*), although they are not introduced on account of Miners.

That with respect to Maize, barley, and other things destined for the subsistence of animals used in the Mines and *Haciendas* of reduction, no excise duty (*Alcabala*) is to be exacted, as this exemption is not considered contrary to what was directed by me, when I was Visitor-General in that kingdom.

That all the aforesaid be put in execution, passing, for that purpose, the proper order to the Direction of *Alcabalas*, in order that, communicating it to all the Administrators, they may observe the said terms in the exaction thereof.

And lastly, that a report be made to His Majesty, setting forth the advantages and conveniences, which the Junta may consider will result to the Royal Treasury, from the absolute and universal exemption from excise duties (*Alcabalas*) of all the effects which are introduced into the *Reales* of Mines, for the use and working thereof, although the introduction be not made by Miners, conceding also the same exemption to all provisions and supplies.

The King being made acquainted with all the aforesaid, with what was represented by the Tribunal of Mining, by the Direction of *Alcabalas*, and by the *Fiscal* of Royal Treasury, and having heard the opinion of the Council, His Majesty has not approved the Resolution of the *Junta*, with respect to extending the exemption from excise duties (*Alcabalas*) to effects and cases not included in the Declaration of the Viceroy, Don Martin de Mayorga, on the 24th of April, 1781, and approved in Opinion of the Council in the said Royal order of January 13th, 1783. And His Majesty has resolved that said Declaration be observed in the manner and form most conducive to the prevention of frauds.

I communicate it to your Excellency, by His Royal order, for your information and observance.

God preserve Your Excellency many years.

Aranjuez, May 9th, 1786.

THE MARQUES OF SONORA.

To the Viceroy of New Spain.

June 17th, 1786.

[From Note to Article 8, Title XIX, Mexican "Ordinanzas de Mineria."]

Exemption of Miners from Militia duty.

Under date of the 17th of June of this year, the Most Excellent Señor Marques de Sonora, by order of his Majesty, says to me as follows:

"In letter of October 27th, of the year 1785, Your Excellency made a report to the King, with authentication of the *expediente* formed at the instance of the Tribunal of

Mining, respecting the exemption of Miners and their operatives from militia duty, so that they may not be impeded in their operations. In view of the contents of the said *expediente*, and having heard the report of the Council, His Majesty has been pleased to declare exempt from service in the militia, the Miners and their operatives, while they both are employed in the service of the Mines, and in time of peace. I communicate, to Your Excellency, this Royal order, for your information and observance."

The which I transmit to Your Honor (V. S.) for your information and observance in the cases which may occur.

God preserve Your Honor (V. S.) many years.
Mexico October 22d, 1786.

His Excellency rubicated it October 25th, ——.

FRANCISCO FERNANDEZ DE CORDOVA.

December 4th, 1786.

[From the "Real Ordenanza de Intendentes, en el Reino de la Nueva-España," dated at Madrid, December 4th, 1786.]

On the powers and duties of Intendants with respect to Mines and Mining.

Art. 150. The Royal due of one-fifth of the gold, silver, copper, and other metals which the *Placeres* and Mines of those dominions and others of America produce, always received the best attention of my glorious ancestors. And whereas, my Father and Lord, Don Philip V, in the year 1723, reduced that contribution, in the kingdom of New Spain, to one-tenth, for the general benefit of the State and the nation; and whereas, I, influenced by the same reasons, have reduced the dues on gold to three per cent., and the price of quicksilver and powder, in favor of the Miners, and have granted to them besides, in order to stimulate their industry and to honor their most useful profession, other favors very appreciable and set forth in the Royal Ordinances which, for the Direction, Regulation and Government of the Important Body of Mining of that Kingdom I was pleased to issue on the 22d of May, 1783;—I will that the Intendants shall direct their principal care, not only to the encouragement and protection of the said Body in the Provinces under their charge, but also to watching, both personally and by means of the subaltern Judges, over the most exact observance of what I have commanded in the said Ordinances; that the *Contadurias* and Royal Treasuries and Administrators of *Estanco* permit no gratuities, gifts or presents to be charged to Miners for quicksilver and powder which they may require, even under the name of fees of the officers or clerks, the offenders being severely punished, in addition to making restitution and paying a fine of double the amount, or quadruple in case of repetition; and that to poor persons employed in Mining there be given quicksilver, in retail, which they may ask, for cash, or on security to pay at the end of the credit which shall be given them, without charging them for it any premium or additional price.

Art. 151. Whereas, it has been directed in Article 13, Title III of the said ordinances of the Body of Mining that the Court of Appeals (*Juzgado de Alzadas*), ordered by it to be erected in each Province, for the objects of its institution, shall be composed of, among others, the Judge best qualified, and appointed by Me, who may be found in it;—I declare it to be my Sovereign will, that it be understood that the exercise of this most important duty, and consequently the Presidency of said courts, with all the authorities and powers granted by the said Ordinances, belong to the Intendants in their respective provinces, except those of Mexico and Guadalajara, because provision is made for both these capitals in the aforesaid Article. And with the just object of avoiding doubts and of preventing the injuries which may result to Mining, if its members should be subjected, by the great distances, to long and consequently expensive proceedings on appeal (*recursos*), I order, that in the Villages (*Villas*) of Chihuahua and Rosario, and in all other places which are very distant from the residences of their respective Intendants, their Subdelegates shall exercise, for them and in the appropriate form, the said office of Judge of Appeals; and, in case of their absence, sickness or vacancy, the oldest officer (*Ministro*) of my Royal *Hacienda* and its respective Treasury (*Tesoreria*.

Art. 152. With the just object of preventing the concealment and fraudulent extraction of gold and silver in bullion (*en pasta*), which the Miners are under the necessity of selling to the merchants and ransomers of these metals, in manifest contravention of the laws, which prohibit the purchase and sale of them, before the duty is taken out; I command, that in all the principal, exterior and subaltern (*Foraneas y Menores*) Treasuries of the Provinces which have Mines actually worked and reduced, there shall always be the amount of coin deemed necessary, for the redemption and full payment of the gold and silver which the Miners shall bring in for sale. And in order that their value at the current price may be promptly satisfied, and with the proper knowledge of the *ley* of the metals, the Intendants will, with the knowledge and approval of the Superior *Junta* of Treasury, through the Superintendent subdelegate, make dispositions, where this has not already been done, to have incorporated in, and connected with, my Royal Treasury, as soon as possible, the offices of Smelter and Assayer of the present Treasuries (*Cajas*), and that they be filled with persons trustworthy, skillful, and salaried, in order that the parcels of silver and gold may be smelted and assayed as soon as their owners present them, in the presence of such owners and of the officers of the Royal Treasury, as is ordered in Law XI, Title XXII, Book IV, of the Recopilacion : observing, that for the said appointments of Assayers, it is indispensable that there first be had the examination and approval prescribed in Law XVII, of the same Title, and Book, and that they must be executed by the Chief Assayer of that Kingdom.

Art. 153. For the direction, administration, and collection of the branches (*Ramos*) of quicksilver, stamped paper, *Medias-anatas y Lansas*, there have been established, in my dominions of the Indies, delegated and special Judges, which practice exists in New Spain. But having considered that these separations of the general management of my Royal Patrimony, in addition to the burthen they impose on its revenues, embarrass, not a little, the knowledge of their legitimate value, and do not prevent notable omissions in their collection ;—I have resolved, that the said branches (*Ramos*) be united to the general body of the administration of the other branches of my Royal Treasury: and I therefore order, that the Superintendent Subdelegate of the Treasury take under his general charge, and the Intendants, in their respective provinces, take under their special charge the directive and administrative cognizance of the said four branches (*Ramos*), subject to the special rules which are indicated in this Instruction, and that they exercise in these matters the contentious jurisdiction which is granted to them, for the other branches of my Royal Treasury, in Article 78.

Art. 154. The quicksilver which, for the extraction of the gold and silver reduced by that means, is carried from this Kingdom, and sometimes from that of Peru, to New Spain, shall be under the authority of the officers (*Ministros*) of the Royal Treasury,—the Auditor-General and Treasurer-General,—who shall keep the account of this branch of my Royal Treasury, through its *Contaduria General* established in Mexico: which must, for the present, continue under the Regulation which, with respect to the number of its members, and the other dependents of the same branch and the pay of both, the *Junta Superior* shall immediately form, according as the actual circumstances and prudent economy of my Royal Treasury may require, putting it in practice and reporting it to me for approval. And whereas, the abundant supply of said ingredient is both useful and indispensable, for the reduction of those ores which are not suitable for smelting, the Superintendent Subdelegate will take care to order the Treasuries of those Provinces, to be provided with such quantities, as the *Reales* of Mines, situated in them, may require; and their Intendants will take care that there is never a want of quicksilver in the respective store-houses, and in other places where it is required; and that the Officers of the Royal Treasury, whether Principals or Exteriors, to whom the duty may respectively pertain, shall give them timely notice, so that they may, in time, notify the said Superintendent Subdelegate, in order that he may opportunely command the necessary remittances to be made.

Art. 155. For the direction, regulation, and government of this branch, there was issued, on the 15th of January, 1709, the Instruction which was then considered proper, and afterward there were successively despatched various *Cedulas* and Royal Orders, for the same object, according as events and time required, under which rules it has been managed till the present time. But as they ought to receive the changes, which necssarily result from what is declared in this Instruction, and as it is, therefore, expedient, in order to avoid confusion, that the precepts which are consequently to govern hereafter, be reduced to a formal Ordinance, I command the Superintendent Subdelegate to immediately proceed to its formation, having before him, for that purpose, all the *Cedulas* and Royal Orders which are indicated, the said Instruction, and the contents of this Instruction on the subject, and of the Ordinances of the Important Body of Mining, which are cited in Article 150, hearing moreover its Royal Tribunal, the *Tribunal de Cuentas*, and the *Contaduria General* of the same branch; and that when so formed, he pass it to the *Junta Superior* in order that, having examined into it with the advice ot its *Fiscal*, it may amend it where deemed expedient, and direct it to be put in practice in the meantime and until, it being reported to me in the prescribed form (*via reservada*), with a certified copy thereof and of all the proceedings on the subject, I may be pleased to approve it according to my Sovereign pleasure.

July 18th, 1789.

[From Note to Article 18, Title I, Mexican "Ordenanzas de Mineria."]

Separation of the office of Director of the Royal Tribunal of Mining from the duties of Fiscal and Defensor.

The King, being made acquainted with your representations of January 27th and March 27th of this year, in relation to the request that the duties of *Fiscal* and *Defensor* be separated from the office of Director of the Royal Tribunal of Mining in that kingdom, which are united by provision of the Royal Ordinance, has been pleased to order me to communicate, as I do this day, to the Viceroy of that kingdom the following Royal Order:

"Most excellent Señor,

"The Director of the Royal College of Mining of that "kingdom, Don Fausto de Elhuyar, has reported to me in "full, in letters of January 27th and March 27th of this "year, the *expediente* which, in entering upon the duties of "his office, he found drawn up in that superior Govern-"ment, with respect to separating from him the duties of "*Fiscal* and *Defensor* of the Royal Body of Mining, on "account of the inconveniences which result from their "being united, as provided in Article 18, Title I, of the "Royal Ordinances, the *Oidor*, Don Baltasar Ladron de "Guevara, as Judge of Appeals of the same Tribunal, gave "opinion that they ought to continue; and that having "asked that the *expediente* be communicated, he set forth in "it his opinion at length, arguing the necessity of separating "from his office those of *Fiscal* and *Defensor*, if there be "desired the good of Mining and of the State, and the circumstances which ought to concur in persons who shall "be appointed to serve them with advantage: which "opinion, although read in the *Junta*, whose business it is "to regulate the affairs of Mining, produced no favorable "result, that body reserving its decision for another time, "which delay was undoubtedly greater than suited the "objects for which Elhuyar and the Germans, whom he "took with him, were sent to that kingdom, considering "that *Juntas* were only held on Saturdays, and that the "business to be transacted in them is very extensive, entangled, and of difficult solution; concluding the whole "with the request that the King be pleased to declare that "the principal object of his duties ought to be the encouragement and perfection of the working of Mines and the "operations of reduction, with the power of making the "journeys which may be required for that purpose, relieving "him from all the occupations which might prevent him "from so doing; for which purpose there be left united "with the office of Director only the duties of the Government of the Mining College, and the scientific and professional part of Mining, with a vote in the Royal "Tribunal General, in matters directive, executive and "economical, and the prerogative of *ex officio* Associate-"Judge of the Court of Appeals.

"His Majesty, in view of all this, and of what Elhuyar "represented in support of his application, which he "deemed to be in accordance with the intentions of the "King in appointing him Director of that Royal Tribunal, "has been pleased to resolve, that, notwithstanding what "the Royal Ordinances of Mining establish and ordain, it "is his Royal will to separate him (the said Elhuyar) from "the offices of *Fiscal* and *Defensor* of the Tribunal, and to "leave him free to exercise his functions as Director over "all Mining matters, with a vote for that purpose in the "said Tribunal, the prerogative of *ex officio* Associate-Judge "of the Court of Appeals, and the government of the min-"eral College, without thereby preventing him from making "such journeys as may be necessary, to the Mines of that "Kingdom which may require his presence, and for the "better regulation of operations, granting to your Excel-"lency the power of appointing and electing *ad interim* the "persons who are to perform the duties of *Fiscal* and *De-"fensor* of the Royal Body, with the salaries which you "may deem proper, and under the condition of first draw-"ing up the *espediente* of instructions which must be trans-"mitted, at the proper time, in order that, on being exa-"mined, it may receive the Royal approval.

"Finally, the delay of the Tribunal in holding its *Juntas*, "has appeared very strange to His Majesty, especially as "by the clear provision of Article 33, Title III, of the "Royal Ordinance, it is commanded that they be held "every day except holydays or mass-days, from eight "o'clock to eleven, and also on extraordinary occasions in "the evening, and even on any day if rendered necessary "by the importance and urgency of business; and in order "to remedy this, His Majesty commands me to direct your "Excellency to issue the proper instruction (*providencia*), "making a compliance herewith a most stringent duty, "and to report to me the results. And I transmit it to "your Excellency for your information and in order that "you may communicate to the Royal Tribunal the contents "of this Royal Resolution, taking notice that they are also "transmitted to Elhuyar, in a Royal Order of this date, so "that he may know them and observe them so far as relates "to himself."

God preserve you many years.

Madrid, July 18th, 1789.

VALDEZ.

To Señor DON FAUSTO DE ELHUYAR.

December 26th, 1789.

Don Carlos III, by Royal Order, November 28th and Cedula of the Council December 26th, 1789.

[Book IX, Title XX, Law II, Novisima Recopilacion.]

Rules for the working of Mines of stone-coal.

In order to remove the difficulties encountered in the use of the Mines of stone-coal, and to simplify the method of working them, without prejudice to the proprietors, and with public advantage; I have been pleased to resolve, declare and command as a general rule and regulation the following:

Art. 1. Stone-coal, not being a metal or a semi-metal, or any of the things included in the laws and ordinances which declare Mines the property (*propias*) of the Royal Patrimony, shall be free for working, and for trade by sea and land, throughout the whole Kingdom, and no impediment shall be placed in the way of its exportation by sea, for trade in foreign countries.

Art. 2. Such Mines should belong to the proprietors of the lands in which they are, by the word *proprietor* the direct owner being understood, and not the lessee or the holder of *enfiteusis;* without its being necessary to ask for a license from any Court or Tribunal, to work, lease, sell or grant them; but if the proprietor, the Mine being once discovered, shall refuse to make use of his ownership, in any of the aforesaid modes, in order to have it effectively worked, my council, and the Intendant of the Province, or the *Corregidor* of the *Partido*, shall have power to grant the right to work it, to the discoverer, he giving to the proprietor the fifth part of its produce.

Art. 3. If there should be Mines of coal in the municipal lands (*Propios*) of Pueblos, they shall belong to the Pueblos, and shall be worked or leased on their account, with the previous permission of the council: and Mines found in the commons, shall be for the use of the citizens, they being distributed among those persons who may wish to work them, or being rented, for the benefit of all; but if found in the municipal or common lands (*propios ó comunes*), and the citizens do not work them, they shall be adjudged to the discoverer, on the same terms as those in the lands of private proprietors.

Art. 4. No one shall be at liberty to dig pits or trial-pits (*calas ni catas*) in the land of another, without the license of its owner, or to take out coal, under the pretext of being the discoverer of the Mine, unless he who does so, shall present some authority for enjoying said Mine.

Art. 5. In order to obviate any doubts with respect to the execution of all this, I repeal, and declare to be of no effect, the Royal *Cedula* of the 15th of August, 1780, and whatsoever decree before or subsequent to that, so far as they conflict with what is herein established.

January 28th, 1790.

[From Note to Article 9, Title I, Mexican "Ordenanzas de Mineria."]

Royal order with respect to the organization of the Royal Tribunal of Mining.

In the Royal Order which, under date of the 28th of January last the Most Excellent Señor Bº. F. Don Antonio Valdes communicated to me, he directed me as follows:

"Most Excellent Señor:

"I have reported to the King the contents of the letter "of your Excellency dated the 28th of May last, No. 1036, "and the *testimonio*, which accompanied it, of the *Expe-"diente* drawn up upon the petition of the *Consultores* and "*Associate-Judge* of Appeals of that Royal Tribunal of "Mining, that there be appointed other persons to fill their "offices, on account of the expiration of the time which they "were obliged to serve, agreeably to Article 15, Title I, of "the Royal Ordinance; as also of the doubt occurring to the "said Tribunal with respect to the individuals who ought "to compose it, at the time when the *Juntas Generales* are "held, in which the new election for said offices were to "take place, during which time, it would have only three "proprietary officers: and His Majesty, informed of all, and "agreeing with what, in that particular, the *Ministros, Ase-"sores* and *Fiscales* of the Superintendency General of "quicksilver and Mines have reported to him, has ap-"proved the opinion of your Excellency in that *Expediente*, "that, in default of the five proprietary members (*vocales*) "of which the Tribunal ought to be composed, when the "*Juntas Generales* are held, there concur in them the "precise number of four votes, viz; that of the Direc-"tor, the two Deputies and one *Consultor*, and in case "of disagreement the *Consultor* who succeeds, in order "to decide it, agreeably to the practice of the Tribunal, "disregarding the petition of the *Consultor*, Don Juan Eu-"genio Santelices, in which he asked, separately, that the "other *Juntas* which should be held in future, should be "composed of five individuals, and that in default of one or "more of the proprietary members of the Tribunal, the cor-"responding *Consultor* or *Consultores* should supply the va-"cancy or vacancies, because the concurrence of many "electors would be injurious, rather than advantageous, "and because, in reference to the fact that Article 2, Title "I, of said ordinances, permits the diminution of them and "expressly prohibits their increase, the Tribunal has not "represented that any prejudice or inconvenience has re-"sulted from the disposition of your Excellency, to whom "I communicate this by Royal order, for your information "and government, and in order that you may communicate "it to the Tribunal and to the said Santelices so that they "may know this resolution."

And I transmit it to your Honor (V. S.), for your information and observance.

God preserve your Honor (V. S.) many years.
Mexico, May 31st, 1790.

El Conde de Revilla Gigero.

To the Royal Tribunal of Mining.

August 18th, 1790.

Don Carlos IV, by Royal Decree, August 18th, and Cedula of the Council, September 15th 1790.

[From Book IX, Title XX, Law III, Novisima Recopilacion.]

On the observance of the law of 1789, *with other declarations for the working of Mines of stone-coal.*

The Director General of Mines having represented the inconveniences which may result from my Royal *Cedula*, issued the 26th of December, 1789 with respect to Mines of stone-coal, in consequence of my Royal Order of the 28th of November of the same year,—I have resolved, that, until the approval of the new general ordinance of Mines which I shall order to be drawn up, with attention to the present condition of this branch, the dispositions of the said *Cedula* shall be in force; with the declaration, that any one shall be permitted to make pits and trial-pits (*calas y catas*), in order to search for Mines, paying the damages to the owners of the lands, if in fact they should cause them any such damages; and that, the Mine being discovered, if the owner of the land desire to work it, he shall be preferred, provided, he does so, according to the regulation, mode, and art, and within six months after he is informed of its discovery, making it produce all the fruit it is capable of producing; and if he shall not desire to do so, or shall not be in condition to do so, then it shall be adjudged to the discoverer, if he have the means of doing it, and if not, to any one who has; paying to the owner of the land, for that which is occupied by the Mine, and for the buildings pertaining to it, which may be necessary, ten per cent. of the coal which may be taken out, the expenses being deducted; or, adjusting the whole amount as so much per annum for the rent of the land, so long as the Mine shall last; and in case no agreement is made between the parties in either of these modes, then, the land shall be appraised for sale, taking into consideration its superficies and what there is on it, and the sale-price shall be paid to the owner, or the interest on that sum, at the rate of five per cent. per annum: that all this is to be understood of Mines of stone-coal which are discovered, or which shall be discovered, subsequent to the date of the said *Cedula*, and not of those previously worked, which must continue upon the same footing as when they were begun to be worked, without any one being at liberty to embarrass or impede the workers, until the new ordinance is made and approved: and also, that this is to be understood of Mines of the same kind which are in the lands of individuals, and not of those found in common lands; which latter shall be immediately adjudicated to the discoverers, compensating the inhabitants of the place, or the council, to whom the usufruct belongs, for the benefit which they would derive from the pasturage, fire-wood, or from any thing else, at a just avaluation.

October 19th, 1790.

Don Carlos IV, by resolution of October 19th, 1790.

[From Book IX, Title XVIII, Law VI, Novisima Recopilacion.]

Jurisdiction of the Superintendent of the Mine of quicksilver of Collado de la Plata.

As well the Commissioner, as the Subdelegate in the commission of the Mine of Quicksilver of the Collado de la Plata in his case, or the Superintendent of the Mine according to various Articles of the Royal ordinances concerning Mines of Quicksilver, has criminal and civil jurisdiction, to take cognizance of causes and matters, both civil and criminal, of the employes and operatives belonging to the Mine, as exclusive Judge thereof, with prohibition to other tribunals than that of the general Superintendency; it being their peculiar care and duty to see that each one complies with his obligation, punishing severely their failure to do so; that the aforesaid Commissioner, and his substitute, or the person who shall hereafter be Superintendent of said Mine, must, in all respects, be subject to the general Superintendency, giving an account of whatever may occur, or be worthy of being brought to its notice; and there shall not be recognized any other jurisdiction than that of the general Superintendency, in administrative and contentious matters of the management and dependence of the Mine and its works, and in those matters which may incidentally arise; so that, obedience shall be given alone to its orders and despatches, and not to any others of other Tribunals, if they are not transmitted through the said Superintendency General; but, nevertheless, due observance must be given to the orders which shall be communicated by my Royal Person: that the jurisdiction of the Superintendent of the Mine, in virtue of the Royal *Cedula* issued in the year 1685, includes wagoners, vehicles, and oxen, bound to serve, and actually employed, in the Mine, having first, with respect to the obligation and license, complied with the formality prescribed by Royal orders: that, consequently, the Judge has exclusive jurisdiction to take cognizance of the excesses which the wagoners, bound as aforesaid, may commit, with respect to pasturage, and the cutting of timber for their implements of trade (*aperos*), and of everything relating thereto and to the Mine, both civil and criminal: that, it shall also be the duty of the Judge to preserve and defend the aforesaid wagoners from every grievance, injury, or violence, which may be done against their right of pasturage and of cutting timber for their wagons, or any other thing pertaining to their business: that the said Superintendent of the Mine is, and shall be in future, the exclusive Judge-Conservator of the Forests and commons which are, or shall be, assigned for the benefit and use of the said Mine and its works; and he shall have exclusive cognizance of all the causes and denouncements relating to the felling, cutting, and burning of trees (*talas, cortas é incendios*), and of other causes and things which may be prejudicial to the Mine: and no other Tribunal, except that of the Superintendency-General shall have cognizance, by way of appeal, or any other legal proceeding (*recurso legitimo*), of the decisions of the said Superintendent, for on this point all are prohibited: that the despatches which the Superintendent may issue, in the exercise of his jurisdiction, (which is territorial and extends to all within the territory designated), for the execution of sentences, the exaction of penalties, and the imprisonment of culprits, being directed to the Subdelegate, *Guarda-mayor*, or other person, and not to the courts, must be obeyed by these courts, and they must render the corresponding aid to the Subdelegate, or special Commissioner, in order to carry such despatches into effect, without interposing any obstacle; under the responsibility of paying all the damages and injuries, which may result from their neglect to chastise and punish evil-doers: that those who are employed with a fixed salary, or who constantly labor in the said Mine, are, and shall be, exempt, in their persons, and horses, from military service and from other drafts, and shall not be made to contribute thereto, nor shall they be subject to be drafted or taken away, in time of war, or be made to pay money in order to procure substitutes; nor shall they be coerced by the courts which assess excise and personal duties (*alcabalas y servicios*), or be made to pay certain Royal and church dues called *moneda forera* and *bulas;* nor shall they be obliged to accept and serve in, against their will, these and other like servile offices (*oficios de servidumbre*) being also exempt from having quartered on them soldiers, men-at-arms, or other military persons; and finally, that in the aforesaid Superintendent, or Commissioner, shall reside power to correct and restrain any of our vassals who may disturb, or, in any manner, impede the proper working of the Mine, as it is also their duty to watch over the observance of these privileges and exemptions; and if any one or more persons shall wish, or attempt, to interfere in this matter, or to cause to any supplier engaged in the service of said Mines and works, vexation, or shall collect from him, under the pretense of excise duty (*alcabala*) or any other taxes, any money (*algunas maravedises*), in addition to their being liable for all the damages and costs which may accrue, the Superintendent shall proceed against them, and, in case of their refusing to make restitution, shall enforce it, fixing the amount thereof: which declarations, as now regulated, and which have their origin in the Royal ordinance of 1735, issued for the government and direction of the Royal Mines of Almaden, and communicated to all the Tribunals of the kingdom for their punctual observance, it is my will, shall form the rule and guide in that of the Collado de la Plata, in order that there may be an end of those doubts and difficulties which only cause conflicts of jurisdictions: and in order to prevent frauds, in the matter of exemptions from municipal charges granted to salaried persons and the regular laborers, the present Sub-delegate will form a list of all of them, and transmit to the Alcaldes of the rural Pueblos a statement of the inhabitants of each who are enrolled for the works: and it is my Royal will, that the Corregidor, as well as the Superintendent Sub-delegate, confine themselves within their appropriate limits, and that, with the greatest harmony, they mutually assist each other in all matters which belong to their respective jurisdictions, for the good of the Royal service and the administration of justice, for otherwise, I shall adopt toward them such serious measures as each may deserve.

December 25th, 1790.

[From "Pandectas Hispano-Mexicanas," No. 221.]

Royal Order, that all Miners be encouraged to construct chapels in Reales of Mines distant from the Pueblos.

Most Excellent Señor:

In view of the doubt which occurred, to the officers of Royal Treasury of that capital, with respect to the exaction which they have made of Don Miguel Francisco de Arzeniega, Miner of the Rancho del Oro, for the license which has been granted to him to build a chapel, in which the operatives can hear Mass on the days of precept, and in view of the *consulta* of Your Excellency's letter of July 27th of this year, Number 18, asking his Majesty's decision, in order that there may be a rule to govern in like cases, the King has decided, that this favor (*gracia*) is not included in those of Article 27 of the *Arancel de Media-Annata*, as such structure is for an object so just, and one in which Arzeniega has no private interest; and, consequently, His Majesty declares him exempt from the payment of such duty; and it is His Royal will that Your Excellency encourage the laudable thought, that all the Miners of that Kingdom may imitate that good example, constructing chapels in *Reales* of Mines which are distant from the Pueblos, so that the operatives may not fail to comply with their most important obligation, and that they may have the spiritual food afforded by our holy religion.

I communicate it to Your Excellency, by Royal Order, for your information and observance.

God preserve Your Excellency many years.

Madrid December 25th, 1790. LERENA.

To the Viceroy of New Spain.

November 12th, 1791.

From Note to Article 16, Title XIII, Mexican "Ordenanzas de Mineria."

Royal Order with respect to excise duties on goods, &c., taken to Mines.

Señor Don Diego Gardoqui, under date of the 12th of November last, communicated to me, in the absence of the Most Excellent Señor Conde de Lerena, the Royal Order which follows:

"Most Excellent Señor:

"Under date of November 26th of last year, No. 179, "Your Excellency directed three *expedientes* of as many "other *testimonios* on the exemption from excise duty (*Al-* "*cabala*) of utensils, goods and fruits, which are introduced "into the *Reales* of Mines of Guanajuato and other places, "of which Your Excellency recommended favorable dis- "position, and that it be extended to all others of the "Kingdom. The King, having ordered the matter to be "passed to the Council for its opinion, in conformity with "its report, has been pleased to resolve, that, as he saw fit "to grant liberty or exemption from excise duty, to the ef- "fects which are mentioned in Articles 2 and 4 of the Cir- "cular Order of September 2d, 1785, the said Miners bring- "ing them into the *Reales*, he has been pleased to enlarge "and extend it to those carried and transported, by any "subject whatever, to the *Reales* of Guanajuato, Fresnillo, "Bolaños, and others of that Kingdom, charging Your "Excellency to take the most active and effective meas- "ures, which your judgment and prudence may dictate, to "prevent the frauds, which, under this pretext, evil intentions may invent, so that, without molestation or trouble, "the fruit of this Royal favor may be enjoyed. And, by "his Royal Order, I convey it to Your Excellency, for "your information and punctual observance."

And I enclose it to Your Honor (V. S.) for your information and satisfaction, in order that you may communicate it to the members of that Body.

God preserve you many years.

Mexico, March 29th, 1792.

EL CONDE DE REVILLA GIGEDO.

To the Royal Tribunal of Mining.

August 24th, 1792.

Don Carlos IV, by Resolution in Council of State, and Cedula of August 24th, 1792.

[From Book IX, Title XX, Law IV, Novisima Recopilacion.]

Free trade in stone-coal; and rules for the working of Mines thereof.

The Council having given its opinion, that the subject of coal Mines has already all the directions and light necessary to definitively distinguish them from all other

Mines; but, that the general welfare of the kingdom, and the sacred right of property, require that the subject be simplified, by abolishing the formalities and onerous regulations which might embarrass it, and by entirely confiding its advancement to the reciprocal interests of the proprietors, of those who work the Mines, and of commerce: I have been pleased to resolve and order as follows:

Art. 1. That notwithstanding the construction (*inteligencia*) which has been, or may be, given to the general laws and ordinances, that every species of Mines, even though not expressly named therein, pertain to the Crown, those of stone-coal shall be of free use, the same as, by ancient custom, are those of iron and of other substances which are extracted from the bosom of the earth.

Art. 2. But the Crown will retain the supreme *Regalia* of incorporating in itself such Mine or Mines as it may require, or as may be deemed expedient for the use of the Royal Marine, foundries, arsenals, or other objects of public service. Those Mines which may be found in ungranted lands (*terrenos valdios*), will be so incorporated, without any recompense; but those found in lands of Municipalities, communities, or private proprietors, shall be paid for at their just value.

Art. 3. The direct proprietary owners of lands in which there are coal Mines, whether they be Municipalities, communities, or individuals, shall be at liberty to discover and work them themselves, or to permit others to do so, by selling or leasing them, at their pleasure, without there being requisite any other license or formality than for working, leasing, or selling the land itself in which they are contained; all this being done by voluntary contracts and agreements, the conditions, time, and price being arranged between the parties themselves, or by public auctions where the lands are municipal, and where the laws prescribe this mode of disposition.

Art. 4. The coal extracted from said Mines may be freely dealt in, by wholesale or retail, within the kingdom, without being subject to Royal or municipal duties of any kind, even the most exceptional and privileged duties; and those which are exported in Spanish vessels from any one part to another of my dominions, and even to foreign dominions, shall also be exempt from the payment of general duties (*Rentas Generales*), but if carried in foreign vessels, they shall be charged and made to pay the duties of *Rentas Generales*, and others which are or may be imposed on the exportation of native products in foreign bottoms.

Art. 5. In order to encourage exports of this kind by sea, Spanish vessels which are induced to carry coal and which belong to places where there are registration ot sailors (*matricula*), will be permitted to have landsmen for one-third of their crews, provided, their owners cannot obtain enrolled sailors (*matriculada*) for the same wages: but the Officers of Marine in the Provinces must form a list of these landsmen (*marineros terrestres*), in order that, without their being required to serve in the Royal *Armada* in ordinary cases, they may be the first to be drafted into such service, on extraordinary occasions, when there shall be a deficiency of enrolled sailors, this being due for the favor which is granted to them in prejudice of the privilege which enrolled sailors enjoy of being the only persons to receive advantages from sea service.

Art. 6. Although, by Article 2 of the Royal Cedula of April 13th, 1790, issued for the encouragement of commerce and the mercantile marine, vessels of less than one hundred tons are excluded from the rewards specified in Article 1; nevertheless, as it is very expedient to encourage, by all means, the exportation of and trade in coal extracted from said Mines, and to cause a coal marine to be formed, particularly on the coast of the ocean; it is declared, that there shall be included in the premium of three hundred *reales* all vessels of Spanish build, and owned by a Spaniard or a person of Spanish domicil, of whatsoever capacity, not under fifty tons, which, within a year, shall make two voyages, with an entire and sole cargo of coal, from any port of the Province to any other out of the Province, but in the Peninsula, including Portugal, or one voyage to a foreign port, not in the Peninsula. Such reward shall be paid by the Administrator of the Custom-House of the ports of embarcation, proof being given of where they are discharged; and the same Administrators will report, at the end of the year, to the Direction General of Duties, the number of rewards, and the amounts paid for them.

Art. 7. In order that the interior and exterior trade in coal may have the increase, of which it is susceptible from the abundance and good quality of the Mines, particularly in the Asturias; it is indispensable to facilitate the transportation, by opening and repairing wagon roads and cross-roads, and improving the navigation of some rivers. With respect to the wagon roads, the Superintendent of that branch will see that those already begun are continued, and that others are commenced, according as the funds destined for that purpose shall permit; also stimulating the Pueblos, for their own advantage, to assist, by putting the cross-roads of their jurisdictions in traveling order. And with respect to the navigation of rivers, particularly of that called Nalon, in the Asturias, the Minister of Marine will have this subject examined into, and he will report on it in a separate *expediente.*

Art. 8. With the same separation, the proper Ministry

will provide that there be established in the Asturias a school of Mathematics, Physics, Chemistry, Mineralogy, and Navigation, in order that there may be diffused, in that Principality, the scientific knowledge which is absolutely necessary for the working and reduction of the Mines and for forming sailing-masters (*Pilotos*) who may direct navigation, for although now, the Mines being new and superficial, coal is taken out of them in abundance, the same will not follow when they are sunk to a great depth, and it will then be impossible to work them without the aid of skill.

Art. 9. By these declarations, the first, second and third of which shall have the force of law, are anulled the laws and ordinances which refer to Mines, and the *cedulas* decrees and orders which particularly treat of those of stone-coal, so far as either of them conflict with what is here established, they remaining, in all else, of full force and effect.

February 5th, 1793.

[From Note to Article 28, Title I, "Mexican Ordenanzes de Mineria."]

Explanations, additions, and modifications of the Mining Ordinances, with respect to jurisdictions, &c.

Art. 1. Most Excellent Señor:

On the 13th of January, 1791, your Excellency reported that, on your entering upon that command, *Juntas* were passing enactments for the regulation of the Tribunal of

Mining, in compliance with the provision of the Royal Order of June 7th, 1786, and that, although the subject was very far advanced, your Excellency, knowing that the multitude of points which were treated of would produce considerable delay and long disputes, directed that the members (*vocales*) collect the necessary information to form a report, each one giving his opinion in writing, and, this having been done in the form expressed in the *testimonios* which were transmitted, your Excellency, knowing the variety of opinions upon the multitude of points treated of, and not agreeing with any of them, transmitted it all for the Royal decision.

Art. 2. This diffuse *expediente* having been examined in the Supreme Council of State, over which the King presided, His Majesty has been pleased to resolve as follows:

Art. 3. That there be made no change in the number of officers of the fund of the Royal Tribunal of Mining,—an Administrator-General, a Director-General, and three Deputies-General,—as that is in conformity with Article 2 of Title I of the Ordinances of Mining, and as it has been proved, by e .perience, that three Deputies are sufficient.

Art. 4. That the office of *Fiscal* continue separate from that of Director, as is commanded by the Royal Order of June 10th, 1791, Don Juan Eugenio Santelices Pablo filling said office of *Fiscal*, with the salary of three thousand dollars which are allowed him, under the obligations and circumstances which are set forth in the *Expediente* which your Excellency transmitted, in your letter of February 7th, of the same year, No. 15.

Art. 5. His Majesty consents that there be in the Tribunal the offices of *Assessor*, Secretary, Factor, two clerks (*oficiales*) of the Secretary, and two Porters, who shall be obliged to serve as bailiffs (*Ministros Executores*), and the

Assessor shall be obliged to assist daily in the Tribunal, for the advantages which are proposed.

Art. 6. The salaries which the said employes are to enjoy are:

The Administrator	$5,000
The Director	4,000
The Deputies, each	4,000
The Factor	2,500
The Secretary	2,000
The first *official* of the Secretary	1,000
The second " " "	600
The first Porter	400
The second "	300

And with respect to the *Assessor*, he will receive a salary of $2,500, being absolutely prohibited from charging any fees, as this mode will facilitate the quicker despatch of business, and there will be a less number *expedientes*, the parties agreeing amicably and without making any figure in court.

Art. 7. The offices of the said Tribunal must be given to persons who are practical, intelligent, and expert Miners, by personal knowledge acquired in that occupation for more than ten years, in exact observance of Article 3, Title I, of the Royal Ordinances; for, as the Body of Miners support these offices, it is just that they should enjoy them; moreover, no one can perform their duties with more intelligence and interest than they themselves.

Art. 8. Strict observance must be given to Article 7 of Title I, with respect to the votes which each Mining place (*Mineral*) ought to have, always inducing those appointed to attend in person, and in default thereof, to give their proxies to other Miners actually engaged in Mining, as is directed in the last words of Article 5, Title I.

Art. 9. The Royal Tribunal of Mining must remain erected into a General Tribunal of Appeals, with the same contentious jurisdiction in second instance, and the same economical, administrative and directive powers which the ordinances give to it, with the right of appeal to the Judge of Appeals in all cases allowed by law; observing, that, as the said Tribunal of Mining and Judge of Appeals have succeeded in place of the *Audiencias*, and as the latter could take cognizance of all causes of the district, on appeal from the judgments of the Judges of Mines and Alcaldes Mayores, so now the Tribunal and Judges of Appeals ought to exercise it in their respective cases; and all those of the territory included by the *Audiencia* of Mexico, must go to Mexico, and those of New Galicia and Viscaya to that of Guadalajara, the Court of Appeals being in fact maintained there, agreeably to the ordinances, and continuing to take cognizance in second and third instances, because there is no Tribunal of Mining there, and the distance of those Provinces is very great, repealing in this respect Article 2 of Title III of the said ordinances, and declaring, that, in proceedings in first instance, the territorial Judge, Judge of Mines and Intendants, where there are such, shall take cognizance with the two territorial Deputies, and exercise, in all cases, contentious jurisdiction, His Majesty in this respect enlarging Article 4, Title III, of the said ordinances.

Art. 10. The King approves the annual expenditures of the Tribunal of Mining with respect to the salaries which were stated and are, $1,000 to the *Oidor Juez de Alzadas*, $100 to the *Procurador*, and $594, as estimated for postage on letters and for other small items, $1,000 also being disbursed for the Associate Judges (*Con-Jueces*) of the Tribunal of Appeals, as it is not just that they pay, out of their own pocket, the fees of lawyers whom they consult on decrees, in order to vote properly in their decisions.

Art. 11. The establishment of a Mining College in that city being a matter of the greatest importance, His Majesty approves the appropriation of $25,000 for its support, and everything else which your Excellency has directed, so that the students may immediately begin their studies in in it; and, in order that this school may be well supplied with professors, books, instruments and other requisites, this Ministry will take the management of the whole, and your Excellency, besides what you set forth in your letter of April 26th, 1790, No. 496, will recommend the works which may be useful for the better instruction, and report, every year, the progress which the students may make.

Art. 12. Because of the burthens already imposed on the Royal Tribunal, it is the will of His Majesty that no change be now made in the consignment ot five thousand dollars in favor of the Academy of San Carlos; for, although a knowledge of civil architecture may not be necessary for Miners, it may be useful for them to know its rudiments, and drawing (*dibujo*) is the foundation of all the arts.

Art. 13. The King has not decided with respect to the suspension of the four thousand dollars appropriated to Señor Don José de Galvez and his posterity, as there is no party interested who claims it.

Art. 14. His Majesty approves the reduction, which your Excellency proposes, in the reward of three thousand and three hundred dollars given to the employes in the Mint, to two thousand six hundred and fifty, and your assigning four hundred to the Superintendent, two hundred to the *Contador*, two hundred and fifty to the Treasurer, and leaving to the *oficiales* their former salaries.

Art. 15. The surplus of the revenues of the Tribunal shall remain at the order and disposal of its members, agreeably to the ordinances contained in Titles XV and

XVI, and to Article 20 thereof, and with the condition, that no Mine can be supplied, nor any funds be withdrawn, without the consent and concurrence of all the *Ministros* and *Consultores*, the direct intervention of the Director, and, especially of its *Fiscal-Defensor*, who, in default of these requisites shall make the corresponding appeals to Your Excellency, and to this Superiority, giving preference to works and Mines which may be most worthy of attention; for, as the fund of the eight grains belongs to the Miners, and to their Tribunal, which represents all those who contribute to it, justice does not permit that they be deprived of their property or of its use; and notwithstanding that some of its Members should have failed to comply with their duties, this measure (*providencia*) shall enure to the benefit of the other employes, who are full of probity, skill, purity, and notions of expedient operations, as is expressed in Article 2, Title XVI, aforesaid, and His Majesty directs that that government shall protect them in everything and shall not perplex them by putting impediments in their way, or by obliging them to form *Expedientes*, and to present the same to Your Excellency for your approval, which can never be granted to them without having the matter examined by the *Fiscal* and referred to the *Assessor*, and going through other formalities which greatly retard the rapid course of business, from which cause originate very grave inconveniences; in view of which, the ordinances impose upon the Tribunal no other obligation than that of reporting to Your Excellency, of obtaining your permission, and of giving notice of its elections and changes, and of reporting to His Majesty, through Your Excellency, at least every year, and whenever an extraordinary case shall occur which requires it to be done,—all in conformity with the provisions of Article 19, Title I, and Article 37, Title III, of the said ordinances, which must be fully observed.

The King does not approve that the Tribunal of Mining form the partnership, which Your Excellency proposes, of one million of dollars with that *Consulado*, each furnishing five hundred thousand dollars, for various reasons of fitness, which render it impossible.

Moreover the Treasurers, who are mentioned in Article 6, Title XVI, for the security and custody of the funds of the Tribunal, must, in all the deposits and disbursements, intervene with the depositors, the *Promotor Fiscal* and the Secretary, and not pay any draft without the signature of the officers (*Ministros*) of the Tribunal, note being taken thereof.

His Majesty approves what Your Excellency proposes, with respect to the demands of the *Fiscal* on the Tribunal; and on all those who have received the rewards and amounts which you have referred to; and also approves Your Excellency's having ordered to be passed to the Tribunal of Mining, only the *expedientes* with respect to the three thousand and nine dollars which have been given to Don Francisco Salesan, that of the fifteen hundred which have been given to the person who asked the payment of fifty thousand dollars from the Royal Treasury (*Cajas*), and that which settles the three thousand and ninety delivered to the *Regidor*, Don Antonio Rodriguez de Velasco, giving an account of the results, and His Majesty declares that there has been no excess in the rewards given to the Viceroy, Don Martin de Mayorga, and to the Director, Don Joaquin de Velasquez, and relieves from all responsibility the *Ministros* of the Tribunal who intervened therein.

It is conformable to the will of His Majesty that, although some of the members (*vocales*) have extended their report to the formation of other Ordinances, Your Excellency has not acceded to it, for power is given to the *Junta* by the Royal Order of June 7th, 1786, only to enlarge or modify those which refer to the regulation, government and administration of the Tribunal, and to elections and salaries, and not all the ordinances indiscriminately.

The King wills, that all the points decided in this declaration be observed with the greatest punctuality, and that Your Excellency cause them to be published, to serve as an addition to the last Royal Ordinances, and that they be communicated to the Royal Tribunal of Mining and to all its *Reales;* I, in his Royal name, giving thanks to Your Excellency for your great merit in moving and concluding this important *Expediente*, with your laudable, active, and appreciable zeal, and with your commendable talent; giving thanks also to the members of the *Junta*, and to the Tribunal of Mining, manifesting to the latter that it has merited and still merits the Royal confidence and protection of His Majesty, with direction that it make this known to all the Miners, for their encouragement and satisfaction, and that it proceed immediately to hold its elections of Administrator, and of other individuals who are required to complete the Tribunal.

All which I impart to your Excellency, for your information and punctual observance.

God preserve your Excellency many years.

Aranjuez, February 5th, 1793.

GARDOQUI.

To the Viceroy of New Spain.

A True copy.—Mexico, June 17th, 1793.

BONILLA.

August 5th, 1793.

Don Carlos IV, by Resolution and Cedula of the Council, August 5th, 1793.

[From Book IX, Title XX, Law V, Novisima Recopilacion.]

Declaration, explanatory of the law of August 24*th*, 1792, *with respect to the working of Mines of stone-coal.*

I have been pleased to resolve, as declaratory of Article second of the preceding Royal Cedula [of August 24th, 1792, Law IV, Title XX, Book IX, Novisima Recopilacion], as follows:

That, although the crown will retain the supreme *Regalia*, which belongs to it, of incorporating in itself any of the said Mines, it will not do this except in case of necessity, paying to the owner thereof their just value, or accepting the cession which he may voluntarily make: That, as declaratory of the third Article, it is to be understood, that the usufruct and enjoyment of Mines of stone-coal ought to belong to the council, parish, place, community, or person, to whom belong the usufruct and enjoy- of all else which the land, in which such Mines are found, produces, without any difference; and that the councils, parishes, or places, cannot sell or alienate their Mines, without authority of the Royal Council, which will grant it, if there be just and reasonable motives; but in case the citizens thereof do not themselves wish to work the Mines, they may lease them, at auction, for a fixed period not to exceed nine years, without any one having a right of perference, or of taking them at the price bid (*tanteo*); the products being expended in things necessary and useful to the public, such as the construction of bridges, and the opening and repairing of roads.

September 11th, 1794.

[From "Pandectas Hispano-Mexicanas," No. 3751.]

Royal Order, respecting exceptions taken to the jurisdiction of courts and Judges in matters relating to Mines.

Most Excellent Señor:

His Majesty approves the order (*Providencia*) given by your Excellency that ordinary and sub-delegate Judges of the *Reales* of Mines may be excepted to in matters relating to Mines, in the same manner as they would be in any other matter; but His Majesty has resolved that the corresponding power of appointing an associate must be understood in the cases of simple exception (*recusation simple*) with the prescribed oath (*juramento de estilo*), and without expression of cause; but not in those where it is made with the clause of inhibition (*con clausula de inhibicion*) offering to prove the cause of enmity, partiality or interest, under the competent bonds, as in that case they must be wholly prohibited from the cognizance, as is practiced by the other Judges, also advising your Excellency that, in case of disagreement between the sub-delegate Judge and his associate, the appointment of a third be always so made that it may devolve upon the person most honored, informed, and suitable, of the respective Pueblo, because this matter interests the administration of justice and the welfare of His beloved vassals, observing the same for the appointment of the associate who takes cognizance of the matter in place of the sub-delegate Judge when the latter is to be entirely prohibited from the cognizance. And, by His Royal order, I communicate it to your Excellency, in answer to your letter of the 28th of February last, number 831.

God preserve your Excellency many years.

San Ildefonso, September 11th, 1794.

GORDOQUI.

November 17th, 1796.

[From Note to Article 8, Title II, Mexican "Ordenanzas de Mineria."]

Declaration of the Superior Government respecting what belongs to the Tribunal, and what to the Government, in erecting Reales of Mines into Deputations.

The reasons given by the Tribunal, for the revocation of the Superior Decree of June 28th, 1791, are better calculated to prove the justice and necessity of requiring that the new territorial Deputations, of which it speaks, shall be erected with the approval of this Superior Government.

The proper *Fiscal* of that body knew this, and so admitted in the answer found in the package (*cuaderno*) which your Honor (V. S.) transmitted with your representation of the 4th of December of the same year, proceeding, not by induction from mere fitness, as that Tribunal says, but by legal inference deduced from the provision in Article 14, Title II, of its ordinances, and from the just obligation which all bodies are under to recognize this Superiority in matters of such high authority.

The inference is, in fact, very natural that, if, for the elections of Deputies, the said Article imposes the absolute necessity of applying for the Superior confirmation, with much greater reason ought this to be so, when the question is to create a new court which is to have the exercise of jurisdiction.

This due acknowledgment interferes, in no respect, with the powers which, by Article 1, Title III, are conferred on that Tribunal, for those always continue in force which belong to it, of examining whether or not the Deputation ought to be elected, and afterward consulting this Superiority, the same as is done in the elections which, after being examined by that Tribunal to see if they have any defect, it transmits for the Superior confirmation.

There is, therefore, not the least reason of difference why the same should not be done with respect to the establishment of new Deputations, and if, notwithstanding the jurisdiction which by the said Article is conferred on that Tribunal for elections, it must indispensably apply for the Superior confirmation of the Government, it necessarily follows, that the antecedent proceedings, i. e., the creation of Deputations, ought to be made in the same manner and with the same requisite of the Superior approval. Nor was it necessary that it should be specially prescribed in the Royal ordinances of that Body, as a thing isdispensable;

for indeed this and other like dispositions may be considered as included in Article 16, Title II, which provides that an account be given every year to this Superior Government of the condition of Mining places (*Minerales*) and Miners, the number of Mines which may be worked, and those which may be discovered in each *Real*, since these particulars have an intimate connexion with the creation of new Deputations, for which purpose an examination must be previously made of the condition of the *Reales* of Mines, to ascertain whether they have the requisite number of Miners, so that these may hold alternately the offices of Deputies and Substitutes, without the necessity of a re-election.

After all, the creation of Courts of Justice is, as that Tribunal says, a matter of great importance, inasmuch as the Deputations, which should exist in the *Reales* or *Haciendas* of Mines which have the prescribed qualifications, acquire from their creation the exercise of jurisdiction, and in these dominions the proper cognizance of these matters corresponds in a great degree to the superiorities of the Viceroys and to their high vice-regal powers.

The other argument which that Tribunal uses, that the election of the temporary officers (*Ministros*) which compose them, properly belongs to the courts or political bodies, is manifestly contrary to its intention, because, if the election of Deputies, although of less importance than the creation of the same territorial Deputations, requires, according to the ordinance, the superior confirmation of the Government, how much more indispensable must that be for establishing new Bodies or Courts which are to administer justice?

It is true, as appears, that the said confirmation was not asked for, in case of the Deputations mentioned in the certificate which that Tribunal transmitted; but that has resulted from not having considered the want of a requisite so essential, it being also in truth very strange that that Tribunal, as is expressed in the said certificate, has not, until now, had any other notice of the Deputations which have been created, than the first elections which have been remitted to it.

This gave a motive for the Director-General to justly protest against the election of the Deputies which the Miners of Hostotipaquillo sent, because they had assembled in Deputations without the proper permission ; and, if the fact that the other Deputations did not apply for its approval, has not served as an embarrassment to this Tribunal in the vindication of its powers, neither can the examples which are alleged prejudice those of the Superior Government.

Much less the opinion which is attributed to the *Fiscal* of civil matters, who was Don Lorenzo Hernandez de Alva, for having said, in his answer of June 30th, 1790, that it appeared to him a proper matter for that Tribunal to determine, whether there should be deputations in Hostotipaquillo, for he expressed this in opposition to what was determined in the matter by the Intendant of Guadalajara; but not in relation to this Superior Government, whose high powers can, at no time, be prejudiced by any expression which the said *Fiscal* may have uttered, through error, ignorance, or other want of information.

That Tribunal cannot fail to so recognize it, and if it be considered odious to treat of restricting the jurisdiction which is granted to it, it is much more so to think of limiting, in these matters, the Vice-Regal power which resides in this Superior Government.

In attention to all, and to what that Royal Tribunal, giving proof of veneration and respect, determines upon in its last representations, saying, that with the provisional determination which it had proposed in its preceding representations, it did not pretend that the power of creating the Deputations be left to it, but that it would leave, with pleasure, the matter to this Superiority, if it should be so determined, with the reservation of the rights which belonged to it: I have declared, finally, at the request of said *Fiscal*, and in accordance with the report of the *Asesor General*, commissioner of the Viceroyalty, that there is no reason for the revocation, asked for, of the superior Decree of June 28th, 1791, and that what was therein determined must be carried into effect: which I communicate to your Honor (V. S.) for your government in the matter, with the information that you may immediately proceed to form the *expedientes* as may appear proper, as well for creating new Territorial Deputations in the *Reales* of Mines, where there concur the essential requisites for taking this proceeding (*providencia*), as for abolishing the Deputations now established in places (*Lugares*) where experience has shown they are not necessary; your Honor (V. S.) giving me an account of those in that condition, reporting all that occurs or may offer, and waiting my superior resolution, in order that they may proceed to the first election of Deputies and Substitutes, and the consequent exercise of jurisdiction by the new Court, with the understanding that this is the best and only mode of reconciling the powers which may belong to that Tribunal, with those of this Superior Government.

God preserve your Honor (V. S.) many years.

Mexico, November 17th, 1796.

BRANCIFORTE.

December 6th, 1796.

[From Note to Article 16, Title XIII, Mexican "Ordinanzas de Mineria."]

Order respecting the exemption, from excise duty, of goods introduced into Reales of Mines.

The most excellent Señor, Don Pedro Varela, communicated to me, by order of the King, under date of December 6th, 1796, what follows:

"Most Excellent Señor:

"The predecessor of your Excellency, Conde de Revilla "Gigedo, in letter of July 29th, 1793, No. 577, reported "with proofs (*testimonio*) the second *Expediente* drawn up in "*Junta* of Royal Treasury, in respect to the doubt which "occurred to the Director of duties on foreign goods (*Alca-*"*balas foraneas*), respecting the exact fulfillment of the Royal "Order of November 12th, 1791, in which the King was "pleased to grant exemption, from this Royal duty, of "goods and effects which are consumed in the *Reales* of "Mines of that Kingdom, which grant the said Director "thought should be understood with the condition that "said effects are to be used by the purchasers in the work-"ing of Mines and in the reduction of ores, proposing, at "the same time, in order to avoid any fraud which, on "account of this exemption, may be committed, that, on "the introduction of these effects into the Mining places, "(*Minerales*), the introducers shall swear that the sale was "made to Miners and that these shall swear that they "purchased them especially for the uses of that profession. "The King, informed of all, and conforming to the opin-

"ion of the Council of the Indies, given in *consulta* of the "10th of November last past, has been pleased to approve "what was determined, on this subject, by the said *Junta* "of Royal Treasury in sessions of the 17th of June, 1792 "and the 18th of June, 1793, commanding that due effect "be given to the resolution contained in the said Royal "Order, with the precautions proposed by the same *Junta*, "with respect that, those oaths, which the Director Gene-"ral of *Alcabala* adopted, would produce no other result "than that of being taken and repeated with falsehood, "especially by those whose conduct is not well regulated, "or who are ill instructed in the obligations which Religion "gives to an oath, and, at the same time, placing others, who "are timid and of good conduct, in a painful agitation of "conscience as to whether they might or might not make "use of that means, unjust in its origin; which incon-"veniences it is proper to prevent.

"Which I communicate to your Excellency, by Royal "order, for the observance and government of the *Junta* "of Royal Treasury."

And I insert it to your Honor (V. S.) for your information and government, and in order that you transmit it, for the same objects, to the territorial Deputations of Mining.

God preserve your Honor (V. S.) many years.

Orizaba, January 17th, 1798.

BRANCIFORTE.

To the Royal Tribunal of Mining.

February 12th, 1797.

[From Note to Article 2, Title III, Mexican "Ordinanzas de Mineria."]

Orders, explanatory of the ordinances of Mining with respeet to judicial proceedings, jurisdictions, &c.

By Royal order of February 5th, 1793, communicated to the Viceroy of New Spain for the regulation and government of the Royal Tribunal of Mining, the King was pleased to command, among other things, in consequence of what was agreed upon in the Council of State over which he presided, that the said Royal Tribunal continue erected into a General Tribunal of Appeals, with the same contentious jurisdiction for proceedings in Second Instance, and to the same extent which the ordinances give it in matters of administrative and directive economy, with appeal to the Judge of Appeals (*de Alzadas*), in all cases which correspond according to law; with the remark, that, as the said Tribunal of Mining and Judge of *Alzadas* have succeded in place of the *Audiencias*, as the latter took cognizance, on appeal, of all causes of the district, from the decisions of Judges of Mines, and the Alcaldes Mayores, so now should the Tribunal and Judge of *Alzadas* take it in their respective cases, and all those included within the *Audiencia* of Mexico, should go to Mexico, and those of New Galicia and Viscaya, to that of Guadalajara, the court of *Alzadas* being maintained there for that purpose, agreeably to the ordinances, and continuing to take cognizance in Second and Third Instances, because there is no Tribunal of Mining there, and because the distance of those provinces is very great, repealing, in this respect, Article 2 of Title III of the said ordinances, and it being declared, that for First Instances, the territorial Judge, Judge of Mines, and the Intendants where there may be any, shall take cognizance concurrently with the territorial Deputies, and exercise contentious jurisdiction in all cases, His Majesty extending, in this respect, Article 4 of Title III of the aforesaid ordinances.

In consideration of the aforesaid, and also of the contents of another *Expediente* moved in the said Kingdom of New Spain, His Majesty has now been pleased to resolve, that, in accordance with what is directed in said Article of said Royal Order, the territorial Deputies of Mining shall proceed, in conjunction with the respective Intendants, to the examination (*formacion*) of all civil and criminal causes, in which class are considered deaths caused in Mines by the falling in of their workings (*labores*), so far as to establish that they did not result from culpable negligence, and that the preliminary investigations (*sumarios*) be drawn up by direction of the *Asesor Letrado* (there being no Royal Judge), with whose assistance is to be made out the declaratory act, as to whether the cause is proper to be continued before the Court of Mining, or to be removed to the Royal Judges, according to the provisions of Article 29 of Title III of the Royal Ordinances given for the government of the important Body of Mining of New Spain. And also, it is his Sovereign will, that, in observance of Article 10 of Title IX of said ordinances, visits be made to all the Mines, as is commanded, issuing, for that purpose, the proper orders to the territorial Deputations, which will transmit, to the Tribunal General, evidence of having done so each year, in order that it may include this circumstance in the annual report which it is ordered, in Article 19 of Title I, to make, and to direct to His Majesty.

I communicate it to your Honor (V. S.), by his Royal order, for your punctual observance in what corresponds to you.

God preserve your Honor (V. S.) many years.

Aranjuez, February 12th, 1797.

VARELA.

June 10th, 1797.

[From Note Article 14, Title I, Mexican "Ordenanzas de Mineria."]

Respecting certificates of pension for Miners and dependants of the Royal Tribunal.

The King has declared that the Royal Tribunal of the important Body of Mining, instead of removing from office, without just causes, the employes who hold the offices of Secretary, Auditor and Treasurer thereof, with fixed salaries, it ought to solicit their promotion, as vacancies occur, and according to their merits, so that they may be induced to faithfully perform their respective duties; and, moreover, His Majesty has resolved, that the incorporation of all the aforesaid persons who enjoy a fixed salary into the *Montepio* offices be taken into consideration; and that there be immediately drawn up and delivered by Your Honors (V. SS.) annually to that Viceroy for transmission to this Secretary of State and of the Dispatch of Treasury of the Indies, the corresponding certificates (*Hojas*) of their service, in the same manner as is provided generally, in the Royal Order of December 18th, 1792, for the other dependants of this Ministry, but entirely separate, including also all the members of the General and special Tribunals, and the owners of Mines, with a statement, in their certificates of service, of the Mines which they hold and work, and what part of their produce is reduced by smelting or amalgamation, the number of dependants and operatives which they maintain, and whatever else may appear worthy of the notice of the King.

And, by his Royal Order, I communicate it to Your Honors (V. SS.), in order that you may take care that it be strictly observed, in the part which belongs to you.

God preserve Your Honors (V. SS.) many years.

Aranjuez, June 10th, 1797.

EL PRINCIPE DE LA PAZ.

February 6th, 1798.

[From Note to Article 18, Title XVIII, Mexican "Ordenanzas de Mineria."]

Respecting the grant of a Patent for Garcés' invention for smelting ores by the use of Tequesquite, and the power of the Viceroy to grant such Patents.

With this date I communicate to the Viceroy of that kingdom the Royal order which follows:

"Most Excellent Señor:

"The King having read your Excellency's letter of "January 27th of last year, No. 923, and the substance of "the *testimonio* accompanying it, which treats of the useful "and advantageous invention discovered by Don Jose "Garcés, for the reduction of ores by smelting with the "use of *Tequesquite*, and of the doubts arising, as to "whether the power of declaring, in the particular cases, "the reward of the exclusive privilege which Articles "18 and 19, Title XVIII, of the Ordinance concede to "authors of inventions useful to Mining, belongs to "your Excellency or to the Tribunal of that important "body, his Majesty has been pleased to approve the new "despatch which your Excellency ordered to be sent to "Garcés, for the enjoyment of the privilege corresponding "to the said invention during his life, just as the said Tri-"bunal had previously granted it to him; and he has "resolved, in order to obviate any new doubts in future, "that the said power of declaring, in each particular case, "the use of the exclusive privilege conceded by the said "Articles 18 and 19, belongs exclusively to your Excellency "and your successors, in virtue of the specific authority "which is granted to them by Article 36, Title III, Articles "15 and 17, Title VI, Article 17, Title X, and Article 1, "Title XI of the said ordinance, provided, that the invention "be previously examined, approved, and tried by the Tri-"bunal of Mining, hearing its Director, agreeably to the "spirit of Article 17, Title XVIII, of the ordinance, the "Masters of the College, and other intelligent persons, "who in each class of inventions may assist, so that the "reward be granted only to those who legitimately deserve "it, and that when these previous proceedings have been "had in the said Tribunal, that body will transmit them "to your Excellency, with its report, in order that, the "*expediente* being instructively substantiated, you may pro-"ceed to declare the exclusive privilege, if it be found to "be deserved, in which case you will immediately put the "author of the invention in possession of its use, issuing "to him the corresponding document, with the requisite "condition of the Royal approval, which your Excellency "and your successors will solicit, without loss of time, in "order to avoid delays and expenses to the parties.

"Finally, His Majesty charges Your Excellency and the "Tribunal of Mining, to respectively protect and aid the said "Garcés, removing any unjust opposition to the use and "application of his valuable invention, taking the proper "measures (*providencias*) to encourage the cultivation of "*Tequesquiterias*, in order that there may be no scarcity of "this very necessary flux, upon the abundance and mod-"erate price of which the advantages of the method of "Garcés depend: that with respect to his offer in the "writing presented by this Tribunal, asking the approval "of his invention, to publish as soon as he shall have ob-"tained such approval, all the rules which are required to "be observed in order to secure a favorable result of the "smelting; Your Excellency will direct, if you have not "already done so, Garcés to comply with his offer, by "drawing up a treatise in which he shall not only explain "circumstantially the mode and rules of reducing his "method to practice, but also the nature of the *Teques-"quite*, of what it principally consists, and the method of "preparing this mineral salt for use in the smelting; be-"cause it is there announced that that which nature itself "produces is not suitable; and that when this work is "concluded Your Excellency cause it to be examined by "the Tribunal of Mining, and, no objection being found "to it, you will order it to be printed, so that, being furnished "to all the territorial Deputations, and through them to the "Miners, both may be made acquainted with the utility "and advantages of the new method, and may profit by "them, as they may deem expedient; Your Excellency "transmitting, with a report, to this Ministry, some copies "of said work and specimens of the raw Tequesquite, and "also of that which is prepared, so that it may ascertain if "the effects of said smelting correspond to what is there "experienced, and whether it may be suitable to commu-"nicate the discovery to the other dominions of the Indies, "where Mines are, or may be worked, in order that its use "may be extended. I communicate to Your Excellency, "by Royal order, all the foregoing, for your information "and punctual observance."

I transmit to Your Honor (V. S.), by the same Royal order, this Royal resolution, in order that being made acquainted with it, you may provide for the observance of the part which relates to you.

God preserve Your Honor (V. S.) many years.

Aranjuez, February 6th, 1798.

SAAVEDRA.

April 27th, 1801.

[From Note to Article 16, Title XIII, Mexican "Ordenanzas de Mineria."]

Respecting the delivery of powder to Miners for mining purposes.

On this date I communicate to the Viceroy the following:

"Most Excellent Señor:

"The *Expediente* moved by that Tribunal of Mining, on "the 28th of January, 1785, asking that there may be "delivered to the Miners the powder which they consume "in working the Mines, under the usual precautions for "the payment of the duty on powder, or others more "rigid, subjecting the deliveries to the certificates of the "respective territorial Deputations, having been presented "to the king, and he having examined what was stated by "the Viceroy, Marques de Branciforte, in letter of Feb- "ruary 28th, 1798, No. 1207, and his successor, in letter of "November 26th of that year, No. 163, recommending the "Body of Mining for the gift of $5,000 which it had just "made for the necessities of the Crown, notwithstanding "its own want of funds: His Majesty has been pleased "to order, that there be issued to the Mining of that "kingdom the powder which it may consume in work- "ing the Mines, at four *reales* per pound, in considera- "tion of its good services and as an inducement for it "to continue them in future, making itself worthy of "such singular favors as this in the present circumstances "of the Royal Treasury. I communicate it to your Ex- "cellency, by Royal order, to the end that you may issue "the corresponding instructions for its due observance."

This Royal order is transmitted to your Honors (V. SS.) for your information, and for the satisfaction of all the Mining of that Viceroyalty.

God preserve your Honors (V. SS.) many years.

Aranjuez, April 27th, 1801.

SOLER.

January 19th, 1805.

[From Note to Article 1, Title XVII, Mexican "Ordenanzas de Mineria."]

Tariff of fees which the professional Experts of Mines and the Reducing Experts are to receive, in their respective operations.

Art. 1. For the examination which the Expert (*Perito*) shall have to make of the vein in the restored workings of old Mines, or where, in those newly opened, inspection is made of the direction, inclination and other circumstances mentioned in Articles 4 and 8 of Title VI of the Ordinance, and for the execution of the exterior measurement and the establishment of boundary stakes which is made at the time of giving possession to the denouncer, he will charge $20, when he is obliged to go the distance of one league to execute the aforesaid; but if the distance be greater, for each additional league, he shall charge $1 for going and another for returning.

Art. 2. For inspections of exterior openings, when there arises any dispute with respect to the boundaries and stakes of the square, if he shall have to make any measurement, he will charge $8; but if it be required to make a complete survey, he will charge $12, and for making a map, $8 more; and with respect to the number of leagues he may have to go, he will observe what is prescribed in the preceding Article.

Art. 3. For inspections of exterior openings, being a simple examination of a Mine, in which there are no measurements to be made, to a hundred varas in vertical depth, he will charge $15, and for each additional hundred varas in depth examined, he will charge $10, including in this any examinations he may have to make of the passages along the thread of the vein, within the limits of one *pertenencia;* but if it be necessary to pass to other *pertenencias* for the purpose of examining them, he will charge $6 for each.

Art. 4. If he shall have to make measurements on the exterior, in addition to the fees allowed in the preceding Article, according to the qualities of the Mines the Expert (*Perito*) will receive one *real* for every vara in length (*cordelada*) measured, it being understood that the measurements shall be made on the shortest line.

Art. 5. If he shall have to make a map of the said measurements, with a horizontal and vertical plan, he shall be paid for it separately, at the rate of one *real* for each vara measured in the Mine, as aforesaid.

Art. 6. If, at the time when the Expert (*Perito*) proceeds to make the measurement, by any dispute between the parties, or by any point arising which requires to be decided, its execution shall be prevented, inasmuch as it was not in his power to make it, and on the other hand his time has been occupied in it, and it would not be right that he be left without remuneration when it so happens, he will charge $5, having to go for the measurement the distance of one league from the *Real;* and if it be for a greater distance he will receive what is assigned for each league going and returning, in addition to the $5.

Art. 7. In meetings for determining or marking out any work, which, agreeably to Article 3, Title IX, must be done with the intervention of an Expert (*Perito*), his pay will be graduated according to what is prescribed with respect to the measurements, as well exterior as interior, of the Mines; and for the inspection which the work must receive for its examination, $10, besides the number of leagues which he shall have traveled; and for the measurement which he may make, the same being necessary for correcting an error which may have been committed in its progress, his pay shall be regulated by what has already been prescribed.

Art. 8. When an appraisement of any Mine is to be made, and it is desired, not only to appraise the pillars and supports which it may have, but also the machines, sheds and other utensils, for every morning or evening which is necessarily occupied in making the appraisement of them, he will charge $3, besides what is prescribed for the number of leagues he may have to travel.

Art. 9. As, for making an estimate of the pillars, supports and other interior parts of the Mines there may be required greater detail, in which the skill and practice of the Experts (*Peritos*) may be specially employed, although in this they may not make an exact and perfect estimate, on account of the variableness and fallibility of the veins and the uncertainty of the *ley* of the ores, but only an approximate calculation; for what they shall do in the interior of a Mine within the boundaries of one *pertenencia*, there will be charged $50, including the examination which they may make of the whole Mine; and if it be necessary to continue the estimate in a part of another *pertenencia*, they will receive the corresponding fee, according to the estimate made, in addition to the fees established for the depth of the Mine, with the understanding that if they shall not be able to finish in one day this operation or any other of those aforementioned in this Tariff of fees (*Arancel*), there shall not be any addition to what is determined aforesaid, and the fees shall be only charged once.

REDUCING EXPERTS.

Art. 10. To the commissioned reducing experts (*Peritos beneficiadores*) in the examinations and other operations which they may have to perform in the *Haciendas*, there shall be paid the sum of $5, for every day they are accupied.

Art. 11. In Places (*Lugares*) where there are no professional Mining Experts (*Peritos facultativos de Minas*), nor reducing Experts (*beneficiadores*), and it shall be necessary to make use of any one of the *prácticos*, most intelligent and favorably known, who may be there, for the execution of the operations pertaining to those offices, as is directed in Article 2 of Title IX, they shall charge only one half of the fees which are assigned to those who are commissioned (*titulados*).

Approved by Decree of the Viceroy January 19th, 1805.

August 22d, 1809.

[From Note to Article 8, Title XIX, Mexican "Ordenanzas de Mineria."]

Miners exempted from militia duty.

On this date I instruct the Commandant of Brigade, that the individuals who prove that they are employed in the working of Mines and in other business connected therewith, are hereafter to be exempt from enrolment in the militia; which I communicate to you for your information, and in answer to your letter of the 18th inst.

God preserve your Honor (V. S.) many years.
Mexico, August 22d, 1809.

El Arzobispo.

January 26th, 1811.

[From Note to Article 22, Title VI, Mexican "Ordenanzas de Mineria."]

Sovereign Resolutions of the general and extraordinary Cortes, and of the Supreme Council of Regency, conceding the full dominion (pleno dominio) *and acquisition of Mines of quicksilver, free trade in their products* (frutos), *and exemption from all kind of duties; and offering rewards to their discoverers and to those who invest their capital in them; communicated to the Royal Tribunal of the important Body of Mining of New Spain.*

Under this date I notify the Viceroy of that Kingdom, that the prerogative of Seigniory, which, from remote times, the Exchequer (*Fisco*) has reserved to itself, with respect to Mines of quicksilver, when it considered them advantageous, after having paid to their owners their just value, has been annulled by general and extraordinary Cortes, in consequence of the resolution and manifestation of the Council of Regency, enacting, at the same time, that the said Mines shall be worked under the same rules and ordinances as those of gold, silver and other metals, and that their possessors shall preserve their proprietorship (*propiedad*) and usufruct, and that in no case shall they be obliged to transfer them (*enagenarlas*) to the State; giving them permission, moreover, to sell their products (*sus frutos*) to any one who will pay the highest price for them. This measure (*providencia*) affirms, in a manner inviolable, the ownership (*propiedad*) and profits (*utilidad*) of this class of real estate (*tales fincas*), and dispels the reasonable fears which prevented individuals from taking them under their care.

The zeal of Your Honor (V. S.), and your love of the public welfare, must cause you to take interest in promoting, among those Miners, the search for and examination of (*cateo*) Mines of cinnabar, to which there can be no more powerful stimulant than the offer of a large pecuniary reward, to be paid from the funds of that Body, to the person who shall discover, and fully prove that he has discovered, a rich and abundant Mine of quicksilver; promising also, that the Council of Regency will recompense, and ennoble, by honorable distinctions, those individuals who shall invest their capital with evident advantage to the said works, and especially those who shall excel in it by extraordinary progress. To serve as an example to all of the necessity of application to this kind of industry, it will be of advantage for Your Honor to undertake methodically the working of a Mine of quicksilver which, of all those of that Kingdom, offers the best prospect, after careful and repeated examinations and exact and well verified reports, establishing a very simple and well regulated Administration, so that the amounts which shall be dedicated to the enterprise may be actually expended in its advancement, and not in unnecessary works and buildings, nor in large salaries, which, without any profit, are commonly charged to such speculations; the accounts of which shall be presented, for examination and approval, at each one of the general *Juntas* held, at the usual time, by that Body, which shall thereupon order what may be most expedient for the management of the affair (*negociacion*).

This same matter of working Mines of quicksilver, your Honor (V. S.) may entrust to the respective territorial Deputations, especially to that of Guanajuato, as it would be a very good opportunity to continue the working of those which were formerly carried on in the *Real* of Tarjea, in the Jurisdiction of San Luis de la Paz, with reference to an economical and continuous system of expenses, founded upon the probable result of regular profits at some future time. Finally, the council of Regency, by whose order I communicate to your Honor (V. S.) these dispositions, expects that you will use the most efficacious means and measures in a matter of such great importance; and that you will direct the same with that prudence, wisdom and knowledge of which your Honor (V. S.), and all your subalterns, have given so many proofs. May God preserve your Honor (V. S.) many years.

Island of Leon, January 26th, 1811.

ESTEBAN VAREA.

To the Members of the Tribunal of the Important Body of Mining of Mexico.

Don Fernando VII, by the Grace of God, King of Spain and of the Indies, and in his absence and captivity, the

Council of Regency, exercising authority *ad interim*, to all those who may see and understand these presents, *be it known:* That in the general and extraordinary Cortes, assembled at the Royal Island of Leon, the following was resolved and decreed:

"The General and Extraordinary Cortes, desiring that the important Branch (*Ramo*) of Mining, in all the dominions of the Indies and Philipine Islands, may receive all possible increase, and considering that the monopoly (*estanco*) of quicksilver established by Law I, Title XXIII, Book VIII of the Recopilacion of the Indies, and the right which the Royal Treasury has reserved to itself, by Article 22, Title VI, of the Ordinances of Mining of New Spain, of taking and working, on its own account, Mines of this kind, whenever deemed expedient, by making an arrangement with the discoverer or denouncer, thus leaving uncertain the interest of the owner, and taking it out of trade, necessarily restrain people from engaging in the useful and expensive undertaking of discovering and working Mines of quicksilver, and also from soliciting it, transporting it, and creating a competition in it (*proporcionar la concurrencia*), as would result from its being made an article of free trade, and exempted from all duties, even including the duty of the Fifth (*quinto*), or the proportion which the Miner is bound to pay, bearing in mind what was proposed and presented to the same Cortes by the Council of Regency on the 26th of December last in favor of the liberty and free trade (*franquicia*) of so necessary an aid to the operations of Mines of gold and silver, and also what, on that subject, has been, by the Deputies of the Indies, asked and solicited of the Cortes, urging with learning and zeal the advantage of annulling said dispositions, and any others which, wholly or in part, conform to them or oppose free trade in said mineral, and the security of absolute and perpetual ownership (*dominio absoluto y perpetuo*) of the Miner, *provided that in managing and working them he observe the general rules established on the subject:*—after mature examination, have decreed and do decree, the aforesaid repeal, and the grant of the said exemptions, commanding, at the same time, that if, in accordance with the former monopoly or otherwise, the Royal Treasury may have sent, or shall send, on its own account, any quantity of quicksilver to be distributed at cost and expenses, as has been done heretofore, for the benefit of owners of Mines, the distribution shall be precisely and exclusively made by the respective Tribunals of Mining, as they are best acquainted with the necessities of each, and with whatever may conduce to the accomplishment and attainment of the object in view; in virtue whereof, it will be their duty to repay the amount into the Royal Treasury, the Cortes trusting to the honor, integrity and zeal of the said Tribunals, that they will justify the high confidence placed in them by a trust so interesting and worthy of their paternal care. Be it known, to the Council of Regency, in order that they may have it printed, published and circulated to whomsoever it may concern, taking care that it be exactly complied with.

Royal Island of Leon, January 26th, 1811.

ANTONIO JOAQUIN PEREZ, *President.*

JOSE ASNAREZ, *Deputy Secretary.*
VICENTE TOMAS TRAVER, *Deputy Secretary.*

To the Counsel of Regency.

And in order that all may know it, the Council of Regency orders it to be printed, and circulated. Let it be understood, and let the necessary measures be taken for its enforcement.

JOAQUIN BLAKE, *President.*

PEDRO DE AGAR,
GABRIEL CISCAR.

In the Royal Island of Leon, January 29th, 1811.

To DON ESTEBAN VAREA.

I transmit it to your Honors (V. S.) for your information and government, so far as concerns you. God preserve your Honors (V. S.) many years.

Royal Island of Leon, February 8th, 1811.

ESTEBAN VAREA.

To the Members of the Tribunal of Mining of Mexico.

The Deputies, Secretaries of the National Congress, say to me under date of the first instant, what follows:

"The General and Extraordinary Cortes, desiring to encourage (*fomentar*) the discovery and working of Mines of quicksilver, with the attention and particularity corresponding to their great importance, have been pleased to provide for rewarding the discoverers, in America, of Mines of quicksilver, and for giving the greatest reward to whosoever shall discover the richest and most productive Mine; they have likewise resolved that the Tribunals of Mining in the Americas be charged with the exact observance of this most important determination, stimulating its progress by all the means deemed conducive to that object, reporting, with punctuality, to the Cortes, through the Council of Regency; and they declare that they will reward the Chemists and Mineralogists of Europe, who may discover or invent the mode of reducing ores with the least expense and the least possible loss of quicksilver."

Which I communicate to your Honors (V. S.), in order that, with the greatest efficacy and punctuality, you will provide for carrying into effect this Sovereign determination: giving me an exact account of what is done and of the progress made in this matter.

Island of Leon, February 2d, 1811.

ESTEBAN VAREA.

To the Members of the Royal Tribunal of Mining of New Spain.

January 13th, 1812.

[From Galvan's "Decretos, etc., de las Cortes de España."]

Order that salt and powder be issued to the Miners of New Spain, at cost and charges.

The Cortes, general and extraordinary, have consented to the proposal made in the memorial presented by the predecessor of your Honor (V. S.), Don Esteban Varea, on the 27th of July last, that there be issued to the Miners of New Spain, at cost and charges, the salt and powder which they may need.

Cadiz, January 13th, 1812.

August 8th, 1814.

[From Note to Article 16, Title XIII, Mexican "Ordenanzas de Mineria."]

Exempting from Excise duties, certain articles used by Miners.

Under date of the 8th of August last, the Most Excellent Minister of the Indies, Don Miguel de Lardizabal, communicated to me the Royal order which follows:

" Most Excellent Señor:

"The king, having examined, with the greatest attention, the different propositions which were formerly "made, and are now repeated, for the encouragement of "the Mining of that kingdom, by Don José Miguel Gordoa, formerly Deputy of the Province of Zacatecas, has "been pleased to resolve as follows:

" Art. 1. That there be again circulated to all Administrations of revenues in the *Reales* of Mines, the list, which was "circulated in 1798, of articles which were to be exempted "from excise duties (*Alcabala*), as being of prime neces-

"sity for all the operations, works, and uses (*atenciones*) of "Mines and owners of *Haciendas* for reducing bullion, in "their difficult and useful labors; and that to those arti- "cles there be added the timbers which are to be used in "the linings and works of Mines and reducing *Haciendas*, "such as beams, scantling, boards, small beams (*Gualdril-* "*las*), and the like, and the bellows for the smelting fur- "naces, and the forge for making and mending the tools "and utensils of the same Mines: with the understanding "that these favors are to be extended to the territories of "the subaltern Administrations and the dependent recep- "tories of the said *Reales* or *Asientos*, whether of gold, or "silver, or copper or any other metal whatever.

"Art. 2. That, as it is the duty of the Tribunal of Min- "ing to make the distribution of quicksilver in general, "the respective Deputation will make it in detail or to indi- "viduals, all in conformity with the special regulation "which the said Tribunal shall form, with the approval of "your Excellency, in which shall be combined a just and "equitable mode of distribution so as to avoid all recla- "mations and complaints, with the greatest security that "the Royal Treasury shall receive its duties. * * * *

[NOTE.—Remainder of this Royal order is omitted in the Ordenanzas.]

October 13th, 1816.

[From Note to Article 10, Title XVI, Mexican "Ordenanzas de Mineria."]

Regulations respecting the exportation of silver from New Spain.

The Superintendent of the Royal Mint having represented to me the notable loss which the said Royal Establishment has suffered in the profits and duties of His Majesty, by the exportation of bullion which, in time of the insurrection of this kingdom, was made through the ports of the south sea, and desiring to put an end to an abuse so injurious, and to prevent, in future, the injurious consequences which this disorder has caused, I have resolved, in conformity with the petition on the subject, and with the opinion of the Assessor, General Commissioner, that the following Articles be observed:

Art. 1. All metals which are found without the corresponding stamps of the Treasury of their place of origin, shall be declared to have incurred the penalty of confiscation, whether found upon the roads or in any other place whatever; of which amount there shall be applied the respective parts to the informer and apprehenders, agreeably to the rule of confiscations of July 16th, 1812.

Art. 2. Metals which leave the places where there are no Treasury and Assayer, shall incur a like penalty, whenever they shall be found on the road, or shall be introduced, without a pass from the respective Courts or a permit (*Guia*) from the Custom Houses, under responsive bond, with which it shall be accredited when presented in the nearest Custom House.

Art. 3. When any bullion shall be exported by sea, there must be given a bond to present the *tornaguia* of its having been introduced into the port of the Peninsula of Spain to which it was directed, as is commanded by Royal *Cedula* of the 30th of December last, with respect to quicksilver which is sold in the Arsenals (*Atarazanas*) of Seville.

Art. 4. All the maritime Custom Houses of this New Spain will transmit monthly, to this Viceroyalty, a report or circumstantial statement of the number of bars which have been embarked in their respective districts, with a statement of their weight and *ley*, and the names of the *Asientos* of Mines from which they came, explaining the names of the persons who brought them, the bondsmen for the return of the *tornaguia*, and finally the names of the ports of the Peninsula, or of this Kingdom, to which they are sent.

And in order that this resolution may come to the notice of all, and receive the most strict and punctual observance, I command that it be published by Proclamation (*Bando*) in this Capital and in the other Cities and Places (*Lugares*) of the Kingdom, to which end it will be circulated to the Intendants, Tribunals, Chiefs, and Officers (*Ministros*), whose duty it may be to watch over its observance.

Given at Mexico, October 13th, 1816.

FELIX CALLEJA.

January 7th, 1818.

[From Note to Article 16, Title XIII, Mexican "Ordenanzas de Mineria."]

Decree, exempting Quicksilver from excise duties.

On this date I say to the Director General of excise duties (*Alcabalas*) as follows:

"In the *expediente* formed for the purpose of the *consulta* "which Your Honor (V. S.) made to this Superiority, on "the 22d of July of last year, with respect to whether "quicksilver, considered now as an article of commerce, "is free from excise duty, I have declared, by decree of "this day, in conformity with the request of the *Fiscal* of "Royal Treasury and the opinion of the *Asesor-General*, "that, the monopoly (*estanco*) of quicksilver being suspend-"ed by Royal Order of December 30th, 1815, and free "trade in it being permitted, with exemption from Royal "and Municipal duties, it is not subject to the payment of "ordinary or permanent excise duty (*Alcabala*); but is sub-"ject to temporary duty, for that is not really *Alcabala*, but "is an impost substituted in place of the Contribution of "War, Convoy, and Squadron, established by the urgent "necessities of the Royal Treasury: and I notify Your "Honor (V. S.) of it, for your information and consequent "effects."

And I transmit it to Your Honor (V. S.) in order that you may know it.

God preserve Your Honor (V. S.) many years.

Mexico, January 7th, 1818.

APODACA.

December 19th, 1818.

[From Note to Article 16, Title XIII, Mexican "Ordenanzas de Mineria."]

Order, exempting quicksilver, for mining purposes, from all duties whatsoever.

Most Excellent Señor:

In letter of the 26th of last June, No. 450, Your Excellency reported, with the the documents, your having declared that quicksilver, although free from ordinary excise duty (*Alcabala*), was subject to pay the temporary duty which had been substituted in place of the Contributions called of War, Convoy, and Squadron, as an impost established in order to supply the urgent necessities of the Kingdom. In view thereof, and considering that the Sovereign will of the King our Lord was very clear and decided, in the Royal Order of December 30th, 1815, that quicksilver should be completely exempt from all Royal and Municipal duties, for the purpose of furnishing the Miners with this semi-metal at the lowest possible price; His Majesty has resolved, in conformity with the report of the *Contaduria General* of the Indies, that quicksilver, which is purchased for the working of the Mines, ought not to pay the temporary Excise duty, nor any other contribution; but that it ought to so pay, if destined for any other purposes in the Kingdom. Which, by Royal Order, I communicate to Your Excellency for your information, and that you may direct its observance.

God preserve Your Excellency many years.

Madrid, December 19th, 1818.

JOSE DE IMAZ.

To the Viceroy of New Spain.

PART V.

MINING LAWS OF MEXICO,

FROM 1821 TO 1853.

PART V.

MINING LAWS OF MEXICO,

FROM 1821 TO 1853.

February 20th, 1822.

[From Note to Article 1, Title XVI, Mexican "Ordinanzas de Mineria."]*

Decree respecting charges made for smelting and refining silver and gold, duties thereon, exemption of quicksilver from duty, and the delivery of powder to Miners at cost and charges.

The Regency of the Empire, provisionally governing in the absence of the Emperor, to all to whom these presents may come and be understood, *know ye:*

That the Sovereign *Junta* provisionally governing, has decreed as follows:

---

*It will be observed that this law is omitted in Galvan's "Coleccion," while one almost identical with it is there published, under date of November 22d, 1821. Arrillaga accounts for Galvan's error as follows: The resolutions passed on the 22d of November, 1821, were in the form published by Galvan, but, on their reference to the Regency, that body objected to Article 11, which was subsequently changed, and the whole became a law on the 20th of February, 1822, in this form.

"The Sovereign *Junta* provisionally governing, which, "from the first moments of its installation, took under "consideration the deplorable and decaying condition of "Mining, and the urgent necessity of affording to this "branch all the means within its power, in order to "contribute to its greater prosperity, on which depends "that of the Empire; having attentively examined the "exposition which the Regency made to it, in consequence "of the resolution passed on the 22d of November last, "and the report which the Committee (*Comision*) of Min-"ing drew up on this very important matter; in the exer-"cise of its powers, and in conformity with what is provided "in Article 11 of Title XI of its Regulation; has been "pleased to decree and decrees:

"Art. 1. The duties of one per cent., tithe (*diézmo*), and "one *real* seigniorage (*de senoreaje*), are abolished.

"Art. 2. The duty of eight *maravedises* on each mark of "silver, which is assessed for the refinement of the bullion "(*pastas*) which is subjected to this operation, is also abol-"ished.

"Art. 3. So likewise the duty of twenty-six *maravedises* "levied on each mark of mixed *pastas* by reason of the "waste of silver in the smelting House (*Apartado*), is "abolished.

"Art. 4. The duty of four *ochavas* on the pieces of sil-"ver, and that of the half *ochava* on pieces of gold, which "are levied, under the name of *bocado*, in the Mint, are "also abolished.

"Art. 5. All duties which were imposed on *pastas* of "gold and silver, and on coin, during the revolution, are "also abolished.

"Art. 6. As an only contribution, there will be levied

" the three per cent. on the true value of silver, and the " same on that of gold, this duty being collected in the " same manner as was formerly that of one per cent. and " the tithe (*diezmo.*)

" Art. 7. In the Mint of the Capital there shall only be " levied two *reales* per mark of silver, and the same per " mark of gold, for the entire expense of coinage of these " metals; and in the other Mints of the Kingdom, inas- " much as they are recently established, there shall be " formed an estimated duty, which shall be enforced the " first year, and correcting it at the end of that time by " the result of the accounts of the entire expenses, this " corrected estimate shall govern for the following year.

" Art. 8. There shall not be charged for the expenses of " refining more than two *reales* per mark of mixed bullion, " instead of the five and a half *reales* which have been here- " tofore exacted; and all the *pastas* which, according to " their *ley* of gold, will pay the expense of the operation, shall " be separated for the introducers thereof. The owners of " mixed bullion are at liberty to perform this operation " themselves, or to get it done wherever they may deem " most convenient.

" Art. 9. In assays for foreigners there shall only be " charged the actual costs of the operations of the assay, " and those of the smelting of the pieces which may re- " quire it, the duty of *bocado* being abolished.

" Art. 10. The payment of the single contribution, de- " signated in Article 6 for *pastas* of gold and silver, having " been made into the national Treasuries, and the stamp, " which shows it, having been impressed upon the pieces " of these metals, the owners shall be at liberty to sell them, " or employ them for any use they please, without any " official establishment of price.

"Art. 11. There shall only be allowed eight and a half "grains deficiency (*de feble*) in the coin of silver, in place "of the eighteen which are now allowed.

"Art. 12. In future, the professional offices (*empleos facul-* "*tativos*) of the Coining and Smelting Houses shall be given "exclusively to persons who have the knowledge of "Physics, Chemistry and Mineralogy, necessary for the "performance of their duties.

"Art. 13. Pure quicksilver remains absolutely free from "duty, whether it come from Europe, or Asia, or is ex- "tracted from the Mines (*Criaderos*) of the Empire.

"Art. 14. The powder which the Miners require for "working the Mines shall be delivered to them by the "Government, at cost and charges.

"The Regency will understand this, and will make the "necessary dispositions for its prompt observance, and let "it be printed, published and circulated.

"Mexico, February 13th, 1822, the Second of the Inde- "pendence of the Empire.

"JUAN JOSE ESPINOSA DE LOS MONTEROS, *Vice President.*

"JOSE IGNACIO GARCIA ILLUECA, *Vocal Secretary.*

"ISIDRO IGNACIO DE ICAZA, *Vocal Secretary.*

"JOSE MARIA DE JAUREQUI, *Vocal Secretary.*

"To the Regency of this Empire."

Wherefore, we command all the Tribunals, Courts, Chiefs, Governors, and other authorities, civil as well as military and ecclesiastic, of whatsoever class and rank, that they keep and cause to be kept, observed and executed, the present Decree in all its parts. Let it be understood for its observance, and let it be printed, published and circulated.

In Mexico, February 20th, 1822, the Second of the Independence of the Empire.

AUGUSTIN DE ITURBIDE, *President.*
MANUEL DE LA BARCENA,
JOSE ISIDRO YANEZ,
MANUEL VELASQUEZ DE LEON.

TO DON JOSE MANUEL DE HERRERA.

March 24th, 1823.

From Note to Article 10, Title XVI, Mexican "Ordenanzas de Mineria."

Order respecting the collection of the duty of one real per mark for the benefit of Mining.

MINISTRY OF TREASURY.

In the *expediente* moved by the Tribunal General of Mining on the imperfect understanding given to the Sovereign resolution of February 13th, 1822, circulated by this Ministry on the 1st of the following March, with respect to its sixth Article, which imposes, as the only contribution, the three per cent. on the true value of the silver, the Emperor has been pleased to resolve, in conformity with the opinion given on the matter by his Council of State;—that the assessment of duties established on silver and gold, be made in the exact terms which are shown in the three annexed copies of Forms, which he has drawn up, under which regulation there shall be made out the accounts in all the Mints and Treasuries of public Revenue of the Empire.

Also His Imperial Majesty, in concert with the said Council, has been pleased to resolve;—that the duty of a *real* per mark for Mining be levied in the said Treasuries, the Mint of this Court ceasing to collect it, and the officers (*Ministros*) of the Treasuries being paid, for the labor and responsibility which is imposed on them, three per cent., which seems sufficient, as they have nothing more to do than to demand a *real* on each mark of those which of the *ley* of eleven *dineros* result from the operations which have to be made for the levy of the National duties, according to said Forms, and because the third part of what is collected is exempt from this charge, they accounting directly with the said Tribunal; with respect to the rendition of the accounts of this branch, the security, expenditure, and transmission of its products, and other matters relative to it, all must conform to that directed by the said Tribunal, which the *Contaduria Mayor* of accounts of this court has approved.

By order of His Imperial Majesty, I communicate it to you, for your information and observance in the part which concerns you.

God preserve you many years.

Mexico, March 24th, 1823, the third of the Independence of the Empire.

MEDINA.

October 7th, 1823.

[From Note to Article 1st, Title VII of the Mexican "Ordenanzas de Mineria." Vide also Galvan's "Coleccion de Ordenes y Decretos," Tomo II.]

Respecting the right of foreigners to acquire and work Mines, and exemption of quicksilver from all duties.

The Supreme Executive Power, provisionally appointed by the Sovereign Mexican Congress, to all who may see and understand these presents:—*know ye*: That the said Sovereign Congress has decreed the following:—

"The Sovereign Mexican Congress has been pleased to "decree:

"Art. 1. Law XII, Title X, Book V, and Law V, Title "XVIII, Book VI, of the Recopilacion of Castile are for "the present suspended; as also, Law 1, Title X, Book "VIII, and those embraced in Title XXVII, Book IX, of "the Recopilacion of the Indies, with Article 1, Title "VII, of the Ordinances of Mining, which required foreign- "ers, in order to acquire and work Mines as their own "(*Minas propias*), to be naturalized or tolerated by express "license of the Government.

"Art. 2. This suspension only enables foreigners to "contract, with the Owners of Mines which require assist- "ance (*habilitacion*), every class of supplies (*avios*), on the "terms which both parties may deem most convenient, "even to the power of acquiring the ownership of shares in "the enterprise (*negociacion*) which they assist, they being

" subject in every respect, to our Ordinances for the working "of Mines and the reduction of ores, and to the other ob- "ligations and charges under which the Nation grants the " ownership of such Real Estate (*fundos*) to every citizen.

"Art. 3. Consequently they are probibited from regis- "tering new Mines, from denouncing abandoned ones, "and from acquiring any share in other Mines than those "which they supply, whatever may be the title by which "they may attempt to sustain their acquisition.

"Art. 4. No change is now made in the matter of Excise "duty (*Alcabala*), and the privilege of quicksilver, which is "expressly exempted from all contribution, the other arti- "cles of consumption in Mining remaining subject to the "variable duty (*Alcabala eventual*) which is now exacted.

"The Supreme Executive Power will understand it, and "have it observed, causing it to be printed, published and "circulated.

"Mexico, 7th of October, 1823.—3rd–2nd.

"FRANCISCO MANUEL SANCHEZ DE TAGLE, *President.*
"JOSE ARCADIO DE VILLALBA, *Deputy Secretary.*
"MANUEL TEJADA, *Deputy Secretary.*"

Wherefore, we order all the Tribunals, Justices, Chiefs, Governors, and other Authorities, Civil as well as Military and Ecclesiastic, of whatsoever class and dignity, to observe, and cause to be observed, complied with and executed, this present decree in all its parts.

Understand it for its fulfillment and cause it to be printed, published and circulated.

Mexico, October 8th, 1823.

MARIANO MICHELENA, *President.*
JOSE MIGUEL DOMINGUEZ,
VINCENTE GUERRERO.

TO DON LUCAS ALAMAN.

October 29th, 1823.

[From Galvan's "Coleccion de Ordenes y Decretos," Tom. II.]

Capital in money sent to Reales of Mines exempted from the duty of two per cent.

The Sovereign Mexican Congress has been pleased to decree as follows:

"Cash funds which leave the Custom Houses, with a per-"mit, for the *Reales* of Mines, are exempt from the duty of "two per cent., the offices being careful to require the *tor-"naguia* corresponding, agreeably to the dispositions on "the matter."

February 13th, 1824.

[From Galvan's "Coleccion de Ordenes y Decretos," Tom. III.]

Admission of quicksilver from whatsoever source.

The Sovereign Constituent Congress has been pleased to decree:

"That, without examining the origin or place of ship-"ment of quicksilver which may enter our ports, there "shall be admitted whatever quantities of the same may "arrive, or may have arrived, under a friendly or neutral "flag, but this determination shall have no effect upon "causes now pending for confiscation of this article."

July 24th, 1824.

[From Galvan's "Coleccion de Ordenes y Decretos," Tomo III.]

Exclusive privilege granted to Don Juan Bautista Binnon, for the period of five years, to work quicksilver Mines in the territory of the Republic.

The Sovereign Congress General of the United Mexican States, having taken into consideration the petition of Don Juan Bautista Binnon relative to granting him the privilege of working Mines of quicksilver, has been pleased to decree:

Art. 1. There is granted to the said Binnon, for the period of five years, the exclusive privilege of working Mines of quicksilver in the territory of the Republic, he conforming in all respects to the Ordinances of Mining.

Art. 2. The said privilege will commence to run two years from the date of this decree, according to the request of the party interested; and he will hold it without prejudice to those who within this time shall acquire Mines of the same ingredient, and to those who have already acquired them.

Art. 3. There is also granted to him permission to erect,

at his own expense, three establishments (*fabricas*) for the reduction of ores of quicksilver, in the places where the Mines may be found.

Art. 4. If, at the expiration of the period mentioned in Article 2, the said Binnon shall not have arrived with the utensils which he proposes, then, *ipso facto* (*por el mismo hecho*) all the concessions mentioned in this law terminate.

Art. 5. The operations in the Mines of quicksilver which Binnon may work, shall be kept up at a weekly expenditure of not less than two hundred dollars.

Art. 6. The quicksilver which Binnon may extract from said Mines shall not be sold by him at more than $40 per quintal.

Art. 7. As soon as the establishments of the contractor produce all the quicksilver which the Mining of the Republic consumes, and it is proved to the satisfaction of the Government, the introduction of foreign quicksilver is prohibited.

Art. 8. If, in addition to what is necessary for the consumption of Mining, and a deposit of one-quarter more, which it shall be the duty of Binnon to keep, he shall accumulate any more, he shall be at liberty to export it from the territory of the Republic, free of duty.

Art. 9. It is permitted to Binnon to cut the wood required for the establishment for extracting quicksilver, he conforming with the dispositions which are in force on the subject.

Art. 10. The nation accepts, at the expiration of the period of this permission, the cession which the said Binnon makes of the business (*negociacion*) and its respective establishments (*fabricas*).

November 16th, 1824.

[From Galvan's "Coleccion de Leyes y Decretos," Tomo III.]

Authorizing the renting of salines.

The Sovereign General Constituent Congress of the United Mexican States, for the better administration and management of the funds (*caudales*) of the Federation, has been pleased to decree the following law:

* * * * * * * * *

Art. 35. The salines of the Federation shall be leased out, at public auction, to the highest bidder, for the number of years which may, by the Government, be deemed expedient.

Art. 36. It shall be a necessary condition of these auction sales, that salt be sold under the regulations adopted by the Public Treasury for the benefit of Mining, and at a quarter part less than the price at which the Treasury may sell it.

Art. 37. There are excepted from the preceding Article, the Salines of the Villas of Refugio and Reinosa, which are ceded for eight years for the common benefit of their inhabitants and those of Camargo, Mier, Revilla y Laredo, under the rules which may be established by the Legislature of the State to which they pertain.

Art. 38. The Salines of private ownership, which until now have paid duties to the Public Treasury, will in future pay one quarter part less than what they formerly paid. These duties will also be rented out, preference under equal circumstances being given to the Governments of the States which may bid.

Art 39. This preference shall also be extended to include the Salines of the Federation.

May 20th, 1826.

[From Appendix to the Mexican "Ordenanzas de Mineria." Vide also Galvan's "Coleccion de Ordenes y Decretos," Tomo IV.]

Extinction of the Tribunal General of Mining, and creation of a Junta to be called "Establecimiento" of Mining.

The President of the United Mexican States to the inhabitants of the Republic, *know ye:* That the general Congress has decreed the following:

Art. 1. The Tribunal General of Mining must cease according to the general Constitution, in so far as it relates to the administration of justice with which it was charged.

Art. 2. It will also cease in so far as relates to the executive, economical and administrative attributes, which were assigned to it by its institution and laws.

Art. 3. The extinguished Tribunal will immediately proceed to liquidate, within such period as the Government shall designate, and which shall not exceed two months, the accounts of the funds which have been under its charge.

Art. 4. The General *Junta* of Miners will designate an individual who, with an accountant appointed by the Government, and an attorney (*apoderado*) of the creditors of the Mining funds, appointed at the time and in the mode which the Government may designate, shall receive and audit these accounts, having first taken charge of the archives, proceedings (*constancias*), &c., pertaining to the Tribunal.

Art. 5. During the period mentioned in the third Article, the members of the Tribunal will receive their salaries.

Art. 6. The accounts, audited as prescribed in Article 4, shall be transmitted to the Government, which, with such report as it may deem proper, shall send them to the General Congress for its approval.

Art. 7. The products of what is called the *Real de Mineria*, and the other available credits of the extinguished Tribunal, shall be applied to the payment of its officers, the support of the College, the payment of interests and redemption of loans, the *Real de Mineria* ceasing as soon as the debts chargeable to the funds of the *Mineria* shall have been extinguished.

Art. 8. Such portions of these funds as have been received by the States, shall be reimbursed to the establishment, within such time as the Government may designate.

Art. 9. The Nation recognizes its obligation for the amounts which have been taken, from the said funds of the Tribunal, for the urgent wants of the State.

Art. 10. The collection of the capital belonging to this fund shall be made by the respective Commissaries, who, under their responsibility, shall remit its products to the

Mint in Mexico, as a special deposit, until this part of the Establishment is regulated.

Art. 11. The distribution of the funds shall be made according to this law, through warrants drawn by the individual appointed by the General *Junta* of Miners, endorsed by the Minister of the Treasury.

Art. 12. This individual shall be regarded as the general attorney of the Body of Miners, and, in that capacity, he may represent to the Government what he may judge expedient for the better execution of this law.

Art. 13. There shall be published printed extracts from the accounts which are rendered by the Tribunal, and, in each successive month, the receipts and expenditures of the funds of the Establishment.

Art. 14. The persons in the permanent employment of the extinguished Tribunal shall continue in the class of salaried officers (*cesantes*), to be paid from the funds of the Establishment.

Art. 15. The Government shall employ the *cesantes* in the labors of the Establishment, if deemed necessary.

Art. 16. The College of Mining shall continue henceforth in the same manner as heretofore, and with the endowment which has been assigned to it, and which shall be drawn from the fund of the *Mineria*.

Art. 17. It shall be under the direction of the individual whom, as is provided in this law, the General *Junta* of Mining shall appoint, exercising with respect to the College the functions which the Tribunal has had, under the direction of the President of the United Mexican States.

Art. 18. The Government will consult Congress with respect to the salary which the Director is to have, and, with its approval, will form a new plan by which the College shall be regulated, according to the intention of its endowment, referring it all to the General Congress, in order that it may determine what is proper.

LORENZO DE ZAVALA,
President of the Senate.

BERNARDO GONZALES PEREZ DE ANGULO,
President of the Chamber of Deputies.

DEMETRIO DEL CASTILLO,
Senator Secretary.

ANTONIO FERNANDEZ MONTJARDIN,
Deputy Secretary.

Wherefore, it is ordered to be printed, published, circulated, and duly observed. And in order that the provisions of the present decree may be most punctually executed, I have directed to be observed, in the meantime, the Instructions (*providencias*) which follow:

Art. 1. In the act of the legal publication of this decree, the individuals who may compose the Tribunal will remain formed as a *Junta*, which shall be called the *Junta Provisional de Mineria.*

Art. 2. This *Junta* will immediately proceed to convoke, in the usual manner, the *Junta General* of Miners, in order that, having assembled in this capital within three months from this date, it may appoint the individual whom it has to elect, agreeably to Article 4 of this decree.

Art. 3. It will also convoke, without the loss of a moment's time, the creditors of the funds, in order that, personally or through their representatives, they may meet in this Capital for the appointment of an Attorney (*apoderado*), which they are to have, agreeably to said Article 4.

Art. 4. This appointment shall be made within three months from this date, and in the same manner as is the custom, in the meetings of creditors, with respect to the *Sindico procurador*.

Art. 5. The provisional *Junta* of Mining will preside over the elections of the individual whom the Miners are to appoint, and of the Attorney of the creditors of the funds of Mining, taking care that they are held in accordance with the provisions of the preceding Articles.

Art. 6. The same provisional *Junta* will be charged with all the directive and administrative powers which the Tribunal exercised, until the individuals mentioned in the foregoing Article are placed in a position to be able to exercise those which respectively belong to them, according to this decree.

Art. 7. As soon as they are in position to perform the duties of their offices (*functionar*), the individual whom the *Junta General* of Miners shall appoint, and the Attorney-in-fact of the creditors, shall present themselves to the Ministry of Relations, in order that, with the Accountant (*Contador*) appointed by the Government, they may immediately begin to form the Establishment which this decree prescribes, the provisional *Junta* of Mining ceasing by the same act.

Art. 8. The provisional *Junta* will notify successively the Ministry of Relations, of the fulfillment of the Articles of this decree, and of the provisions of the Government in the matters which correspond to it.

Art. 9. The extinguished Tribunal will proceed, within two months from this date, to liquidate the accounts of the funds which have been under its charge.

Art. 10. The amounts which the States have taken from the funds of Mining, must be repaid to the Establishment within six months from this date.

Palace of the Federal Government of Mexico, May 20th, 1826.

GUADALUPE VICTORIA.

TO DON SEBASTIAN CAMACHO.

September 15th, 1829.

[From Appendix to the Mexican "Ordenanzas de Mineria." Vide also Galvan's "Coleccion de Ordenes y Decretos," Tomo V.]

Revocation of so much of the Decree of May 20th, 1826, as requires the Commissary Generals to collect Mining duties, and directing them to be made by the officers of the Establishment of Mining.

The President of the United Mexican States to the inhabitants of the Republic,—*know ye:*

That, the Establishment of Mining having represented the great inconveniences which have been experienced in the collection of the endowment funds by the Commissary Generals of the States, to whose care it was committed until said Establishment should be organized, which has already been done in accordance with the law and regulation of

May 20th, 1826, asking, in consequence, that in future the collection of the duties ordered to be levied, be made by the employes appointed by the Members (*señores*) who compose the said Establishment, with which employes they can settle directly, and thereby have more control over the funds, so as to secure their appropriation to the important objects for which they were designed; in the exercise of the extraordinary powers which are conferred on me, I have been pleased to command:

Art. 1. That the Commissaries General cease the collection of the funds of Mining.

Art. 2. That this be done hereafter by the persons, whom, under its responsibility, and suitable securities being previously given, the said Establishment shall appoint in each one of the places where the duty of Mining is to be collected.

Art. 3. That the said Establishment assign to the collectors a compensation in proportion to the sum they shall collect, observing always the greatest economy of the funds.

Art. 4. That the Commissaries General present to the Establishment the general accounts of the produce of the duty of Mining, during the time its collection was in their charge, with a statement of the amounts remitted to the Mint, and of what remains in favor of Mining (*la Mineria*).

Art. 5. That the foregoing Articles be substituted for Articles 10 and 11 of the said law of May 20th, 1826.

Wherefore I order it to be printed, published and circulated for its due observance.

Given at the Federal Palace of Mexico,
September 15th, 1829.

VICENTE GUERRERO.

To Don LORENZO DE ZAVALA.

April 14th, 1832.

[From Galvan's "Coleccion de Leyes y Decretos," Tomo. VI. Vide also Arrillaga's "Recopilacion de Leyes" &c., 1832–3.]

Exclusive privilege granted to Juan Andrés Velarde to use the new system of amalgamation and cupellation invented by him for the reduction of ores.

The Vice-President of the United Mexican States, exercising the Supreme Executive power, to the inhabitants of the Republic, *know ye:*

That the General Congress has decreed as follows:

Art. 1. There is granted to citizen Juan Andrés Velarde, for the period of ten years, the exclusive privilege to use the new system of amalgamation and cupellation which he has invented for the reduction of ores.

Art. 2. Before using this privilege, he will present to the Government an exact description of his system, accompanied with drawings, models and whatever else he may deem necessary for the explanation of his invention; and the Government will issue to him the corresponding patent to secure his right.

Art. 3. The patent will contain an exact copy of the documents and drawings which Velarde may have presented, and the descriptions of the models.

Art. 4. The period of the ten years having expired, there will be published circumstantially, in the periodicals of the Republic, the new system of Velarde.

Art. 5. In case that, in the opinion of Velarde, there may

be political or commercial reasons which require his new invention to be kept a secret, the Government will cause to be copied, in his presence and by the hand of a person having his confidence, the said description into a special register, which shall be sealed up and remain so for the time which the privilege is to continue; placing upon it or upon its envelope the name of Velarde, the date, and the reasons why the package is sealed up.

May 22d, 1833.

[From Galvan's "Coleccion de Leyes y Decretos," Tomo. VII. Vide also Arrillaga's "Recopilacion de Leyes," &c., 1833.]

Grant of salines, &c., to the State of Tamaulipas extended for the term of 10 *years.*

CIRCULAR OF FIRST SECRETARY OF STATE, IGNACIO MARTINEZ, ETC.

Art. 1. The grant made to the *villas* of the north of the State of Tamaulipas by Article 17, [37?] of the Decree of November 16th, 1824, is renewed for 10 years, for the benefit of that State.

Art. 2. The foregoing grant is made to extend to the other Mines (*Criaderos*) of Salt which exist within the territory of the said State.

May 22d, 1833.

45

July 2d, 1833.

[From Galvan's "Coleccion de Leyes y Decretos," Tomo VII. Vide also Arrillaga's "Recopilacion de Leyes," &c., 1833.]

Products of the Real de Mineria collected in Zacatecas applied to the expenses of the war.

CIRCULAR OF FIRST SECRETARY OF STATE, IGNACIO MARTINEZ, ETC.

* * * * * * *

"Art. 1. Exercising extraordinary powers, and during "the present circumstances, I direct that the products of "the *Real de Mineria*, which are collected in the State of "Zacatecas, be paid into the Commissariat-General of the "same, and be applied to the expenses of the war, issu- "ing to the Governor such sums as he may ask, for that "purpose.

"Art. 2. With the remaining incomes of the branch, "the other objects of the Establishment of Mining will con- "tinue to be attended to without change.

"ANTONIO LOPEZ DE SANTA ANNA."

And in order, &c.

July 27th, 1833.

From Galvan's "Coleccion de Leyes y Decretos" Tomo VII. Vide Also Arrillaga's "Recopilacion de Leyes," &c., 1833.]

The products of the Real de Mineria of Guanajuato applied to the expenses of the war.

CIRCULAR OF FIRST SECRETARY OF STATE, IGNACIO MARTINEZ, ETC.

"Art. 1 Exercising extraordinary powers, and during "the present circumstances, I direct that the products of "the *Real de Mineria* which are collected in the State of "Guanajuato will be paid into the Commissariat-general of "the same, and be applied to the expenses of the war, "issuing to the Governor such sums as he may ask for that "purpose.

"Art. 2. With the remaining incomes of the branch, the "other objects of the Establishment of Mining will con-"tinue to be attended to without change.

"VALENTIA GOMEZ FARIAS."

And in order that, &c.

November 18th, 1833.

[From Galvan's "Coleccion de Leyes y Decretos," Tomo VII. Vide also Arrillaga's "Recopilacion de Leyes, &c.," 1833.]

Repeal of the decrees of the 2d and 27th of July, 1833, applying the products of the Real de Mineria in Zacatecas and Guanajuato to the expenses of the war.

The President of the United Mexican States to the inhabitants of the Republic,—*know ye:*

That having issued the decrees of the 2d and 27th of July last, for the time the circumstances of that period should continue, and they having ceased, using the extraordinary powers, I decree:

Art. 1 The decrees of the Governmont of the 2d and 27th of July last, in which it was commanded to pay into the Commissariat-general of the States of Zacatecas and Guanajuato the products of the *Real de Mineria* collected in the same, are repealed.

Art. 2. The collection of said *Real de Mineria* will be re-established in its former condition and for the ends it was intended.

November 18th, 1833.

May 12th, 1838.

[From Arrillaga's "Recopilacion de Leyes, Decretos, &c.," 1838.]

A reward to those who, during the time specified, shall import quicksilver, not French property, in national or neutral vessels.

Art. 1. During the blockade and six months thereafter, those who shall import quicksilver in neutral or national vessels, shall receive a premium of $5 for each quintal which may be introduced by the ports or frontiers of the Republic, which shall be allowed in the respective custom-houses on account of any duties which may be there due.

Art. 2. If the quicksilver which shall be presented in the ports of the Republic be French property, it shall not be considered as within the provisions of the preceding article, but on the contrary shall be subject to the penalty of confiscation

November 22d, 1838.

[From Arrillaga's "Recopilacion de Leyes, &c," 1838.]

Authorizing the alienation of the Salines of the Peñon Blanco, &c.

The Government is authorized to alienate, for cash and on terms most advantageous to the public Treasury, the

Salines of the Peñon Blanco of San Luis Potosi, and the edifice of the Ex-Inquisition, reserving the amount necessary to cover the charges and incumbrances (*gravamenes*) upon them, and inserting the conditions, in the bidding for the Salines, that the purchaser shall continue selling to the Mines the *sal-grano* and the *sal-tierra*, on the same credits and at the same prices, as they have heretofore been sold.—(Circular of the Ministry of Treasury.)

December 11th, 1838.

[From Arrillaga's "Recopilacion de Leyes," &c., 1838.]

Authorizing the alienation of the Salines of Zacoalco y Zayula, &c.

Art. 1. The Government is authorized to alienate for cash the salines, saltpetres and *tequesquites* of the beach of Zacoalco and Zayula in the Department of Jalisco, consulting as much as possible the greatest advantage of the Treasury, and making as a condition in the sale or bidding, that the purchaser shall not change the usual prices.

Art. 2. It is also authorized to sell the copper which is in the Mint of this capital, acting in this matter in concert with the parties interested in these metals. (Circular of the Ministry of Treasury.)

October 19th, 1841.

[From Lara's "Coleccion de Decretos y Ordenes," No. 56.]

Creating a general Junta of Mining, Coin and Trade.

MINISTRY OF FOREIGN RELATIONS AND GOVERNMENT.

His Excellency the Provisional President has been pleased to take into consideration the regulation, which the branch of Mining requires, in order to avoid, as much as possible, the evils which attend the working of the Mines and the reduction of their ores, and has also desired that these operations may be carried on in such a manner that, with more economy of men, of time, and of expenses, better results may be derived from this important branch of public wealth.

Money is the object intended for the exchange of articles of commerce; in coining it, the nation exercises a sovereign right, and in impressing its stamp upon it, it becomes responsible for its fineness and weight.

The Banks of Europe have observed, in ours, essential differences in both respects, and the stamp of some makes it appear the coin of different powers, to the injury of trade and of the public.

Trade suffers most of these grievances, considering the restrictions which the different laws have imposed on it, and the effects of bad roads, causing such expensive freights on land, and it feels still more, in the interior, the other most grievous evils which our agriculture suffers.

Wherefore, and in order that the important branch of Mining may increase its products to the advantage of

Miners and the public, that our coin may be made as perfect as possible, by being uniform in weight, fineness and stamp, and that trade may have a greater increase, by the removal of all obstacles, His Excellency the President has resolved to create a General *Junta* of Mining, Coin and Trade, on account of the connection which exists between these branches, like that mentioned in law VII, Title I, Book IX, of the Novisima Recopilacion, of which a copy is annexed. This *Junta* will immediately be occupied in proposing what may be expedient for the progress of the said branches; for that of coin, it will be composed of the Superintendent, Don Bernardo Gonzales Angulo, the Administrator, Don José Davila, and the second *sellador*, Don Luis Cadena; for that of Mining, of Don Andres Quintana Roo, Don José Maria Sardeneta, and General Don Pedro Terreros; and for that of Trade, of Don Miguel Muños, Don Francisco Iturbe, and Don German Landa.

These individuals will meet in the Mint of this capital, and will elect a President from their number; and when organized, the *Junta* will have for its Secretary, Don Manuel Maria Sorondo, with the pay of the salary of his present office, which he will continue to hold, in the said Mint. The *Junta* will have its correspondence free, and the establishment of Mining will transfer to the Secretary $50 per month, for a clerk, and for the clerk's expenses.

November 27th, 1841.

[From Lara's "Coleccion of Decretos y Ordenes," No. 183.]

Order, directing the formation of new Ordinances of Mining.

MINISTRY OF JUSTICE AND PUBLIC INSTRUCTION.

His Excellency the President, considering the great utility and the interest of the re-establishment of the special tribunals of Mining, with the modifications which the present system of government requires, and those which experience has indicated, has been pleased to decree, that a *Junta* be formed, composed of Your Honor, and Don Jose Francisco Robles, Don Lucas Alaman, Don Tomas Ramon del Moral, and Don Felipe Neri del Barrio, in order that, having before them the old ordinances, they may form and present, as soon as possible, a new project of organization and proceedings for said tribunals.

His Excellency hopes, from the well known patriotism of Your Honor, that you will accept this commission, and that you will perform its duties with the efficiency which characterizes you.

TO LICENCIADO DON JOSE MANUEL ZOZAYA BERMUDEZ.

March 11th, 1842.

[From Lara's "Coleccion de Decretos y Ordenes," No. 325.]

Rights of foreigners to acquire and hold real estate in Mexico.

Antonio Lopez de Santa Anna, General of Division, *Benemerito* of the country, and Provisional President of the Mexican Republic, to all the inhabitants thereof,—*know ye:*

That after mature and most careful examination into the advantages which will result to the Republic from permitting foreigners to acquire property (*propiadades*): having heard the opinon of the council of representatives which has most scrupulously examined this subject: the reports of the several Departmental *Juntas*, of many well-informed persons, and what has been said *pro* and *contra* by the press: having seen the various projects of a law which for this purpose have been presented; convinced, moreover, that liberal policy and well considered interest require that there should be no longer delay in a concession which may tend to the aggrandizement of the Republic, by increasing its population, by the extension and division of property, which consequently makes greater the national wealth: taking also into consideration, that by this means, the safety of the nation may be more and more secured, since foreign proprietors will be so many more defenders of national rights, when they are interested in the common prosperity: considering also the impulse which will be given to agriculture, industry, and commerce, which are the sources of public wealth; and lastly, that the opinion generally expressed is in favor of said concession,—I have been pleased, in the exercise of the powers conceded to me by the seventh of the Bases established at Tacubaya and sworn to by the representatives of the Departments, to decree as follows:

Art. 1. Foreigners, who are domiciled and residing in the Republic shall be at liberty to acquire and possess urban and country property, (*propiedades urbanas y rusticas*), by purchase, adjudication, denouncement, or any other title whatsoever, established by the laws.

Art. 2. They shall also be at liberty to acquire in property (*en propiedad*) Mines of gold, silver, copper, quicksilver, iron and coal, of which they may be discoverers, agreeably to the Ordinance of that Branch.

Art. 3. No one individual foreigner shall be at liberty to acquire, more than two country estates (*fincas rusticas*) in the same Department without license of the supreme Government, and only under the boundaries which they now have independently of each other.

Art. 4. In the acquisition of urban estates (*fincas*) in the cities, *Villas* and *Pueblos*, as well as in lands contiguous thereto, in which it is wished to construct new buildings (*fincas*), the occupants shall have the right of preference, under the same circumstances and conditions.

Art. 5. Foreigners who may, in virtue of this law, acquire property (*propiedad*) will, in respect thereto, be absolutely subject to the existing laws, or those which may be in force in the Republic, respecting their transfer, use, preservation, and the payments of taxes, without that they shall be at liberty to allege any right of alienage with respect to these matters.

Art. 6. Consequently, all questions of this nature which may arise shall be determined in the ordinary and usual manner by the national laws, to the exclusion of all other intervention, whatever it may be.

Art. 7. Foreigners who may acquire property (*propiedades*) in the country, in cities, or in Mines, and foreigners who may work in them as servants, operatives or day-laborers, shall not be obliged to bear arms, except it be in the police: but they must pay the taxes which are levied for militia purposes.

Art. 8. If the foreign proprietor shall absent himself for more than two years with his family from the Republic, without obtaining permission from the Government, or if the property shall pass by inheritance, or by any other title, into the power of a person not resident in the Republic, he shall be obliged to sell it within two years, counted from the day when the absence or transfer of ownership (*dominio*) took place. If he shall not do this, the official sale shall be proceeded with, observing all the legal formalities, and of the proceeds, the tenth part shall be given to the denouncer, leaving the remaining nine-tenths, in safe deposit, at the disposal of the owner. This shall always be done when it is proved that the owner of the estate (*finca*) resides out of the Republic, and that he who calls himself proprietor is nothing more than in place (*en lugar*) of the absentee.

Art. 9. These provisions do not include the Departments bordering on, or bounding with (*limitrofes o fronterizos*), other nations, with respect to which Departments, special laws of colonization will be enacted, without that property may ever be acquired in them by foreigners, unless by express license of the Supreme Government of the Republic.

Art. 10. In the Departments which are not boundary or frontier, and which may have coasts, only within five leagues of them may foreigners acquire country property.

Art. 11. In order that foreigners, who have acquired property in the Republic, may be citizens thereof, it is sufficient that they prove, before the political authority of the place of their residence, that they are proprietors (*propietarios*), that they have resided two years in the Republic, and that they have conducted themselves well. The *Expediente* drawn up in this manner shall be sent to the respective Ministry, by which the letter of citizenship shall be despatched.

Art. 12. Foreigners shall not be at liberty to acquire public or vacant lands in all the Departments of the Republic without contracting for them with the Government, which possesses this right in representation of the dominion (*dominio*) of the Mexican Nation.

Wherefore I order it to be printed, published, circulated, and duly observed.

Palace of the National Government,
Mexico, March 11th, 1842.

ANTONIO LOPEZ DE SANTA ANNA.

JOSE MARIA BOCANEGRA,

Minister of Exterior Relations and Government.

July 12th, 1842.

[From Lara's "Coleccion de Decretos y Ordenes," No. 414.]

Restorers of abandoned Mines to have the same rights as Discoverers.

Antonio Lopez de Santa Anna, General of Division, *Benemerito* of the country, Provisional President of the Mexican Republic, to all the inhabitants thereof,—*know ye*:—

That, the Decree of the 11th of March of this year which enables foreigners to acquire real estate (*bienes raices*) in the Republic, on the terms which the same Decree expresses, having been published, some doubts have been raised with respect to the true intendment of Article 2, and disputes, resulting from the different interpretations which have been given to said Article, have been sent up to the Supreme Executive. In view of all which, and having before me the respective provisions and ordinances, I have been pleased, in exercise of the powers conferred on me by the seventh of the Bases established at Tacubaya and sworn to by the representatives of the nation, to declare the following:

Natives or foreigners who shall fully prove that they have been restorers of ancient Mines, decayed or abandoned, shall be considered as discoverers, and consequently are enabled (*habilitados*) by Article 2 of the Decree of the 11th of March of this year to acquire property (*propiedad*) in Mines.

Wherefore I order it to be printed, published, circulated, and duly observed.

National Palace, Mexico, July 12th, 1842.

ANTONIO LOPEZ DE SANTA ANNA.

JOSE MARIA BOCANEGRA,
Minister of Exterior Relations and Government.

August 31st, 1842.

[From Lara's "Coleccion de Decretos y Ordenes," No. 452.]

Law of October 7*th* 1823 *not repealed by that of March* 11*th* 1842.

Antonio Lopez de Santa Anna, &c., &c., *know ye:*

That, in exercise of the powers conferred on me by the seventh of the Bases established at Tacubaya and sworn to by the representatives of the Departments, I have been pleased to decree as follows:

The law of the 11th of March of this year which enabled foreigners to acquire real estate (*bienes raices*), did not repeal that of the 7th of October, 1823.

Wherefore I order it to be printed, published, circulated, and duly observed.

National Palace, Mexico, August 31st, 1842.

ANTONIO LOPEZ DE SANTA ANNA.

JOSE MARIA BOCANEGRA,
Minister of Exterior Relations and Government.

October 18th, 1842.

[From Lara's "Coleccion de Decretos y Ordenes," No. 496].

Respecting the delivery of powder to Miners.

Antonio Lopez de Santa Anna, &c., &c., *know ye:* That in exercise of the powers confered on me by the seventh of the Bases established at Tacubaya, and sworn to by the representatives of the Departments, I have been pleased to decree the following:

* * * * * * * *

Article 27. The powder which is sold for consumption the Mines will be accredited in the accounts by the certificates given by the Deputations of Mining; and where there are none, the Miners will ask it of the political authority, and only with this document will the administrators, ware-house keepers, and trusties (*fieles*) issue it at the price which is prescribed for the said Miners.

October 26th, 1842.

[From Lara's "Coleccion de Decretos y Ordenes," No. 520].

Exemption of iron, &c., from duties.

Antonio Lopez de Santa Anna, &c., &c., *know ye:*—That, the protection and encouragement of the national industry being the object of my most constant anxiety, and taking into consideration that the Departmental *Juntas* of Durango, Aguascalientes, San Luis Potosi, Guanajuato, Vera Cruz, Puebla, and Zacatecas have asked, that iron which is produced from the metalic veins of the Republic, be declared exempt from all kind of duties and taxes, in compliance with their petitions and using the powers conferred on me by the Bases established at Tacubaya and sanctioned by the nation, I have decreed as follows:

Art. 1. Iron, which is produced from any vein whatever in the Republic, is declared free, for ten years, from every species of duty.

Art. 2. The machines, cylinders, refractory stones and brick, which are imported into the Republic for the use of iron-works, shall also be free from all duty.

Art. 3. In order that the owners of iron-works may enjoy the exemption which is granted to them by Article 1, they will put upon their manufactures a mark cut into the same, which shall contain, in visible letters, the name of the place where the manufactory is situate, and the cipher of its owner.

Art. 4. The permits (*guias*) which shall be applied for, in order to transport the iron from the place where it is manufactured, shall express the number of bars or pieces which are exported, and the weight of each one respectively.

Wherefore I order it to be printed, published, circulated, and duly observed.

December 2d, 1842.

[From Appendix to the Mexican "Ordenanzas de Mineria." Vide also, Lara's "Coleccion de Decretos y Ordenes," No. 549.]

Regulations of the Junta of Encouragement and Administration of Mining.

Nicolas Bravo, General of Division, *Benemerito* of the country, and President Substitute of the Mexican Republic, to the inhabitants thereof,—*know ye:*

That, considering the necessity and at the same time the importance of encouraging the interesting Branch of Mining, and remembering that although there are various dispositions in relation to this public benefit, they have either fallen into disuse or have been entirely forgotten; without dilating upon the great importance, especially to the Republic, of preserving one of the elements most necessary for its prosperity and greatness, I have been pleased to decree, in the exercise of the powers granted by the seventh of the Bases established at Tacubaya and sworn to by the representatives of the Departments, the following:

REGULATION FOR THE JUNTA OF ENCOURAGEMENT AND ADMINISTRATION OF THE BODY OF MINING.

Title I.

Of the Junta of Encouragement and Administration of Mining, its formation, renewal, and attributes.

Art. 1. There shall be a *Junta* which shall be denominated *de Fomento y Administrativa de Mineria.* It shall be composed of an agent (*apoderado*), of the Miners, another of the creditors of the endowment fund, and a person commissioned by the Supreme Government. Its mode of election and powers will be detailed in the following Articles:

Art. 2. The Presidency of this *Junta* shall be held in turns by the three individuals who compose it, alternating each year. In the first period the person commissioned by the Supreme Government shall be President.

Art. 3. As soon as the present decree is published the Supreme Government and the creditors of the funds of the Establishment will proceed to appoint their respective Commissioners, and the Miners resident in this capital shall also on their part appoint one *ad interim.*

Art. 4. The Governor of the Department of Mexico will assemble and preside over the creditors and the Miners in separate *Juntas,* in order that each class may appoint their respective Commissioner, conforming in voting to the laws which are in force and observed in the present Establishment.

Art. 5. The *Junta* being formed it will require of the individuals who go out of office, the delivery by formal and exact inventory, of everything which may belong to the Establishment, as also the rendition of accounts for all the time which has elapsed since the last were presented, and it will cause that the same be laid before the tribunal for the revision of accounts, within the period of three months. It will hold its sesssions in the place which is designated in the edifice of the same Establishment, and its office shall be there also.

Art. 6. Three months after the publication of this decree, the special agents (*apoderados*) who may have been appointed in the Mining places (*Minerales*) by the *Juntas* of Miners, shall meet in this capital, under the Presidency of the Governor of the Department, in order to elect the person who is to represent them in the *Junta* of Encouragement, and this election having taken place, and the proprietary Commissioner having entered upon his office, the one appointed *ad interim* will retire. There will also be elected three Substitutes, who, as well as the Commissioner, must be Miners or suppliers of Mines (*Aviadores*), who will act in place of the proprietors in the order of their appointment, and who will also act as Advisers (*consultores*) in cases where the *Junta* ask their opinion. If from any places, the particular attorneys of the *Juntas* cannot come, on account of the great distance, or their inability to bear the expenses of traveling and sojourn in Mexico, it will be sufficient that the *Juntas* of the *Asientos* of Mines send authority and sufficient instructions to some person of their confidence and who resides in this capital.

Art. 7. On the 31st of December, 1844, this *Junta General* of Miners shall be renewed, in order to make a similar election, in the same manner as the creditors shall do, in regard to their agents, and both may re-elect the persons whose terms expire, the renewal taking place successively every three years.

Art. 8. Each one of the members of this *Junta* shall

receive the annual salary of $3,000, which shall be paid out of the fund which this law establishes, and there shall be paid to the Substitutes half salaries, when they shall enter on duty on account of any legal impediment of the proprietor and whenever the occupation shall exceed fifteen days.

Art. 9. The *Junta* will propose the reforms which it shall deem expedient in the office of Secretary and the other offices of the establishment. In the Regulation, of which mention will be made in the next Article, the salaries of all the employes, who may be considered necessary for the office, shall be fixed, and in their appointment, under equal circumstances, preference shall be given to those retiring (*cesantes*) who receive a salary from the treasury.

Art. 10. The attributes of this *Junta* shall be those which include an economical and faithful administration of the funds mentioned in this Decree, in conformity with the Regulation which it shall draw up and transmit to the Supreme Government for its approval. In this Regulation there shall, moreover, be determined: 1st, the manner in which quicksilver shall be obtained, distributed, and sold to those who reduce ores, fixing the cases and mode in which the working of quicksilver Mines in the Republic is to be supplied, rewarded, or in other ways stimulated and protected: 2d, everything relating to the redemption of the debt of the endowment fund, according to what may be directed in the respective Title: 3d, the regulation and direction of the *Junta* itself; and, finally, it shall be an attribute and object of its most efficacious solicitude, to promote the encouragement of the branch (*ramo*), its funds, and its school.

Art. 11. The *Junta*, after hearing the Director and Professors of the School, will propose to the Supreme Government for its approval, the reforms which it may deem expedient in the ordinances of said establishment.

Art. 12. The *Junta*, in conjunction with the Advisers (*Consultores*), will propose to the Supreme Government a ternary list for the appointment of the Director of the School, the individuals named in the ternary list having the qualifications detailed in Article 13 of Title I of the Ordinances of Mining; it being understood that the Supreme Government may return the ternary list in order to have a new one formed, if that should be deemed expedient; the President of the *Junta* having in the votings the casting vote.

Art. 13. In the charge of the *Junta* of Encouragement shall be all the capital which produce the funds mentioned in this law, that arising from the quicksilver which may be sold, and the stock of said quicksilver on hand. The responsibility for everything aforesaid in its administration, preservation and safe-keeping shall be jointly in the members of the *Junta*. That of the management and distribution shall belong to an Auditor-Treasurer (*Contador Tesorero*), whom the Supreme Government shall appoint on the submission of a ternary list by the *Junta*, with a salary which shall be fixed in the regulation, and shall be paid out of the fund of the Establishment, he giving the bonds which shall be designated in the same regulation; it being obligatory upon the said Auditor-Treasurer to draw up and present the amounts of each year, and to be accountable for the quicksilver which the establishment may keep on hand. For the custody and security of the capital, there shall be an iron chest with four keys, each of the three Commissioners keeping one, and the Auditor-Treasurer the other. The regulation which the *Junta* shall form for the management of the capital, shall be founded upon these bases.

Art. 14. The *Junta* cannot expend the sums which may come into its coffers for any other purposes than those prescribed by the present law, or those for which it may obtain the previous authority of the Government. For the ordinary repairs of the building, or other extraordinary expenses, it shall only have power to expend without such permission a sum not to exceed $250 per annum.

Art. 15. The *Junta* will transmit to the Supreme Government monthly statements of its cash accounts, and each year it will transmit another of its general receipts and expenditures; the *Junta* being required to publish the first every four months, and the second every year at the proper time.

Art. 16. The *Junta* shall have power to settle, with the parties interested, the business which was left pending by the extinguished Tribunal of Mining, and which the Establishment may now have, these transactions being submitted for the approval of the Supreme Government.

Art. 17. There is granted to the same *Junta* the privilege that the Mines which the Establishment supplies at Tasco shall not be denounced during the period of two years, the Articles of the Ordinance of Mining relating thereto being suspended for that case.

Title II.

Of the quicksilver fund.

Art. 18. A fund shall be created destined for the acquisition of quicksilver, which the *Junta* shall manage, distributing it exclusively among the reducers of ores at cost and charges. Said fund shall bė formed from two-thirds of the increase of duty of import imposed on linens and fabrics of foreign cotton, according to the decree of this date.

Art. 19. The *Junta* may issue the quicksilver, with the corresponding guaranty and on the terms which may be established by the regulation, being authorized, for the effectual collection of what may be difficult, to have recourse, by itself or its agents, to the Judges of the Treasury, in order to compel those who contract to fulfill their engagements, and to coerce payment from those who refuse to make it.

Art. 20. Every four months the *Junta* shall publish a statement of distributions of quicksilver, expressing the quantities remitted to each Mining-place (*Mineral*), and the names of the reducers who shall have received it: and in case of any complaint of partiality or injustice in the distribution, the Supreme Government will decide: the said statement shall also include an account of the quantities purchased by the *Junta* and the cost thereof.

Art. 21. When the *Junta* shall have succeeded in collecting a fund equivalent to 24,000 quintals of quicksilver, it shall cease to apply to this object the part assigned of the products of the increase of duties on linen and fabrics of foreign cotton, according to the appropriate decree of this date.

Title III.

Of the endowment fund, and of redemption.

Art. 22. The proceeds of the so called "Real de Mineria," the active credits of the former Tribunal, and those of the Establishment which is reorganized by this law, shall continue liable to the charges and obligations which are designated by Article 7, of the Decree of May 20th, 1826, and which shall continue to be scrupulously complied with.

Art. 23. Every four months there shall be made by the *Junta* a partial redemption of the debts, with the amounts which shall have been collected, taking care as far as possible to reconcile the preference which some debts are justly entitled to, with the greater advantages which other creditors of said fund may offer.

Title IV.

Of the Administration of Justice in matters of Mining.

Art. 24. The Governors of Departments, in concert with the Departmental *Juntas* and with the previous approval of the Supreme Government, will establish in each of them the number of Courts of First Instance which are required within their limits.

Art. 25. Each Court shall be composed of three territorial Deputies elected in the manner which is prescribed in the old Ordinance of Mining; and of these three individuals, the first shall be the President of the Court, and the other two Associates (*Colegas*).

Art. 26. Each one of these Courts shall exercise, within its territory, the executive and economical powers given by the old Ordinance of the branch, (*ramo*) and in judicial matters they shall conform to the same Ordinance, so far as relates to the simplicity and brevity of their proceedings.

Art. 27. In contentious matters, in which the Tribunal may be under the necessity of consulting with some professional person (*letrado*), it shall consult with the Judge of First Instance of the respective *Partido.*

Art. 28. Each tribunal of First Instance shall elect a secretary, and the subordinates it may deem necessary for the despatch of the business of the secretary's office and of the court, designating the salaries which they are to receive, which it will report to its respective Departmental Governor, in order that, with his report, it may receive the determination which may be the pleasure of the Supreme Government.

Art. 29. From the final decisions of these courts, provided the amount in dispute does not exceed $500, there shall be no recourse of appeal, nor shall there be admitted any recourse of revision (*suplica*) when the decision in Second Instance shall conform to that in First Instance, and the amount in litigation does not exceed $2,000.

Art. 30. Besides the three individuals who are to compose the tribunal of First Instance, there shall be appointed three others, who shall serve as Advisers (*consultores*) in all executive matters in which the said Tribunal may desire to hear their opinion, and they shall supply the places of the regular Judges, in case of their inability to act, or of their being excepted to.

Art. 31. The Second and Third Instances which may take place in matters of Mining, and the extraordinary appeals (*recursos*) which may be made, shall be tried and determined in the Superior Tribunals of Justice in each Department respectively, and in the tribunals designated by the laws, or which may hereafter be designated.

Wherefore I command it to be printed, published, circulated, and duly observed.

Palace of the National Government, Mexico, December 2d, 1842.

NICOLAS BRAVO.

JOSE MARIA TORNEL,
Minister of War and Marine.

January 14th, 1843.

[From Lara's "Coleccion de Decretos y Ordenes," No. 574.]

Order permitting territorial Deputations to continue till the Courts of First Instance can be organized.

MINISTRY OF WAR AND MARINE.

This day I say to the President of the Junta of Encouragement and Administration of Mining as follows:

"His Excellency, the President Substitute, to whom I "gave an account of Your Excellency's official letter of "the 11th instant, and of the exposition which accompanied "it, and in which the Miners of Tasco petition that that "territorial Deputation may continue exercising its func- "tions until the Courts of First Instance, mentioned in

"Article 24 of the Law of the 2d of December last are "established; has been pleased to determine, in conformity "with the opinion of that Junta, that it may do so; this "disposition being extended to the other Departments in "which they were formerly established, for which purpose "let proper orders be communicated to the Most Excellent "Señores, the Governors thereof, and to Your Excellency I "say this in reply, for your information and other ends."

Communicated to the respective Governors.

January 31st, 1843.

[From Appendix to the Mexican "Ordenanzas de Mineria." Vide also Lara's "Coleccion de Decretos y Ordenes," No. 579.]

Decree explanatory of the Law of December 2d, 1842.

Nicolas Bravo, General of Division, *Benemerito* of the country, and President Substitute of the Mexican Republic to its inhabitants,—*know ye:*

That whereas the *Junta* of Encouragement and Administration of Mining has expressed some doubts respecting the interpretation of the law of December 2d, 1842, which gave a new organization to the Establishment, and whereas it is important to solve them, in exercise of the powers which were conferred on me by the seventh of the Bases established at Tacubaya and sworn to by the representatives of the Departments, I have been pleased to decree what is contained in the following articles:

Art. 1. In Article 7 of the said law, in place of 1844, read 1845, that having been a typographical error.

Art. 2. The *Juntas* of the creditors and of the Miners will meet immediately, under the presidency of the Governor of the Department, in order to each elect a substitute for the present agents, and this shall be also done in future when there is an election of principals.

Art. 3. The substitute agents (*apoderados suplentes*) will receive full pay when the principal agents are absent fifteen days for their own convenience, but nothing when this absence is caused by the sickness of the principal.

Wherefore I order it to be printed, published, circulated, and duly observed.

Palace of the General Government in Mexico,
January 31st, 1843.

NICOLAS BRAVO.

JOSE MARIA TORNEL,
Minister of War and Marine.

February 11th, 1843.

[From Appendix to the Mexican "Ordenanzas de Mineria." Vide also Lara's "Coleccion de Decretos y Ordenes," No. 588.]

Decree for carrying into effect Article 28 *of the Law of December* 2*d*, 1842.

Nicolas Bravo, General of Division, *Benemerito* of the country, and President Substitute of the Mexican Republic, to its inhabitants,—*know ye:*

That in conformity with what is represented by the *Junta* of Encouragement and Administration of Mining, in exercise of the powers conferred on me by the seventh of the Bases established at Tacubaya and sworn to by the representatives of the Departments, and for the purpose of expediting the fulfillment of Article 28 of the Law of December 2d, 1842, I have been pleased to decree what is contained in the following Articles:

Art. 1. Each Court of First Instance which shall be established agreeably to the provision of Article 24 of the Decree of December 2d, 1842, will form, hearing the Miners of the district (*comprension*), the tariff of duties which shall be charged by the territorial Deputies who compose it and by its respective secretary.

Art. 2. When the tariff mentioned in the preceding Article has been formed, they will transmit it to the Departmental Government, which, with its report, will direct it to the Supreme Government of the Republic through the *Junta* of Encouragement and Administration of Mining, in order that it may receive the proper resolution.

Art. 3. In the Mining establishments (*Minerales*) where there is a Notary Public, he shall be elected secretary of the Court of First Instance.

Art. 4. The specification of the duties to be charged, being made in the tariff referred to in Article 1st of this decree, it will be a full compliance with what is provided, with respect to pay, in Article 28th of the organic law.

Wherefore, I order it to be printed, published, circulated and duly observed.

Palace of the General Government in Mexico, February 11th, 1843.

NICOLAS BRAVO.

JOSE MARIA TORNEL,
Minister of War and Marine.

February 17th, 1843.

From Appendix to the Mexican "Ordenanzas de Mineria." Vide also Lara's "Coleccion de Decretos y Ordenes," No. 592.

Decree authorizing the Mining Junta to contract a loan of two millions.

Nicolas Bravo, General of Division, *Benemerito* of the country and President substitute of the Mexican Republic to its inhabitants,—*know ye*:—

That, whereas the high price at which quicksilver is now purchased is injurious to Mining,—the most important branch of our wealth; and it is therefore the duty of the Government to provide for the procurement of an article of so much interest, at the lowest possible price: whereas there are reasons to hope that the Government of S. M. C. may enter into a negotiation by which the interests of both countries will be provided for; whereas the law of December 2d, 1842, has created a fund for this identical object; and whereas in fine the *Junta* of Encouragement and Administration of the Branch of Mining, is worthy of the confidence of the Government and of that of its constituents:—therefore in exercise of the powers conferred on me by the seventh of the Bases established at Tacubaya and sworn to by the Representatives, I have been pleased to decree what is contained in the following Articles:

Art. 1. The *Junta* of Encouragement and Administration of the Branch of Mining is authorized to contract a loan of two millions of dollars, in the Republic or out of it, on the lowest possible terms, for the purpose of procuring the contract of the quicksilver of the Mine of Almaden.

Art. 2. This contract is subject to the approval of the Supreme Government.

Art. 3. The said *Junta* is authorized to enter into negotiations with the Spanish Government for a contract of the said mineral to the number of quintals mentioned in the said law of December 2d, 1842.

Art. 4. There is set apart for the security of the contract, the fund which the same law created, and also two per cent. of the three which on the value of the gold and silver is imposed as a contribution in Article 6 of the law of February 20th, 1822, which shall go into the funds of the Branch of Mining as soon as this contract of quicksilver is entered into, with the condition of repaying the public Treasury with the same products which are derived from its sale to the Miners.

Art. 5. The contract which the *Junta* of Encouragement and Administration of the Branch of Mining may form with the Government of S. M. C. for the acquisition of quicksilver, shall be submitted to the approval of the Government of the Republic.

Wherefore I order it to be printed, published, circulated and duly observed.

Palace of the National Government,
Mexico, February 17th, 1843.

NICOLAS BRAVO.

JOSE MARIA TORNEL,
Minister of War and Marine.

April 7th, 1843.

[From Appendix to the Mexican "Ordenanzas de Mineria." Vide also Lara's "Coleccion de Decretos y Ordenes," No. 639.]

Decree assigning certain duties on foreign goods to the benefit of Mining.

Antonio Lopez de Santa Anna, General of Division, *Benemerito* of the country, and Provisional President of the Mexican Republic, to the inhabitants thereof,—*know ye:*

That in the exercise of the powers conferred on me by the seventh of the Bases established at Tacubaya and sworn to by the representatives of the Departments, I have been pleased to decree the following:

Art. 1. During the war which the Nation is now carrying on against the revolters of Texas and the non-conformists of Yucatan, there will be an increase of twenty per cent. on the duties on imports which are now levied by the Tariff of April 30th of last year.

Art. 2. Linens and cotton fabrics, white, brown, and colored, mentioned in the Decree of the 2d of December last, will only pay the duty (*cuota*) which is there set apart, for the encouragement of the Branches of Mining and Industry, the collection of the said *cuota* being made from the date which is specified by the subsequent Decree of the 24th of the same December.

Art. 3. The increase spoken of in Article 1 will take effect four months after the publication of this Decree in the Capital of the Republic, with respect to the cargoes which shall arrive in the ports of the Gulf of Mexico, and six months, for those which shall arrive in the ports of the South Sea, the Gulf of Californias, and the sea of Upper California.

Art. 4. Both the amount of the increase of twenty per cent. directed by Article 1, and the difference which there may be between the *cuotas* which the Tariff imposes on linens and cotton fabrics, and those which the said Decree of the 2d of December last designates, shall be satisfied in bills of exchange payable on time, as directed in the Tariff for duties on imports, drawn in favor of the General Treasury, where they will be kept as a special deposit at the disposition of the Supreme Government.

Art. 5. The Maritime Custom Houses for the fulfillment of what is prescribed, will increase the amount of duties as shown in the accounts of issue (*hojas del despacho*), according to the said Tariff, by the twenty per cent. which the first Article of this Decree establishes; and for the collection of the difference which results between the *cuotas* of said Tariff and those of the said Decree of December 2d with respect to linens and fabrics of cotton, they will place it in a separate column.

Art. 6. The collection of the duty of one per cent. on imports mentioned in the law of the 31st of March, 1838, as well as that of the two per cent. of the duty of average (*averia*), shall be made in proportion to the increase made by this Decree and that of the 2d of December of last year.

Art. 7. The Direction General of excise duty (*alcabalas*) and direct contributions, will issue the proper orders for the levy of the duties of consumption in the Maritime Custom-Houses, according to the import duties of foreign goods, and in the land Custom-Houses, on their circulation, it will be governed by the dispositions contained in this Decree.

Wherefore I order it to be printed, published, circulated and duly observed.

Given at Tacubaya, April 7th, 1843.

ANTONIO LOPEZ DE SANTA ANNA.

IGNACIO TRIGUEROS,
Minister of Treasury.

May 23d, 1843.

[From Lara's "Coleccion de Decretos y Ordenes," No. 657.]

Patent to Don Joaquin Martinez for the exclusive use of his apparatus for distilling quicksilver.

Antonio Lopez de Santa Anna, &c., &c., *know ye*:

That desiring to recompense useful discoveries which are made in the working of quicksilver, as well as to give to this branch, so very interesting to Mining, due protection, exercising the powers which are conferred on me by the seventh of the Bases established at Tacubaya and sanctioned by the nation, I have been pleased to decree as follows:

There is granted to Don Joaquin Martinez the exclusive privilege, for ten years, to alone construct the apparatus for distillation of quicksilver, like that which he has established in the mountain of Tapalpa in the Department of Jalisco.

Wherefore I order it to be printed, published, circulated and duly observed.

May 24th, 1843.

[From Appendix to the Mexican "Ordenanzas de Mineria." Vide also Lara's "Coleccion de Decretos y Ordenes," No. 659.]

Decree conceding certain exemptions to quicksilver.

Antonio Lopez de Santa-Anna, General of Division, *Benemerito* of the country, and Provisional President of the Mexican Republic, to the inhabitants thereof,—*know ye:*

That, in accordance with my intentions to encourage whatever may contribute to the national aggrandizement and wealth, and considering as one of the means most suitable for that purpose, the granting of rewards and exemptions to the important branch of Mines of quicksilver, so necessary for the reduction of the precious metals, the most important branch of the industry of the Republic, without which the others can make no progress; having heard the report of the *Junta* of Encouragement of Mining;—in exercise of the powers granted to me by the seventh of the Bases established in Tacubaya and sanctioned by the Nation, I have been pleased to decree the following:

Art. 1. The Royal Orders of January 13th, 1783, of November 12th, 1791, of December 6th, 1796, and of August 8th, 1814, with respect to exemption from excise duties (*alcabala*), granted to articles of consumption in Mining, will be punctually observed with respect to Mines of quicksilver in the Republic.

Art. 2. No general or municipal import shall be levied upon quicksilver extracted from the Mines (*criaderos*) of the Republic.

Art. 3. Quicksilver may be sold throughout the Nation without permits, passes or other Custom-House papers.

Art. 4. A premium of $25,000 is granted to each one of the first four operators (*empresarios*) who shall extract in one year, from the Mines of the Republic, 2,000 quintals of liquid quicksilver.

Art. 5. There shall be paid, for the term of three years, for each quintal of quicksilver, which shall be derived from the aforesaid source, the sum of $5.

Art. 6. The operatives in Mines of quicksilver are exempt from all military service and from personal taxes.

Art. 7. The *Junta* of Encouragement and Administration of Mining, will draw up the proper Regulation for the distribution of the said premiums, they being paid, when due, from the fund which is designated by the 2d Article of the decree of December 2d, 1842, and Article 4th of that of February 17th of this year.

Wherefore, I order it printed, published, circulated and duly observed.

Palace of the National Government in Tacubaya, May 24th, 1843.

ANTONIO LOPEZ DE SANTA ANNA.

JOSE MARIA DE BOCANEGRA,
Minister of Exterior Relations and Government.

July 5th, 1843.

[From Appendix to the Mexican "Ordenanzas de Mineria." Vide also Lara's "Coleccion de Decretos y Ordenes," No 698.]

Decree empowering the Mining Junta to work, supply and protect Mines of quicksilver in the Republic.

Antonio Lopez de Santa-Anna, General of Division, *Benemerito* of the Country and provisional President of the Mexican Republic, to its Inhabitants makes known: That desiring to encourage and give a greater impulse to the important branch of Mining, which forms the principal wealth of the Republic, in use of the power which is given to me by the 7th of the Bases agreed upon in this city and sanctioned by the Nation, I have been pleased to decree what is contained in the following Articles:

Art. 1. The *Junta* of Encouragement and Administration of Mining is empowered to work, to contract for supplying, and to protect Mines of quicksilver in the Republic.

Art. 2. The amounts which the *Junta* may advance, with said object, to those who undertake the working of Mines of quicksilver, besides being secured to its satisfaction, will bear an annual interest of five per cent. which will go to the funds of that branch.

Art. 3. The *Junta* will not contract for the supply of any Mine, without obtaining the necessary data which prove, as far as possible, the goodness of the *negociacion*.

Art. 4. The *Junta* will not work, on account of the funds which it administers, Mines which do not offer known advantages, or, at least, the preservation of the capital which is to be invested in them, and the interest which is exacted of those which are supplied by it.

Art. 5. The *Junta* will receive, at the price which may be agreed upon with the parties interested (and which cannot exceed the market price) in payment of the respective capitals and interests, liquid quicksilver, and will distribute it among the Mines of gold and silver, in a proportional manner and according to the provisions of the Regulation which, for the useful distribution of this fund, it will draw up and pass to the Government for its approbation.

Art. 6. The *Junta* will establish in the Mining departments, depôts of liquid quicksilver, and will distribute it as is provided in the foregoing Article.

Art. 7. The said *Junta* will purchase and construct, on account of the funds which it administers, the flasks required to hold the quicksilver, and will distribute them in the Mining departments, in order that it may be transported with security.

Art. 8. The *Junta* is empowered to send intelligent persons to search for good Mines (*criaderos*) of cinnabar, to cause those already discovered to be examined, and to dictate such partial measures as experience may suggest, in order that the production of quicksilver in the Republic may be efficaciously encouraged.

Art. 9. In order that the *Junta* may accomplish the object of this law and that of the 24th of May last, it will use the funds which are designated by Article 2 of the law of the 2d of December of last year, and by Art. 4 of the law of the 17th of February of this year.

Art. 10. The said funds are immediately at the disposition of the *Junta* of Encouragement and Administration of Mining, those respectively charged with their collection, delivering them to it on its order.

Wherefore I order it to be printed, published, circulated and duly complied with.

Palace of the National Government at Tacubaya, July 5th, 1843.

ANTONIO LOPEZ DE SANTA ANNA.

JOSE MARIA DE BOCANEGRA,
Minister of Exterior Relations and Government.

July 8th, 1843.

[From Lara's "Coleccion de Decretos y Ordenes," No. 703.]

Order respecting the raising of funds in the Department of Jalisco for the working of Quicksilver Mines.

MINISTRY OF TREASURY.

The Most Excellent Señor President Provisional of the Republic, following the principles which he has always manifested, of inciting and promoting all the elements of the wealth and power which the nation possesses and which are calculated to elevate it; appreciating at its true value and importance the encouragement which the working of quicksilver requires, and especially when the high price at which it is now purchased makes some enterprises very expensive and renders others useless, causing an obstruction to Mining, placing it at the will of those who, on a change of increased advantages, may desire to profit by it;—finally, in order to emancipate it from this grievous pupilage, His Excellency being convinced that a generous and voluntary national impulse will be sufficient to extract the quicksilver necessary for the reduction of the large existing quantities of ores, the wealth of which is paralized; the First Magistrate of the Republic has directed that Your Excellency, with the wisdom and prudence corresponding, open in that Department a subscription, by which a gracious gift or voluntary loan may procure the necessary fund.

Your Excellency will proceed to form a *Junta*, composed of the owners of quicksilver Mines, and to regulate it as you may deem expedient, in order that it may organize its treasury, and be charged with receiving all the sums or donations which are destined for the encouragement of the undertaking (*empresa*), and those which shall be collected by taxes for this object in the Department, and in order also that the proper office may repay the loans when the resources of the establishment shall be sufficient for that purpose, and satisfy the lenders, or the person or persons with whom they have to arrange to make the delivery, and the mode and terms in which they are to be repaid.

His Excellency also directs that, at the head of the citizens who may wish to subscribe, his name be placed, with the sum of $5,000, thus presenting an example which he wishes others to imitate for the public good. He confides to the zeal of your Excellency the care of carrying into effect the desires of the Supreme Government, because on your activity it will depend whether the Mines of quicksilver of that Department receive the desired impulse, and thence the advantages which are reasonably expected. His Excellency desires to know the advances which this project may make, and while, omitting to overvalue, to that Government, the zeal of the Chief Executive, I must not fail to remind it, that upon the prompt organization of the subscription, and upon the removal of obstacles which are presented, will depend, in a great measure, the success of the undertaking (*empresa*).

July 11th, 1843.

[From Appendix to the Mexican "Ordenanzas de Mineria." Vide also Lara's "Coleccion de Decretos y Ordenes," No. 708.]

Regulation respecting the transportation, to Mining places, of goods exempt from duty.

Antonio Lopez de Santa Anna, &c., &c. I have been pleased to decree the following:

* * * * * * * *

Art. 32. The favors granted to Mining by the laws in force, will continue; but in order that the respective exemptions from duties may take place, the goods must be transported with a permit (*guia*) and responsive bond; and besides, the political authority of the Mining place (*Mineral*) to which they are carried shall give a certificate to the conductor or consignee which shall show that the goods were introduced into the Mine to which they were consigned, without which proof, the *tornaguia* shall not be issued till the duties are first paid. When all or a part of the goods are sold on the road, or do not arrive so as to be introduced into the Mining places (*Minerales*), the duties which correspond will be required, and the Administrators will take care to collect them in their executive capacities.

July 14th, 1843.

[From Lara's "Coleccion de Decretos y Ordenes," No. 717.]

Decree authorizing the Departmental Junta of Jalisco to make a loan for the encouragement of Mines of quicksilver.

Antonio Lopez de Santa Anna, &c., &c., *know ye:*

That, desirous to promote the national prosperity, by encouraging those Branches which, with the greatest facility and promptitude, should produce this beneficial result, and convinced that one of the most important is the Mines (*Criaderos*) of quicksilver, determined to extend to it all protection that may be possible, by uniting the co-operation of the authorities with the efforts of individuals, in exercise of the powers which are conferred on me by the seventh of the Bases established in this city and sanctioned by the nation, I have been pleased to decree what is contained in the following Articles:

Art. 1. The Departmental *Junta* of Jalisco is authorized to raise in that Department a loan, and in case it cannot be negotiated, to impose an impost upon the Branches which it may consider most proper, to produce the sum of $100,000, within the period which it may designate and which may be most expedient for the object.

Art. 2. The $100,000 mentioned in the preceding Article must be devoted to the encouragement of Mines of quicksilver in that Department.

Art. 3. The said Departmental *Junta*, fifteen days after having received the present Decree, will regulate the distribution of the loan, and, in case of the impost mentioned, the mode which may make effective the collection and application of said sum.

Art. 4. The Governor of the Department of Jalisco will form a *Junta*, over which he will preside, composed of the owners of Mines of quicksilver, and will regulate it as may be most expedient, creating and organizing its treasury in such a manner that it shall have charge of receiving all the amounts which this law determines, and those which shall be collected by donation, as well as their reimbursements, when the resources of the establishments (*negociaciones*) which shall be assisted by this means, shall be sufficient for the purpose.

Art. 5. These concessions which are made to the Department of Jalisco, are extended to all the Departments which shall work Mines of quicksilver.

Wherefore, &c.

July 28th, 1843.

[From Appendix to the Mexican "Ordinanzas de Mineria." Vide also Lara's "Coleccion de Decretos y Ordenes," No. 726.]

Circular respecting the payment of the premium of $5 per quintal of quicksilver extracted from Mines of the Republic.

MINISTRY OF EXTERIOR RELATIONS AND GOVERNMENT.

Most Excellent Señor:

In conformity with the Report of the *Junta* of Encouragement and Administration of Mining, the Most Excellent Señor President Provisional of the Republic has been pleased to decree, that every individual who shall extract any quicksiver from the *criaderos* or Mines of this production, and shall desire to avail himself of the benefit of the premium of $5 per quintal granted by law, will prove before the Government of the respective Department the fact of having extracted the quantity for which he may demand premium, presenting, for that purpose, certificates of the Justice of the Peace, of the Parish Curate and of the respective Mining Court, or its Delegate, which proof shall be remitted, for the corresponding ends, to the said *Junta* of Encouragement. And I communicate to Your Excellency this supreme order, so that you may be so good as to give it due publicity.

Communicated to the Governors of the Departments.

August 24th, 1843.

[From Lara's "Coleccion de Decretos y Ordines," No. 756.]

Explanation of the Decree of July 14*th*, 1843.

Antonio Lopez de Santa-Anna, &c., &c., *know ye:* That doubts having arisen, whether the loan which the Junta Departmental of Jalisco was authorized to raise by Article 1 of the Decree of the 14th of July last is voluntary or forced, I have been pleased to declare as follows:

The loan of $100,000 which, for the encouragement of Mines of quicksilver in the Department of Jalisco, was established by Article 1 of the Decree of the 14th of July last, is voluntary; and with this understanding they will proceed to its collection, and to carry into effect the other provisions of the said Decree.

September 6th, 1843.

[From Lara's "Coleccion de Decretos y Ordenes," No. 768.]

Circular respecting the establishment of Mining Courts.

MINISTRY OF EXTERIOR RELATIONS AND GOVERNMENT.

It being very expedient that, with the greatest brevity, that Government, in concert with the Departmental Assembly, proceed to designate the places in which there are to be established Courts of First Instance of Mining, which are mentioned in Article 24 of the law of the 2d of December last; His Excellency, the Provisional President of the Republic, has been pleased to direct your Excellency to proceed immediately to advise the Supreme Executive of the designation of places in which they are to be established in that Department, giving, as reasons, the reports which, on this subject, that Government may make, in concert with the said Assembly, and with the understanding that as soon as they are established, they themselves will designate the district of their respective jurisdiction, specifying the names of the *Asientos* of Mines which they may include.

September 25th, 1843.

[From Appendix to the Mexican "Ordenanzas de Mineria." Vide also Lara's "Coleccion de Decretos y Ordenes," No. 786.]

Decree to encourage the discovery and working of Mines of Quicksilver.

Antonio Lopez de Santa Anna, General of Division, *Benemerito* of the country, and Provisional President of the Mexican Republic, to the inhabitants thereof,—*know ye:*

That, desiring to render effective the benefits which the Government proposed to confer on Mining, in the authority conferred upon the *Junta* of encouragement of that Branch, by Decree of the 5th of July last, to assist and encourage the working of *criaderos* of quicksilver, I have decreed the following:

Art. 1. The *Junta* of encouragement of Mining will appoint one Commission at least, in each Department of the Republic, to explore and examine all the *criaderos* of cinnabar which may be there.

Art. 2. The examination which these Commissioners are to make shall be scientific, and moreover they shall be charged to report on the following points: 1st, whether in the respective Departments there are, or have been, Mines of quicksilver which are now worked, or which have formerly been worked; 2d, what is their present condition; 3d, what ones are most susceptible of being worked; 4th, what works are necessary to put them in operation, and the expense required to put them in order; the *ley* which they have, and the produce which is known ; 6th, the cost of its extraction and reduction.

Art. 3. The *Junta* of encouragement, in view of the said reports, will determine what places shall be assisted in preference, and the sum with which they may be supplied.

Art. 4. Within six months from this date, the said examinations must be concluded, and within seven months from the same date, the supplies (*avios*) of the Mines shall be decreed, it being permitted to grant, before that time, some assistance (*habilitaciones*) to Mines which notoriously deserve them.

Art. 5. Out of the funds which are set apart for supplies (*avios*) of Mines of quicksilver, and out of those which this decree designates, the assistance (*habilitaciones*) mentioned in the preceding Articles shall be taken.

Art. 6. In furnishing the supplies mentioned in this decree, the *Junta* will adopt one of two modes ; 1st, to furnish the necessary money, in the character of a loan, at an interest of six per cent. per annum ; 2d, to constitute itself a supplier (*aviador*), sharing in the losses and gains, the same as in ordinary contracts of supplies (*avios*).

Art. 7. When money is advanced on interest, security must be given to the following effect ; that the money be returned within the term which is agreed upon ; that the capital and interest be secured with guaranties to the satisfaction of the Establishment ; that it be invested necessarily and exclusively in the business (*negociacion*) contracted for, and for this purpose an inspector may be appointed by the Establishment and paid by the owner of the Mine ; and that these loans shall only be made in favor of operations (*negociaciones*) which the said Establishment has ascertained to be worthy of being supplied, according to the examinations which this Decree prescribes.

Art. 8. If the supply (*avio*) shall be made, by the Establishment becoming itself the supplier (*aviador*), the following rules will be observed: 1st, That the supply be made to a Mine which is worth working, as shown by the examinations which are directed by this Decree; 2d, That the sum which is to be furnished, be regulated according to the estimates formed by the Commission which may have examined the Mine; 3d, That it be stipulated that one-half at least, of the profits shall belong to the supplier (*aviador*); 4th, That the exclusive direction shall be under the charge of the supplier (*aviador*), with the right on the part of the owner of the Mine to place there an inspector; 5th, That every four months there shall be a settlement of accounts and a division of profits, if any there be; 6th, That the Establishment, under its responsibility, shall have the making out (*glosa*) of the accounts; and 7th, That the profits which result shall be first applied, as a sinking fund, to pay the capital supplied, and, until this has been entirely provided for, there shall be no division of profits among the parties.

Art. 9. The Establishment will draw up a Regulation for supplies (*avios*), on the basis of the two preceding Articles, submitting it for the approval of the Government.

Art. 10. The funds which are set apart for the supplies (*avios*) which may be decreed, are, 1st, The one per cent. duty laid on coin which is carried from one Department to another; 2d, The $130,000 which have been decided to belong to the Mining out of the fund created by the Decree of the 2d of December last.

Art. 11. The one per cent. duty shall be collected by the Establishment of Mining, for which purpose it may appoint and distribute the Commissioners, as shall be deemed expedient.

Art. 12. The sum of $130,000 shall be paid by the maritime Custom Houses of Vera Cruz and Tampico, the first furnishing $80,000 per annum, and the second $50,000, payable by monthly instalments, which shall be remitted in bills of exchange in favor of the *Junta* of Encouragement.

Art. 13. The *Junta* shall apply, of the said funds destined to the working of Mines of quicksilver, the $15,000 which were set apart by the Decree of August 18th of this year, to the endowment and annual expense of the Seminary of Mining.

Wherefore, I order it to be printed, published, circulated and duly observed.

National Palace at Tacubaya, September 25th, 1843.

ANTONIO LOPEZ DE SANTA ANNA.

MANUEL BARANDA,
Minister of Justice and Public Instruction.

October 3d, 1843.

[From Lara's "Coleccion de Decretos y Ordenes," No. 812.]

Decree regulating the organization of the College of Mining.

Antonio Lopez de Santa Anna, &c., &c.—*know ye:* That in view of the project of a decree proposed by the Director of the College of Mining, in accordance with the provision of Article 9 of the law of the 18th of August last,* I have been pleased, exercising the powers with which I have been invested by the Nation to decree as follows:

Art. 1. In the College of Mining there shall be studied the professions (*carreras*) of Land-Surveyor (*Agrimensor*,) Assayer, Refiner (*Apartador*) of gold and silver, Reducer of ores, Engineer of Mines, Geographer and Naturalist.

Art. 2. The preparatory studies for these professions shall be as follows: In the first year, logic, idealogy, Spanish grammar, and drawing: In the second year, pure mathematics in the branches of arithmetic, elementary geometry, plain trigonometry, algebra through equations of the second degree, the French language, and drawing: In the third year, geometry, analytical and descriptive, its application to exterior and interior measurements, the theory of perspective and shadows of bodies, stereometry, spherical trigonometry, the general principles of infinitesimal calculus, the French language, and drawing.

Art. 3. For the profession of Land Surveyor (*Agrimensor*) four years of study are required: the three years of preparatory studies mentioned in Article 2, and in the fourth year that of the following subjects:—elements of rational mechanics, theory of caloric, electricity and magnetism; elements of optics, acoustics and meteorology; the English language and delineation. The students shall practice for this profession in the most convenient place, under the direction of the respective professor of mathematics, and at the end of the third year.

* The law of August 18th, 1843, is published at length in Lara's "Coleccion," No. 753. Article 9 is as follows:

"The study of the Natural Sciences shall be organized, as well in the duration and matters of the preparatory studies, as in the division into two or three different professions (*carreras*), and the duration of the course of instruction, by a special Decree, which the Director of the College of Mining shall propose to the Government."

Art. 4. The profession of Assayer shall require a course of study of five and a half years: in the first four the subjects prescribed in Articles 2 and 3; in the fifth, the elements of general chemistry and the application of inorganic chemistry to the Art of Assaying and metallurgy; including as much as possible the national and foreign practical methods; delineation and English. The practice for this profession shall take place, during the remaining half year, in the Assay office of this city and in the laboratory of the College, being devoted to chemical analysis.

Art. 5. The course of study for the profession of Refiner or Separator (*Apartador*) of gold and silver shall continue six years, five and a half the same as for that of Assayer, and the remaining half year, practicing in some office of Refiner.

Art. 6. The profession of Reducer of ores shall require seven years: the first three, in the preparatory studies; the fourth and fifth, in those prescribed in Articles 3 and 4; the sixth and seventh in practice, which shall be six months in the laboratory of the college, and a year and a half in Guanajuato, where there shall be established a school of practice.

Art. 7. The Engineer of Mines shall pursue a course of nine years: the first five, in the subjects designated for the profession of Reducer; the sixth shall be employed in the study of mineralogy, geology, the working (*explotacion*) of Mines, and the German language; the seventh, eighth and ninth in practice, divided as follows:—the first half year to mechanics applied to Mining and chemical analysis in the college; the next year and a half in the School of Guanajuato; and the last year in some other well known Mine, during which time, he shall devote himself to metallurgical construction, in which he shall be examined.

Art. 8. The Geographer shall require eight years course of study; the first four the same as for a land surveyor;

the fifth and sixth in cosmography, geodesy, uranography, and geography; the last two, in practice with the Geographical engineers of the Government, in the class of Adjuncts to the commissions which the said officers may perform.

Art. 9. The Naturalist shall have seven years course of study: the first six the same as those required for the Engineer of Mines; and the seventh shall be employed in the study of botany and zoology.

Art. 10. The examinations for the preparatory studies, shall be made by the respective professors. That of Assayer shall be made by the Chief Assayer and the Professors of Chemistry and Physics. That of Refiner of gold and silver, by the Chief Refiner and the Professors of Chemistry and Physics, it being requisite, to be received in this profession, that the individual who pretends to it shall have been approved as Assayer, or that, at the same time, he be examined in both professions, in which case there shall be associated with the examiners, the Chief Assayer and one of the Professors of Mathematics. That of Reducer of Ores, by the Professors of Chemistry, Physics, and one of those of Mathematics. That of Engineer of Mines, by the Professors of Geology, of Mineralogy and Working of Mines, of Chemistry, of Physics, and one of Mathematics. That of Geographer, by the Professors of Geography, Geodesy, Physics, and two of Mathematics. That of Naturalist, by the Professors of Mineralogy, Zoology, and Botany.

Art. 11. The Diplomas of all these professions shall be issued by the Director of the College, as President of the *Junta facultativa*, and shall be certified with the signature of the Secretary of the said *Junta*.

Art. 12. The Director of the College will exercise his

functions in civil, moral and religious matters, by means of the Rector chaplain: in scientific matters, by the *Junta facultativa*, and in economical matters, by the Steward (*Mayordomo*).

Art. 13. In case of the resignation or death of the Director, the *Junta* of Encouragement and Administration of Mining will immediately elect one *ad interim*, and, within a month, will make to the Government the corresponding ternary nomination. In his default or temporary absence, which may not exceed a month, the Director will appoint one of the Professors in his place; but if it shall exceed that time by obtaining the permission of the Supreme Government, the said *Junta* will make the appointment.

Art. 14. The *Junta facultativa* will be composed of the Director who will preside, and of four Professors, who shall be annually elected by all of them, on the day when the College courses are opened. It will appoint one of its members as secretary.

Art. 15. For the aforesaid course of studies, the professorships will continue as they now are in the College; but that of cosmography and delineation shall be divided into two, with the following changes—one of cosmography, geodesy and uranography, and the other of delineation.

Art. 16. There shall be established professorships of Logic and Ideology, of Geology, of Geography, of Zoology, and of the German language.

Art. 17. There shall be a professor who shall direct the study of the application of mechanics to Mining, and another for chemical analysis.

Art. 18. There shall also be a Prefect of studies, and two substitute assistants of professorships (*de cátedras*).

Art. 19. The new employés shall be appointed by the

Government, on nomination, this time, of the Director, and in future, all of them by competition.

Art. 20. The present Professors of the College may change, this once, their respective professorships with the approval of the Director, and each one may have more than one chair, provided that they shall serve in them the hours which shall be fixed in the Regulation, and with the modifications which this law establishes; but, in the last case, they shall not receive out of the fund of Mining any more than that which results from the endowments of the united chairs (*cátedras.*)

Art. 21. Those who remain only with the chairs which they now have, will keep their old allowance, even though it may be less than that given in the new Regulation.

Art. 22. The office of Vice-rector is abolished, and he who now holds that employment shall receive the half of his salary, it being deducted from the salary which is assigned to the Prefect of studies.

Art. 23. In place of the twenty-five students of endowment, there will hereafter be supported ten of this class, and thirty of half-endowment; but these shall be supported only while they remain in the College in this city. This provision will take effect as there may be vacancies in the present places.

Art. 24. In future no student of endowment shall be admitted under the age of 16 years.

Art. 25. As the expenses of the college are increased by this Decree, and in order that they may be provided for, there will be appropriated from the fund of quicksilver $2,000 monthly, instead of the $15,000 assigned in the law of August 18th of this year.

Art. 26. The funds mentioned in Articles 82 and 86* of the said law, shall be distributed in the following manner:

Salary of the Chair of Spanish Grammar, ideology and logic,	$1,000
Salary of the Chair of French	600
" " " " English	600
" " " " German	600
" " " " Drawing	700
" " " " Delineation	600
" " " " 1st course of Mathematics	1,200
" " " " 2d " " "	1,200
" " " " Physics	1,500
" " " " Mechanics applied to Mining	600
" " " " Chemistry	1,500
Salary of the Professor of Analysis in the Laboratory	500
Salary of the Chair of Mineralogy and Working of Mines	1,500
Salary of the Chair of Geology	1,300
" " " " Cosmography, Geodesy and Uranography	1,200
Salary of the Chair of Geography	600
" " " " Botany	1,200
" " " " Zoology	1,200
Salary of the Director of the Museum	1,200

* EXTRACTS FROM THE LAW OF AUGUST 18th, 1843.

"Art. 82. What is said of funds and endowments does not apply to the College of Mining, to which, in addition to what it already has, are assigned $15,000 annually, out of the fund created for quicksilver by the law of December 2d, 1842, and which shall be first paid.

"Art. 86. The National Museum and Cabinet of Natural History and the Chair of Botany which now exist, shall be annexed to the College of Mining, and shall form a part of its establishment, for which end there shall be transferred to the said College the appropriations which the said establishments now have."

Decree of October 3d, 1843.

Salary of the Prefect of Studies	$ 1,200
" " Rector and Chaplain	800
" " two substitutes, $500 each	1,000
" " Steward	800
" " Porter—Dispenser	400
" " Cook	216
" " five Servants	960
" " ten Students @ $300	3,000
" " thirty of half endowment @ $150	4,500
" " two for Rector and Prefect	300
" " *Semanarios* of ten Students	130
Pay for mending and washing	120
" " Public exhibitions	1,000
" " practice of the Students, estimated,	2,000
" " books, instruments, laboratory and printing	3,000
Pay for lights and extras	1,574
Assigned to Museum and Cabinet of Natural History	3,000
Drawing materials and stationery	1,400
Difference of salaries of present Professors	2,300
Clerk *Archivero* of the direction and *Junta facultativa*	300
Total	$46,800

Art. 27. The Director will draw up the Regulation of the College and submit it to the Government, for its approval, with the understanding that it shall go into effect, January 1st of the coming year.

December 26th, 1843.

[From Appendix to the Mexican "Ordenanzas de Mineria." Vide also Lara's "Coleccion de Decretos y Ordenes," No. 865.]

Circular, respecting the organization of courts of Mining.

MINISTRY OF JUSTICE AND PUBLIC INSTRUCTION.

Most Excellent Señor:

Under date of the 22d instant, this Ministry said to the Government of the Department of Guanajuato that which I copy:

"Most Excellent Señor:

"In view of the advice which that government was pleased to ask, in note of the 9th of November last, with respect to the mode in which the periodic renewal of the members (*Vocales*) of that Court of Mining is to be made, His Excellency the President *ad interim* of the Republic has been pleased to determine in council of Ministers, that this year there will only be renewed the President of said Court (*Jusgado*), leaving the Associates, and that next year these will be changed, leaving the President. Which I have the honor to communicate to your Excellency for the effects corresponding."

And His Excellency having resolved that this instruction (*providencia*) be observed as a general rule by all the Courts of Mining of the Republic, I have the honor to communicate it to your Excellency for your information, and the corresponding effects in the Department under your command.

Communicated to the Departmental Governments. }

December 30th, 1843.

[From Lara's "Coleccion de Decretos y Ordenes," No. 880.]

Decree regulating the officers, &c., of the Junta of Encouragement and Administration of Mining.

Valentin Canalizo, &c., &c.—*know ye:*

That, having examined the Regulation proposed by the *Junta* of Encouragement and Administration of Mining, agreeably to the provisions of Article 9 of the law of December 2d, 1842, I have been pleased to decree, in cabinet meeting, as follows:

Art. 1. For the performance of its functions, the Junta will have an office, divided into sections—of secretary, auditor and treasury (*secretaria, contaduria y tesoreria*). The first shall consist of a secretary, a first, and a second officer (*oficial*), and two scriveners (*escribientes*). The second, of an auditor, a first and a second *oficial* and one scrivener. And the third of a treasurer and cashier (*contador de moneda*). Besides, the office shall have a keeper of archives (*archivero*), and an office servant.

Art. 2. The salaries of these officers shall be as follows:

Secretary	$2,500
First *oficial de Secretaria*	1,200
Second " "	800
Archivero	700
First *escribiente de Secretaria*	500
Second " "	400
Auditor	1,600
First *oficial de Contaduria*	1,000
Second " "	700
Escribiente de Contaduria	500
Treasurer	2,000
Office servant	300

Art. 3. The Government shall have the appointment of the employes of said office, on ternary nominations of the *Junta*. Its secretary shall be the person who was secretary of the former Establishment of Mining.

Art. 4. The Junta will immediately transmit to the Government the corresponding nominations, having reference, in making them, to the employes of the Secretary's office of the said Establishment, and observing the final clause of said Article 9.

Art. 5. Promotion by rank in their respective section is declared to the employes of the office. Vacancies or temporary absences of any of them shall be filled by those immediately below. All shall have the same consideration and be subject to the same laws which govern those of the public Treasury.

Art. 6. The treasurer shall give bonds for the performance of his duties (*su manejo*) with three bondsmen of $4,000 each, or four bondsmen of $3,000 each, to the satisfaction of the Junta, with the understanding that he shall appoint, under his own responsibility, and shall pay, out of his own salary, the cashier.

Art. 7. That part of Article 13 of the law of December 2d, [1842], which establishes an auditor-treasurer, is repealed.

Art. 8. According to Article 2 of the aforesaid law, the first period of the Presidency of the *Junta* belongs to the Commissioner of the Supreme Government; the second, which should commence the first of January next, corresponds to the agent of the Miners; the third to the agent of the creditors; and so on successively.

Art. 9. The Government retains the power to regulate the laws in respect to Mining, dictating such measures as it may judge expedient for its encouragement.

Wherefore, &c.

December 30th, 1843.

[From Lara's "Coleccion de Decretos y Ordenes," No. 882.]

Regulations of the College of Mining.

Valentin Canalizo, &c., &c., * * *know ye:* That, the Regulation of the National College of Mining, formed by its Director, agreeably to Article 27 of the Decree of the 3d of October last, having been submitted to the Supreme Government, I have been pleased to approve it, in concert with the Junta of the cabinet and exercising the power conferred on the Supreme Government by the seventh of the Bases established at Tacubaya, in the following terms:

CHAPTER I.

Of the Director.

Art. 1. His duties are, *first*, to watch over the strict observance of the laws, decrees and regulations relating to the College, which have been or may hereafter be issued. *Second*, to distribute the labors of the College in the most convenient manner, making such reforms as experience may require. *Third*, to transmit, at the end of the year, to the Ministry of Public Instruction, a report on the condition of the studies in the College, their reforms and improvements. *Fourth*, to examine the receipts of the collections which the steward shall make, as also those of extraordinary expense. *Fifth*, to be present, at least at the end of each third of a year, at the inspection of the cash of the steward. *Sixth*, to visit, at least once every month, the museum, the cabinet of natural history, and the botanical garden. *Seventh*, to transmit, every third of a year, to the *Junta* of Encouragement and Administration of Mining, the accounts of the College.

Art. 2. The attributes of the Director are: *First*, to propose to the Government the legislative measures which may be deemed expedient for the improvement of the said college. *Second*, to call a *Junta* of the faculty when he may deem expedient, &c., &c., &c.

[NOTE BY THE TRANSLATOR.—Of the remainder of this law, which mostly consists of minute details, only a brief summary is given.]

CHAPTER II.

Of the Rector.

This officer is charged with the moral and religious instruction of the students, their manners, recreations, &c.

CHAPTER III.

Of the Junta of the Faculty.

This Body is to meet once every month, or oftener if called by the Director. It regulates the course of studies, the examinations, &c., and the purchase of books, apparatus, instruments, &c. It may be called upon for reports, by the President or by the Junta of Mining.

CHAPTER IV.

Of the Professors.

This chapter gives in detail the duties of the Professors of the College.

CHAPTER V.

Of the Prefect of studies and substitute Professors.

The Prefect has charge of the students during the hours of study, and the *sustitutos de catedras* assist the Professors, and in their absence supply the vacancies.

CHAPTER VI.

Of the Chiefs of Sections.

One out of every ten students is elected as the Chief of the Section, whose duty it is to assist the rector and prefect.

CHAPTER VII.

Of the Librarian.

This person has charge of the library and is responsible for the good order and condition of the books.

CHAPTER VIII.

Of the Examinations.

This chapter contains special directions for the several examinations, and for the issuing of diplomas in each branch, in accordance with the provisions of the law of October 3d, 1843.

CHAPTER IX.

Of the Students, and qualifications for admission.

There are four classes of students, those of endowment (*de dotacion*), of half endowment (*de media dotacion*), boarders (*pensionistas*) and externes (*externos*). Those of the first two classes, for admission, must be descendants of Miners, 16 years of age, able to read, and acquainted with the first four rules of arithmetic. Those from the Departments are to be preferred, and the selections to be so made as to distribute them equally over the Republic. The boarders must be 15 years of age, but for the externes no other qualification than good conduct is required. The College course is estimated at $150 per annum. The endowment and half endowment students are required to remain and complete the full course, for the first, till they graduate as Engineers of Mines, and for the second, till they graduate as Reducers of Ores, being obliged, if they leave before having so completed the course, (except in case of sickness), to repay to the College the expenses of their previous instruction.

CHAPTER X.

Of Rewards and Punishments.

The rewards consist principally in special privileges and appointments, and the punishments in privations of amusements and recreations, confinement and expulsion.

CHAPTER XI.

Of the distribution of time, &c.

This chapter regulates in detail the distribution of the time from 5½ o'clock in the morning to 9½ in the evening.

CHAPTER XII.

Of the Steward.

This officer collects monthly the amounts due the College from the parents or guardians of the students, and from the *Junta* of Mining, the sums appropriated from the endowment fund and from quicksilver. The amount allowed each student for clothing, books and other necessities must not exceed $150 per annum. He is required to give bonds in the sum of $4,000.

CHAPTER XIII.

Of the Porter, Dispenser, Cook and Servants.

CHAPTER XIV.

Of the Conservator of the Museum of Antiquities and the products of Industry.

CHAPTER XV.

Of the Professor of Botany, and Director of the Garden and the Cabinet of Natural History.

April 30th, 1844.

[From Appendix to the Mexican "Ordenanzas de Mineria."]

Decree and Regulation respecting the manufacture of powder in certain Departments, and its sale to Miners for Mining purposes.

MINISTRY OF THE TREASURY,
Section Fourth.

The Most Excellent Señor President *ad interim* has been pleased to issue the Decree which follows:

Valentin Canalizo, General of Division and President *ad interim* of the Mexican Republic, to the inhabitants thereof,—*know ye:*

That the National Congress has decreed and the Executive has sanctioned the following:

Art. 1. The government will regulate the manufactories of powder existing in Mexico and in Zacatecas, so that the current price thereof, for working Mines, shall not exceed two and a half *reales* per pound.

Art. 2. It will also establish a manufactory in the Department of Guanajuata, another in Chihuahua, another in New Leon, arranged in such a manner that powder for the Mines may be delivered to the Miners, when highest, at the aforesaid price.

Art. 3. If, within the term of one month from the publication of this law in the Capital, the Government shall not be able to collect from the existing revenues (*rentas*) the necessary funds, both for putting the present manufactories in order, and for the establishment of new ones, mentioned in the foregoing Article, he will enter into contracts for both, with private individuals or corporations, agreeably to the following bases: 1st, An equitable price will be fixed for the rent of each manufactory, to be paid in money, or in war-powder, on which no bidding (*postura ni puja*) shall be admitted; 2d, The bids (*posturas y pujas*) shall be on the price of the ordinary Mining powder, the maximum being fixed at two and a half reales per pound, and the bid of the individual or corporation to supply it at a less price, shall be considered as a better bid (*mejor postura*); 3d, These contracts shall be limited to a moderate period, not to exceed five years, at the end of which time, new bids (*pujas*) shall be received, in the manner prescribed for the first; 4th, In case of the same bids by an individual or any corporation, and by the Body of Miners in general, or the Tribunal of Mining of the Departments in which the manufactories are to be established, that of the Body of Miners or of the respective Tribunal of Mining shall be preferred; 5th, In case of competition between the Body of Miners in general and the Tribunal of Mining of any of the Departments, the latter will be preferred under equal circumstances.

Art. 4. For the making of these contracts the Government will advertise for bidders for the period of thirty days, and the notice will be posted immediately after the expiration of the month mentioned in the preceding Article.

Art. 5. Whether or not the existing manufactories are put in order, and whether or not the new ones mentioned in this law are established, the Government will sell to the Miners, the ordinary powder which is already manufactured, at the price of two and a half *reales* per pound, in each

Mining place (*Mineral*), at the expiration of the month specified in Article 3.

RAFAEL ESPINOSA,
President of the Chamber of Deputies.

JOSE CIRILO GOMEZ ANAYA,
President of the Senate.

JOSE LUIS DEL HOYO,
Deputy Secretary.

BERNARDO GUIMBARDA,
Senator Secretary.

Wherefore I order it to be printed, published, circulated and duly observed.

Palace of the National Government, Mexico, April 30th, 1844.

VALENTIN CANALIZO.

IGNACIO TRIGUEROS,
Minister of Treasury.

And for the better observance of the present Decree of the National Congress, His Excellency the President, in accordance with the report issued by the Council of Government, has been pleased to direct the following instructions to be observed.

First. The day preceding that on which the period of one month from the publication of this law in this Capital expires, there shall be made a deposit of this article in all the revenue warehouses, where Mining powder is expended, with the intervention of the highest local political authority, remitting to the principal administrations, and these to the general direction, statements of the amounts which the powder on hand will produce, at the new price of two and a half *reales* per pound, which price shall commence the day after the deposit.

Second. The Deputations of Mining of each Department will give to the principal administration of the respective monopoly revenues (*rentas estancadas*), a list of the names of the Mines in operation, which shall be transmitted to the Direction, leaving a certified copy. With respect to the sale of powder for the use of the Mines, they will observe what is prescribed in Article 27 of the Regulation of October 18th, 1842.

Third. In the powder expended for the working of Mines, the cost of the casks is not included, but he who takes them shall pay for them separately; and in that which is sold to individuals, the price shall continue as prescribed in Article 19 of the said Regulation of October 18th.

Fourth. The price which is paid by the Treasury for saltpetre being very high, and also disproportionate to the convenience and cheapness introduced in the process of extraction, it will hereafter be purchased at the prices fixed in the Tariff which goes herewith, that which was issued by the Ministry of Treasury on the 16th of July, 1825, being hereby annulled.

Fifth. The Governors of the Departments will transmit to the Ministry of Treasury, as soon as possible, and afterward at the beginning and middle of each year, notice of the Mines which are worked and denounced, with an approximate calculation of the amount of powder which will be required in six months, and how much of it the new manufactories ought to supply, in order to arrive at a knowledge of the work which corresponds, and to lay in a timely supply of the component ingredients.

Sixth. If it should not be easy for the Government to establish manufactories in all or any of the points mentioned in Article 2 of this law, they will be contracted for in the terms prescribed in Article 3.

Seventh. The price for the rents shall be in Guanajuato, $4,700 ; in Chihuahua, $1,400 ; in New Leon, $900 ; and in Sinaloa, $1,350.

Eighth. Any favorable conditions whatever, whether with respect to the revenue or with respect to the Miners who are not versed in the price of rents, shall be reputed as better bids (*por mejoras en la postura.*)

Ninth. The contractors shall enjoy all the rights which pertain to the Public Treasury with respect to the monopoly, without excluding that of adjusting and giving licenses for the extraction and purchase of sulphur and saltpetre.

Tenth. The contractors will only have the right of monopoly within the limits of the Department in which the manufactory is.

Eleventh. The notice (*convocatoria*) for making the bids, will be issued by the Government, on the day fixed by the present law.

And I communicate it to your Honor (V. S.) for your information and the corresponding ends.

God and Liberty, Mexico, April 30th, 1844.

TRIGUEROS.

June 28th, 1852.

[From Navarro's "Legislacion Mejicana," Tomo de 1852.]

Decree, for the Administration of the Fund of Mining.

Mariano Arista, General of Division and President of the United Mexican States, to the inhabitants of the Republic,—*know ye:*

That, for the purpose of regulating the Administration of the Fund of Mining, which, in virtue of the law of November 30th, 1850, was incorporated in the general revenues of the Federation, I have been pleased, in use of the powers which are conferred on me by the law of the 21st of May last, to decree as follows:

Art. 1. The Fund of Mining will be administered by an officer whom the Government will appoint, and under the inspection of the Ministry of the Treasury.

Art. 2. The organization and pay of the employes of the administration will be as follows:

An Administrator, to be paid per annum	$3,000
A Secretary, who will have charge of the archives	1,600
A Clerk	600
A *Contador*, who will keep the books	1,600
A Clerk, who will act as cashier	600
A Messenger	360
A Watchman	180
Office expenses	250
Total	$8,190

Art. 3. The creditors, in virtue of the agreement made with the Government on the 15th of February, 1851, will continue to appoint their agent (*apoderado*), who will intervene in the administration of the fund, and take care that the interest is paid according to the stipulations of said agreement. This agent will be paid by the creditors.

Art. 4. It shall be the duty of the Administrator of the fund:—1st, to collect and manage the funds spoken of in the present decree, appointing for that purpose, under his own responsibility, provincial collectors: 2d, to pay the expenses of the branch (*ramo*), they being previously approved by the Government, every month; and this approval shall also be requisite for any extraordinary expenses which may occur: 3d, to transmit his monthly balance to the Ministry of Treasury and to the Treasury-General, to which he will also send his annual account, one month after the end of the fiscal year: 4th, to give security, for the performance of his duties, to the satisfaction of the Ministry of Treasury: 5th, to pay monthly the expenses of the College of Mining, on the previous order of the respective Ministry in charge of Public Instruction, communicated by the Ministry of Treasury: 6th, to pay triannually as due, and by an order on the Treasury-General, the interest to the creditors of Mining: 7th, to promote with efficatious solicitude the encouragement of the important branch (*ramo*) of Mining and its College: 8th, to have the accounts kept in the office by the system of double-entry.

Art. 5. On the 1st day of next July, at which time the fiscal year begins, the present *Junta* of Mining will deliver to the Administrator who is appointed, conformably to the present Decree, whatever it may have on hand, the furniture, utensils and office fixtures, and whatsoever belongs to the fund; making a detailed inventory of every thing, at which there shall be present the Chief of the directive section of the Ministry of Treasury, to whom the duty belongs.

Art. 6. The employes of the organization which is prescribed in this Decree, are subject to the provisions of the law of May 21st of this year.

Wherefore, I order it to be printed, published, circulated, and duly observed.

Palace of the National Government,
Mexico, June 28th, 1852.

MARIANO ARISTA.

September 2d, 1852.

[From Navarro's "Legislacion Mejicana," Tomo de 1852.]

Circular, respecting the disposition of the surplus of the products of the Mining fund.

MINISTRY OF THE TREASURY,
Section Fourth Directive.

His Excellency the President, having observed that the fund designated by the Decree of the 16th of June last to cover the estimates of the Chambers and the *Contaduria Mayor*, is not sufficient for the purpose, has been pleased to direct that it be increased by the twenty-eight thousand dollars which are paid into the public treasury from the products of the *Real de Mineria*, after paying the expenses of the College, the interest due to the creditors, and the administration of the fund, which sum the *Junta* of Public Credit will distribute monthly. He also directs that the proper person collect the 8 per cent. of products of the Maritime Custom-Houses, mentioned in said Decree, in the same manner as the 3 per cent. assigned to the interior debt, which person shall be the one who administers the fund in the Chambers and pays monthly their estimates, accounting directly to the Presidents of the committees of police (*policia*), in compliance with what the said Chambers may hereafter enact.

And, by Supreme Order, I communicate it to your Honor (V. S.) for your information, and the corresponding effects.

God and Liberty, Mexico, September 2d 1852.

ESPARZA.

June 1st, 1853.

[From Navarro's "Legislacion Mejicana, Tomo de 1853.]

Miners permitted to manufacture powder for Mining purposes.

Antonio Lopez de Santa Anna, *benemerito* of the country, General of Division, Knight of the grand cross of the Royal and distinguished Spanish Order of Carlos III, and President of the Mexican Republic, to the inhabitants thereof—*know ye:* That, using the powers which the nation has been pleased to confer on me, I have decreed as follows:

Art. 1. Miners are permitted to manufacture powder which may be necessary for the working of their ores.

Art. 2. For the said manufacture of powder, it shall be indispensably necessary to have the previous permission of the Commandants-General of the respective States, or the Chiefs of the Territories, who, on granting it, will inform the Government through the Ministry of the branch (*ramo*).

Art. 3. The rules under which said powder may be manufactured, shall be prescribed by the Ministry of Treasury.

Art. 4. Powder for other purposes, as also its ingredients, sulphur and saltpetre, will continue to be a Government monopoly.

Art. 5. By the Ministry of Treasury there will be issued the regulation, with detailed instructions for the supply of these articles for the public use, the manufactories of powder of the Republic continuing under the charge of the corps of Artillery.

Wherefore, I order it to be printed, published, circulated, and duly observed.

Palace of the National Government,
Mexico, June 1st, 1853.

ANTONIO LOPEZ DE SANTA ANNA.

June 28th, 1853.

[From Navarro's "Legislacion Mejicana," Tomo de 1853.]

Circular, respecting the payment of dues to the Mining Fund.

MINISTRY OF GOVERNMENT.

On this date I say to the Administrator of the Fund of Mining as follows:

I have reported to His Excellency the President of the Republic, the official letter of Your Honor (V. S.) of the 25th instant, representing that the individuals who ought to pay the duty of the *Real de Mineria*, refuse to do so, on the ground that, in the Decree for the classification of revenues, of the 29th ultimo, it is declared that what is derived from the said duty is not a national tax.

His Excellency, in view of what is represented, and what is reported by Your Honor (V. S.) for the information of the Government, has been pleased to direct as an answer, that the law of said classification of taxes does not extinguish the obligation of the Miners to pay the *real* per mark which forms the Fund of Mining, but that it restores it to the condition in which it was prior to the law of public credit, and makes it subject to the old laws by which it was regulated.

God and liberty, Mexico, June 28th, 1853.

AGUILAR.

Circulated to the Governors, Political Chiefs and Ministries.

July 16th, 1853.

[From Navarro's "Legislacion Mejicana," Tomo de 1853.]

Circular, respecting the collection of the Mining duty.

OFFICE OF SECRETARY OF STATE AND GOVERNMENT.

His Excellency the President of the Republic having ordered that the respective assayers must not permit the withdrawal of bullion, until it has paid, as formerly, as well the duties which belong to the public Treasury as those called of the *Real de Mineria*, I have the honor to communicate it to your Excellency, by Supreme Order, so that you may issue the necessary instructions for the most exact fulfillment of this resolution; with the understanding that the product of the *Real de Mineria* must remain at the disposition of the Administration of the branch (*ramo*).

God and Liberty, Mexico, July 16th, 1853.

AGUILAR.

To the Governor of the District.

July 30th, 1853.

[From Navarro's "Legislacion Mejicana," Tomo de 1853.]

Decree, establishing a school of practice for Mining.

Antonio Lopez de Santa Anna, *benemerito* of the country, General of Division, Knight of the grand cross of the Royal and distinguished Spanish Order of Carlos III, and President of the Mexican Republic, to the inhabitants thereof,—*know ye*:

That, in use of the powers which the nation has been pleased to confer on me, I have been pleased to decree as follows:

Art. 1. There will be established a practical School of Mines and Metallurgy, which, for the present, will be situated in the Mining place (*Mineral*) of Fresnillo, under the immediate direction of the College of the same branch in the Capital.

Art. 2. All the students of the College who are intended for engineers of Mines and reducers of ores will be obliged to practice in said School.

Art. 3. Those who, without having pursued the theoretical studies in the College of Mining in Mexico, may desire to enter the practical School, will first be examined in the said College in all the matters which are there taught with respect to the branches of Mines and Metallurgy.

Art. 4. The course of practice will continue for two years and a half, the first year being devoted to the working of Mines, the second to Metallurgy, and the remaining six months to visiting other mineral districts, and to the practice of Geodesy and Topography in that capital and its suburbs.

Art. 5. There shall be three Professors, to take charge of these courses of instruction; two of them charged respectively with instruction in the working of Mines and the reduction of ores, and the third to preside over the expeditions to other districts, to direct the geological examinations, and to draw up a statistical statement of the Mining of the Republic.

Art. 6. In addition to the practical study in the Mines and *Haciendas*, the said Professors will give, successively, theoretical courses of review to the practising students in their several branches, and as far as may be compatible with the main object.

Art. 7. The Professors will also establish, as soon as possible, academies of instruction for the liners or drainers (*ademadores ó paleros*), pumpmen (*bomberos*), carpenters, machinists, etc.

Art. 8. They will form collections of rocks, minerals and metallurgical products, as well as other kinds of indications (*pintas*), of the Mining districts visited by the Professor who takes charge of these excursions, which collections will be in duplicate, one set to be deposited in the practical School, and the other in the cabinets of the College in Mexico.

Art. 9. The three Professors, or, in the absence of one on the said excursions, the other two, will take charge, alternating weekly, of the studies and good order of the School.

Art. 10. The administration of the funds of the School

will also be under the charge of the three Professors jointly, who will every third of a year render account of their use and management to the direction of the College.

Art. 11. At the end of the first month after the organization of the School, they will draw up Regulations for its interior government, and submit them to the *Junta* of the faculty of the College; it, moreover, being their duty to present annually a programme of the improvements which experience may prove to be necessary for the advancement of the School.

Art. 12. The appointments for the chairs of the three Professors of practice, shall be by competition (*oposicion*) which shall be presided over, this time, by the general *Junta* of the Professors (*catedraticos*) of the theoretical college, with the exception of those of languages and drawing: this *Junta* will propose to the Government the persons who are approved by it, and in future the vacancies will be filled by competition, which shall take place in the practical school, before an examining *Junta*, composed of the two remaining Professors of practice, and those of the working of Mines, Metallurgy and Geology, of the College in Mexico, one of whom shall be authorized by the Director to take his turn, observing, in said competition, the provisions of a special Regulation which will be immediately drawn up by the three Professors who are this time appointed. Said examining *Junta* will propose to the Government, through the Director of the College, the person who may be qualified for the place.

Art. 13. Immediately after the publication of this law, a notice (*convocatoria*) will be given for the competition, which shall take place precisely within a month, counted from such publication of this law, it not being required, this time, that the competitors have the diplomas (*titulos*) of engineers of Mines or reducers of ores, provided, it be proved at the competition that they have sufficient knowledge, both theoretical and practical, in their respective branches.

Art. 14. A chemical and metallurgical laboratory shall be established, and shall be under the charge of the Professor of this last branch.

Art. 15. For the study of the reduction of ores (*beneficio de patio*) there shall be provided for the students 250 or 300 piles (*montones*) of ore selected by the Professor of Metallurgy who will bargain for them with those who carry on the works (*empresas*) of the Mining place (*Mineral*) where the School is established, paying the amount of bullion which falls short of the estimated *leys* as determined by professional assays, after making the discount of the difference per cent., as well as those reduced by quicksilver after the deduction of the consumption and loss which experience in reducing has proved to be usual.

Art. 16. There shall also be deposited in the practical school, collections of Mineralogy and Geology, formed from the duplicate specimens which can, in the opinion of the respective Professors, be spared from those now in the College.

Art. 17. From the duplicate works now in the library of said College, there shall be taken those which are necessary for the practical school, also subscribing for those periodicals which are taken by said College, and for others which may be deemed necessary.

Art. 18. There will also be formed a collection of models, for the use of the School, of machines, furnaces, and different utensils used in the operations of working Mines and in Metallurgy.

Art. 19. The School will also be provided with all the instruments necessary for the measurement and examination of Mines, as also those proper for the drawing and delineation of plans, machines, &c.

Art. 20. Each of the three Professors of the School will receive a salary of three thousand dollars per annum, besides their board, and the keeping of a horse.

Art. 21. The annual endowment (*dotacion*) of the students will be five hundred dollars,—$150 for board, $150 for clothing, $100 for keeping a horse, and $100 for traveling expenses, including those of their return to the Capital.

Art. 22. The students who are *portionistas* and of half endowment, will pay the same amount of $500 per annum, every four months in advance, the price of a horse and equipments; all other expenses will be on the account of the School.

Art. 23. For the expenses of the first establishment, there will be assigned the sum of $5,000, to be distributed as follows:

For instruments	$1,200
For the establishment of the laboratory	1,000
For preparing the edifice, purchase of horses, equipments and furniture	2,300
For supplying a chapel	500
Total	$5,000

Art. 24. The fund which is intended for the practical School is $18,000, which, as well as the $5,000 mentioned in the preceding Article, will be taken from the fund called of Mining (*de Mineria*), it being distributed as follows:

For the pay of the three Professors, ($3,000 each)	$9,000
Their board, ($150 each)........................	450
Keeping of their horses...................................	300
Endowment of eight students at $500 each..........	4,000
Salary of chaplain..	500
Repairs of the building, and extraordinary expenses...	926
Payment of subscriptions, purchase of books, &c...	300
Laboratory expenses...	300
To pay for the losses in reduction of ores as mentioned in Art. 14.....................	1,000
Pay of porter...............	300
Pay of two servants...	384
Pay of two hostlers...	240
Pay of a cook..	300
Total...	$18,000

Art. 25. The surplus and savings which may result from the fund of the practical School, will be applied to the improvement of the same.

Art. 26. The funds, which this law establishes, being independent of those now enjoyed by the College, the *Junta Facultativa* thereof is authorized to form, for every two students of endowment who go to the practical School, another new place of endowment with what shall be given to the students of half endowment, or *porcionistas*, who shall have made advances, and are of known application and good conduct, and are absolutely without the means to pay for their instruction in the School.

Art. 27. For each one of the places which are created by the preceding Article, the College will pay the School, monthly in advance, the amount corresponding thereto, and also that prescribed for the horses and equipments.

Art. 28. The expenses of the journey from Mexico to

the School of practice shall be paid by the theoretical College, as heretofore, for all the students of endowment or free students who may go to said School.

Art. 29. All dispositions which conflict with the present decree are repealed.

TRANSITORY ARTICLES.

1st. As soon as the School is established there will go to it the students of endowment who are found practicing in the different Mining places, and they will remain there the time required to complete the period of two years and a half, devoting themselves to the working of Mines, or to the reduction of ores, or to both branches, according as their instruction may have fitted them.

2d. The *Junta Facultativa* may excuse, from practice in the School, any of the students of half-endowment who may now be in the College, if they shall not have the means necessary to pay the fee of $500, and if they can practice with greater convenience in any other Mining place; but all the students of this class, who in future enter the College, will be subject to what is prescribed in Article 2.

3d. The three Professors of the school of practice will be appointed from those who enter the competition, immediately after that takes place, in order that they may commence the work of organizing the School.

Wherefore, I order it to be printed, published, circulated, and duly complied with.

Given at the National Palace of Tacubaya, July 30th, 1853.

ANTONIO LOPEZ DE SANTA ANNA.

APPENDIX.

APPENDIX No. I.

SPANISH LAW AND REGULATION OF MINING,

1849.

[From "Ley de Mineria, Madrid, 1854."]

LAW OF MINING, APRIL 11, 1849.

Doña Isabel II, by the grace of God and the constitution of the Spanish Monarchy, Queen of Spain, to all to whom these presents may come and be known,—*know ye:*

That the Cortes have decreed and we have sanctioned the following:

CHAPTER I.

Of the objects of Mining.

Art. 1. All inorganic substances which are susceptible of being worked, whether metalic, combustible, saline or precious stones, whether found in the interior of the earth or on its surface, are the special object of the branch of Mining.

Art. 2. The property (*propiedad*) of the substances mentioned in the preceding Article belongs to the State, and no one shall work them (*beneficiar las*) without concession from the Government in the form prescribed in this law.

Art. 3. Mineral productions of an earthy nature, as silicious stones and those for building, sands, argillaceous and magnesian earths, and calcareous stones and earths of all kinds, will continue, as heretofore, to be of common or private profit, according as may be the lands in which they are found. The working of these substances in the lands of another will not be permitted without the consent of the owner. Nevertheless, when these materials are applied to pottery, the manufacture of earthenware or porcelain; fire-bricks, crystal or glass works, or any other branch of fabrile industry, or for structures of public interest, authority may be granted by the Government, an *expediente* being previously drawn up by the political Chief, a hearing being given to the owner, to the Engineer of Mines and to the provincial Council. If the owner binds himself to work them within the period of six months, he will be preferred; but in the case of structures of public interest, the Government will fix the term. In no case can the working be commenced without having first indemnified the owner of the land for its value and one-fifth more, unless he prefer the value of the damages which shall be done to him. This class of concessions become extinct (*caducard*) when the conditions prescribed in the Regulation are not observed. The substances to which this Article has reference are not subject to the conditions of this law with respect to the manner of working: nevertheless they are subject to the vigilance of the Administration, with respect to the rules of police, whenever shafts and subterranean galleries are constructed.

CHAPTER II.

Of the exploration and concession of Mines.

Art. 4. The auriferous sands and all other mineral productions of rivers and *placeres* are of free use, without the necessity of a license or other formality, unless these operations are carried on in permanent establishments. This disposition is applicable to the use of minerals of iron, for whose working, shafts and galleries are not requisite.

Art. 5. Concession of a *pertenencia* of a Mine cannot be made, unless the deposit or mineral has been discovered, and it must be preceded by an *expediente* drawn up in the form which the Regulation may determine, the corresponding section of the Royal Council being heard. To the grantees there shall be issued a title in property (*de propiedad*) by the Minister of the branch. In it shall be expressed the conditions, which, in the opinion of the Government, the special circumstances of the undertaking or the public convenience may require. These cannot be other than the general conditions, or some of the special conditions which the Regulations provide. A condition being refused by one company (*empresa*) or individual, a grant of that *pertenencia* to another company or individual cannot be made without the same condition, unless the first abandons his right of preference, for which he will be invited. The Regulation will determine when silence is to be considered as an abandonment.

Art. 6. The grants of *pertenencias* of Mines are for unlimited time, so long as the Miners comply with the conditions of this law and with those of the grant. They may also dispose freely of their products, subject to the laws. Quicksilver and common salt are excepted, so long as they are articles of government monopoly (*generos estancados*), and their products must be delivered into the store-houses of the State, at the price which is, or shall be, established.

Art. 7. All Spaniards or foreigners may freely make explorations and examinations for discovering the minerals mentioned in Article 1, whether in Royal lands, commons, or municipal lands, or in lands belonging to individuals, provided that these operations are limited to mere trial-pits (*calicatas*). These shall not exceed four *varas* of superficies, by one in depth. When the trial-pits have to be made at a less distance than fifty *varas* from an edifice, or in gardens, orchards, vineyards, enclosed or irrigated lands, or in public servitudes (*servidumbres*), they shall not be commenced unless with the permission of the owner or of his representative, and on his refusal, with that of the political Chief, who shall not give the permission without hearing the owner, and on report of the provincial Council, an examination being previously made by an expert (*facultativo*). The explorer is obliged to indemnify the proprietor of the land for the damages and injuries which he may, in any way, cause; and in default, as in case of insolvency, he shall be regarded, for all legal purposes, as a trespasser (*dañador voluntario*).

Art. 8. If, within the space which shall be designated for one *pertenencia*, two or more persons shall open trial-pits, preference shall be given in the concession of the Mine, to the first of them who shall discover the mineral, and he may include other trial-pits within his demarcation. If two or more persons shall discover the mineral at the same time, there being land free and suitable for the concession of one *pertenencia* to each of the discoverers, it shall be so granted to them. When there be not space or proper ground, all those who shall have first discovered the mineral, will have equal rights, and one *pertenencia* shall be adjudged to them in common. In all these cases, if the land be owned by an individual, such owner shall have the right, if he demand it, to enter into partnership with the discoverers, for the tenth part of the profits and expenses. The demand must be made within two months after he shall have received notice of the discovery.

Art. 9. When, from not finding any mineral in the trial-pits, the explorers shall desire to continue their investigations by means of pits and galleries, they must ask the permission of the political Chief of the province, in writing, of which record shall be made in a formal register which shall be kept for that purpose. The permission cannot be refused, provided the petitioner give proper security for indemnification of the damage and injury which he may cause, and the fulfillment of the other obligations which the concession may impose on him. No pits or galleries shall be opened within the radius of 1,500 *varas* from fortified places and points, without the previous permission of the Minister of War. Nor shall pits or galleries be opened within a radius of 100 *varas* of settlements (*poblaciones*), without the previous license of the Minister of that branch. In rural settlements (*poblaciones*), the license to work Mines in the intermediate space, may be granted by the political chief. Whenever pits or galleries are to be opened in lands of the class mentioned in the first paragraph of Article 7, it shall be indispensable to have the *expediente* and license mentioned therein, if they have not been made and issued.

Art. 10. To the first person who shall solicit the permission of the political Chief to open a pit or gallery, there shall be reserved, for the period of one year, the land necessary for a *pertenencia*, which he shall designate within the period of three months from the date of the permission. If, at the expiration of the one year, he shall have proceeded with activity, and shall have constructed works of importance, the political Chief, on hearing the provincial Council, and an examination being previously made by the engineer, will extend it for as long as the Mine shall be worked (*poblada*). If the investigator shall discover the mineral, in soliciting the grant he may vary the direction of the *pertenencia*, provided, there be vacant land, and that he do not occupy the land included in the reserve of another legally authorized explorer. The owner of the land in which there may be discovered a mineral deposit by means of pit or gallery of more than a *vara* in depth, has no right to participate in the Mine.

Art. 11. Each *pertenencia* of a Mine constitutes a solid of a rectangular base 300 *varas* long by 200 wide, measured horizontally in the direction designated by the party interested, and of an indefinite vertical depth, without including the surface. There cannot be granted upon any one vein (*criadero*), to one single person, more than two contiguous *pertenencias*, and three if there be a company of four or more persons. In Mines of coal, lignite, or turf, each *pertenencia* shall be 600 *varas* long by 300 wide, and there may be granted to the amount of four *pertenencias*. The discoverer of an unknown vein, layer (*capa*), or pocket (*bolsada*), shall have a right to one more concession than those specified in the different cases of this Article.

Art. 12. The demarcation of a Mine which has only one *pertenencia*, is indivisible. If the primitive concession shall include two or more *pertenencias*, these may be separated, with the authority of the Government.

Art. 13. The space between two or more *pertenencias*, which cannot conveniently form another that shall contain a rectangle, at least equivalent to two-thirds of the space of an ordinary *pertenencia*, shall be adjudged as a surplus (*demasia*) to the adjoining Mines, it being divided in proportion to the lines of contact.

CHAPTER III.

Of the workings and enjoyment of Mines.

Art. 14. The use of the water found inside of a Mine belongs to the owner of the Mine, so long as he retains that ownership; but it will be his duty to make indemnity for the damage and injuries which may be done to a third party, in its opening, conveyance and incorporation into rivers, creeks or drains. The owners of Mines are also responsible for all the damage and injuries which, on account of the working, may be caused to a third party.

Art. 15. Every Miner is obliged to indemnify his neighbor, for the injuries which may be done to such neighbor, by the water accumulated in his workings, if, on being required so to do, he shall not drain it off within the period which the Regulations shall prescribe. The Miners shall also be obliged to contribute, proportionally to the benefit they may receive from the draining of the Mines near to them. The same shall take place, when, by the authority of the Government, which must always be preceded by a professional report and a hearing of the parties interested, general galleries shall be constructed for drainage or transportation, for a group of *pertenencias*, or for a whole Mining district (*Comarca.*)

Art. 16. The minerals which, in making the adits or general galleries for drainage or transportation, and their pits and ventilating shafts, shall be discovered in vacant land, shall be subject to concession of *pertenencias* in favor of the contractors (*empresarios*). If the minerals shall be found within known *pertenencias*, they shall belong, one-half to the owners of these *pertenencias*, and one-half to the contractors for the adit, the latter paying all the expenses until they are brought to the surface. In these lands the contractors can never go outside of the line and dimensions of the land marked out for the adit.

Art. 17. The owners of the *pertenencias* which an adit for drainage or transportation shall cross, cannot work the mineral which the walls of the adit may contain within a space of three varas, without strengthening them according to rule, at their own expense, and according to the judgment of the engineer of the branch.

Art. 18. Adits and general galleries of investigation cannot be opened without the authorization of the Government, and the consent of the owners of the *pertenencias* which have to be crossed. The rights of the contractors, with respect to the minerals which are found in the conceded *pertenencias*, shall be those which are agreed upon with the owners of the *pertenencias*; and with respect to what may be found in vacant lands, those which, in like cases, are granted in Article 16 to the contractors for adits of drainage.

Art. 19. The Miners and reducers (*beneficiadores*) of minerals, shall be considered as citizens (*vecinos*) of the Pueblos in which are situate their Mines, works, or reducing offices, in respect to the use of waters, woods, commons (*dehesas*), pastures, and other common advantages in relation to their business (*industria.*)

Art. 20. The same may obtain, from the owners and legal administrators, the lands which they may require for the mouths of their Mines (*bocaminas*), their ventilating shafts (*lumbreras*), edifices, store-houses, reducing offices, the deposit of their rubbish and scoria, washing-places and other dependencies, servitudes, and roads, and which shall not exceed half a league, whether public or common. In case there be no agreement between the parties, the matter will be submitted to the conditions and proceedings which are established by the law of forced expropriation for the sake of the public good. When the manufactory or office of reduction shall require the use of combustible vegetable matter, or of any fall of water, the permission of the political Chief, with a hearing of the provincial Council shall be requisite for its construction. The same authorization shall be requisite for opening roads of more than half a league, if there be opposition on the part of the Pueblos or owners of the lands which are to be crossed.

Art. 21. The Mines shall be worked according to the rules of art; their owners and workers shall submit to the rules of police which the Regulations shall designate. Violations of these shall be punished by fine, of from 400 to 2,000 *reales*, to be doubled in case of repetition. If any other crime be committed, it shall be punished according to law. In every case, there shall be indemnity for damages and injuries, if such shall be caused.

Art. 22. No Mine shall be considered as worked (*poblada*) or in activity, if there be less than an average of four workmen continually employed for each *pertenencia.*

Art. 23. The works of a Mine cannot be suspended, with the intention of abandoning it, without first giving notice to the political Chief, in order that an engineer of the branch may examine it and ascertain whether the works are in a secure condition. If not secure, they shall be made so, at the expense of the owner. The infractions of this provision shall be punished by a fine of from 400 to 2,000 *reales.*

CHAPTER IV.

Of the cases in which the ownership of Mines is forfeited, and of the denouncement of Mines.

Art. 24. The right to a Mine is forfeited, and it will be denouncable by any one, in the following cases: 1st, when there is a failure to perform the conditions of the grant: 2d, when six months, from the date of the concession, shall have expired, without the works being commenced: 3d, when, the operations having been commenced, the Mine shall not be worked (*poblada*) for four consecutive months, or eight interrupted months, in the course of one year: 4th, when, by mal-direction of the works, ruin is threatened, and the owner, on being required, shall not secure it within the time designated to him: 5th, when, by an avaricious manner of working, the subsequent enjoyment of the mineral is made difficult or impossible. In the second, third and fourth cases, superior force, which impedes the work, will constitute an exception, it being proved in due form.

Art. 25. A *pertenencia* being abandoned, the edifices dependent thereon will continue to be the property of the owner to whom they belong, unless he has also abandoned them. Mining edifices are to be considered as abandoned, when they are in such a ruinous condition as not to serve the end for which they were intended. After ten years of abandonment of a Mine or reducing works, without their being denounced by another, the lands of the edifices and servitudes will revert to the person who was the owner of the soil when they were established.

Art. 26. A Mine, or reducing works, or *pertenencia* of scoria which has been abandoned, may be denounced by any person before the political Chief: if there be any opposition, the point will be investigated before the provincial Council, with a hearing of the old owners. The abandonment being declared by due sentence, and also the reasons for the denouncement, the grant will be made in the form established in Article 5, although the mineral shall not be apparent.

CHAPTER V.

On the concession of the use of the old slags (escoriales), *and rubbish-heaps* (terreros).

Art. 27. The slag and rubbish-heaps resulting from old abandoned Mines, are declared subject to denouncement, except those which are found within *pertenencias* legally granted, and which have not been denounced prior to the said *pertenencias*. There are also excepted the slag and refuse-heaps belonging to establishments reserved by the State, especially all those found within the radius of four leagues from that of Almaden.

Art. 28. For the concession of refuse and slag-heaps, there shall, as a general rule, be observed the same requisites as for concessions of Mines, but abbreviating the proceedings (*tramites*), according to the difference between Mines and slag-heaps, there being always previously made, by an Engineer, an examination, plan and report. The Regulation will determine the proceedings which are to be observed for the formation and completion of the aforesaid *Expediente*.

Art. 29. In old slag-heaps, and in recent ones which have been abandoned, and are in vacant land, the *pertenencias* will be conceded in the form of a rectilinear polygon which the applicant may designate, provided, its extent does not exceed 80,000 superficial *varas*.

Art. 30. In order that a refuse-heap or slag-heap shall be considered as worked (*poblado*), it must be occupied by at least four workmen.

Art. 31. The right to a slag-heap is forfeited in the following cases: 1st, When it is not worked (*poblado*) according to the provision of the preceding Article: 2d, When no beginning is made to the working within the period of 8 months from the date of the concession: 3d, When the operations of reduction are interrupted for more than two months, without the intervention of superior force.

CHAPTER VI.

Of Mines belonging to the State.

Art. 32. There remain reserved to the State the following Mines: The quicksilver Mines of Almaden; the copper Mines of Riotinto; the lead Mines of Linares and Falset; the zinc Mines of San Juan de Alcaraz, in which only the direct dominion (*dominio directo*) belongs to the State; the sulphur Mines of Hellin and Benamaurel; the graphite or plumbago Mines included within the judicial *Partido* of Marbella; the iron Mines which, in Asturias and Navarra, are destined to supply the necessary mineral for the national manufactories of arms and ammunition at Trubia, Orbaiceta and Eugui; the coal Mines situate in Asturias in the councils of Morcin and Riosa, registered by the Director of the manufactory of Trubia for supplying it with fuel. The extent of *pertenencias* of the aforesaid Mines of the State will be what they now have. For those which have no boundaries expressly marked out, the Government will assign them. Within the perimeter or demarcation of the Mines of the State no one shall make borings (*calas*) or trial-pits (*catas*), or explorations, except by the order, and on account, of the Government, nor can there be made concessions of *pertenencias* of Mines or slag-heaps. Minerals which are not intended to be worked by the State, .are excepted, provided the trial-pits be made at a distance of at least six hundred *varas* from the works and offices of the State. The slag-heaps resulting from the Mines or works of the State belong to the same, and cannot be worked by a private person, even though they be without the demarcation of the Mine or the jurisdiction of the works. The State shall not, in future, alienate or acquire Mines or slag-heaps, unless the Government be so authorized by a special law.

CHAPTER VII.

Of the Tribunals which are to take cognizance of Mining matters.

Art. 33. The provincial Councils will take cognizance, with appeal to the Royal Council: 1st, of oppositions to the denouncements of Mines and slag-heaps, and of reducing establishments, on account of abandonment or because of the concession becoming extinct (*caducada*) as provided in Articles 24 and 31: 2d, of Mining affairs in which the state has a direct and immediate interest, and in any questions which may arise between the Administration and the Miners. For the hearing and decision of these matters, the engineer of Mines, of the highest rank in the province, will assist as a special member (*vocal*) with the right of voting.

Art. 34. The Royal Council will have contentious cognizance (*en via contenciosa*): 1st, of the reclamations which may be made against concessions of Mines, *pertenencias*, and other things which belong to the Government: 2d, of reclamations which are made in opposition to the conditions imposed by the Government for the concession: 3d, of reclamations which are made by resolution of the Ministry against those conditions which result from said remedy.

Art. 35. The ordinary tribunals will take cognizance of all suits between individuals, and of the crimes and offenses which may be committed in the dependencies of Mining.

Art. 36. Of causes which arise from fraud in the mineral products, cognizance will be taken by the tribunals competent to take cognizance of cases of fraud against the public Treasury.

Art. 37. The tribunals cannot, in any case except bankruptcy, decree the suspension of the working of Mines or of reducing establishments, or issue executions against the former and the articles necessary for their supply, but they may against their net products, or their products in kind.

CHAPTER VIII.

Of the Corps of Engineers of Mines and its Schools.

Art. 38. There shall be a Corps of Engineers of Mines, charged with the direction of the works of the Mines of the State, and of the other duties which may belong to it in Mining, and which the regulations shall designate.

Art. 39. There shall be a School of Mines for the education of the students of the Corps of Engineers of Mines. There shall also be practical schools in Almaden and in Asturias, for the engineers, masters, and overseers (*capataces*) of Mines.

TRANSITORY DISPOSITIONS.

1st. The concessions which have already been made, will continue as heretofore: nevertheless, if the grantees wish, they may have the dimensions of their *pertenencias* increased to the 300 *varas* in length by 200 wide, measured horizontally, as prescribed by Article 11 of this law, provided, there be vacant ground for that purpose, in either way. The grantees will continue to enjoy the rights which they may have acquired, according to the laws and provisions which have heretofore been in force.

2d. The same thing is to be understood of Mines of iron which may be of common enjoyment, and they shall not be liable to denouncement, except in case their working cannot be continued in any other mode than by subterranean works.

3d. After the promulgation of this law, reducing works cannot be established by means of high furnaces in which vegetable fuel is employed, or *catalan* forges, without the authorization of the government, with the previous report of the political Chiefs, who will give such permission after hearing the *Ayuntamientos* of the Pueblos where the charcoal is to be made, and the Commissioners of forests of the district.

4th. Matters pending in the inspections, and in the superior tribunal of the branch or Direction of Mines, whose special jurisdiction is abolished by this law, will be transferred, according to their condition and nature, to the tribunals which are competent, conformably to the provisions of said law.

5th. The Government will publish, as soon as possible, the Regulations necessary for the execution and observance of this law, the execution of which, will, in the meantime, remain suspended.

6th. Lastly, a special and probitory law will fix the taxes on Mines and their products, and in the meantime the existing taxes will be enforced.

Wherefore, we command all the tribunals, courts, chiefs, governors, and other authorities, civil as well as military and ecclesiastic, of whatsoever class and rank, to observe, and cause to be observed, kept and executed, this present law, in all its parts.

Given in the Palace, April 11th, 1849.

I, THE QUEEN.

JUAN BRAVO MURILLO.

Minister of Commerce, Instruction and Public Works.

July 31st, 1849.

Royal Decree and Regulation for carrying into effect the Mining Law of April 11th, 1849.

Having heard the Royal Council, I have approved the annexed Regulation for the execution of the Law of

Mining of April 11th, 1849, which my Minister of Commerce, Instruction and Public Works has presented to me.

Given at San Ildefonso, July 31st, 1849.

Rubric of the Royal Hand.

JUAN BRAVO MURILLO,
Minister of Commerce, Instruction and Public Works.

CHAPTER I.

Of the ownership of Mines,—Rights and obligations of the Administration in matters of Mining,—General Dispositions.

Art. 1. The ownership (*propiedad*) of Mines, by Article 2 of the Law of Mining of April 11th, 1849, belongs to the State, and in consequence, and in conformity with what is prescribed in Article 43 of the Constitution of the Spanish Monarchy, the administration of said property belongs to the Government. Wherefore, it is competent for the Government, 1st, to grant the ownership of Mines to individuals or companies (*empresas*) who may offer to work them to advantage, and in the form which the said Law prescribes, and with the previous proceedings which are established by this Regulation; 2d, to authorize, agreeably to Article 3 of the Law, the working of the mineral productions of an earthy nature, which are mentioned therein.

Art. 2. The branch of Mining being one of national industry, the Government exercises the administration thereof by means of the Ministry of Commerce, Instruction and Public Works, which is charged with the protection of Industry.

Art. 3. The Ministry of Commerce, Instruction and Public Works will perform the administrative part of the branch of Mining, through the Direction of Industry. In the provinces it will be represented by the political Chiefs, with the powers which the Law gives them.

Art. 4. The Corps of Engineers of Mines, organized by a special Regulation, as is prescribed in Article 38 of the Law, will assist the Government and its administrative agents, in the professional part of the branch.

Art. 5. The Government and the political Chiefs, by means of administrative acts, will declare rights in the matter of Mining, certain proceedings being previously had. These rights will be acquired by individuals, on their petition, and in order to declare them, those who may be interested in what is to be granted or refused, must be summoned by means of notifications.

Art. 6. For administrative acts, in matters of Mining, no fees will be charged; and the terms of the proceedings, which will be fixed in this Regulation, will always be reckoned from the day succeeding that of the notification: the notifications are also administrative acts.

Art. 7. By administrative notification is to be understood that which, without any fees being due, is executed in the name of the Government, by an agent of the administration, or in the name of that agent by one next below. It must be made to the party interested, or to his representative, by exhibiting to them the communication in which it is ordered to be executed; and, for its fulfillment, the party notified will sign it, or the notification must be certified in writing by the party who made it, and by a witness.

Art. 8. The priority of petition in matters of Mining, will give, in like cases, the better right to the concession.

A failure to perform any of the conditions upon which it is made, or an abandonment of the working, will cause the extinction (*caducidad*) thereof, which will be declared by the administration. Wherefore, the political Chiefs, as evidence of that priority, and of the observance of all the proceedings, will be obliged: 1st, To note immediately on every petition for a grant, the day and hour of its presentation. The chronological order, for the acquisition of rights resulting from petitions, shall be fixed, not by the respective date of each petition, but by the day and hour in which the endorsement states that its presentation took place: 2d, To give to the party interested a certificate (*resguardo*) or receipt of it, as a foundation of his right thereafter. This *resguardo* shall consist of a certificate expressive of the act and its circumstances, conformably to Form No. 1. If, at the time of making out the certificate, (*resguardo*), the political Chief knows that there has been presented another petition, asking for the same thing, it shall be stated therein. The Secretary of the political Government will authenticate this certificate, with the indorsed approval (*visto bueno*) of the Chief, and the seal of the political Government. 3d, To cause to be kept in the Secretary's office the following books: 1st, a diary of Mining of the province; 2d, a book of registries; 3d, a book of denouncements of conceded Mines.

Art. 9. These books must be paged and rubricated by the political Chief; they must be bound and stitched (*á pliego metido*), they must not have alterations or erasures; and any correction which is to be made in them must be done by writing it out in full in the books.

Art. 10. The Mining Diary (*Diario de Mineria*) of the province shall contain, in the order of dates and without any blank space, all proceedings relative to the branch (*ramo*). The entries shall be made in the way indicated in Form No. 2.

Art. 11. The book of Registries and that of Denouncements ought to contain one registry or one denouncement on each leaf, in the order of their dates, so that there shall be no leaf in blank. In it shall be noted all the proceedings which may be had, from time to time, respectively, in the *expedientes*, up to the decision upon the concession in the first, and of the extinction (*caducidad*) in the second. Both books will have their corresponding index, with reference to the name of the Mine, and to that of the person who registers or denounces it. For due uniformity, these books will be arranged according to Forms Nos. 3 and 4.

Art. 12. The political Chiefs as well as those officers specially appointed to assist them in these matters, will proceed, in matters of Mining, with the greatest activity. They will also scrupulously observe both the proceedings and the periods of time which are prescribed for them. When, by unforseen circumstances or on account ot insuperable difficulties, an act cannot be performed within the period which is fixed for it, there will be stated in the proceeding the cause of the delay. The superior authority will determine its importance. In proceedings which have no period prescribed, their nature not permitting it, they will proceed with all possible brevity, with the understanding that they will thereby show their zeal for the service of the State.

Art. 13. The delay of a term shall work prejudice to no individual, when it results from the omission of a functionary, provided, he makes reclamation against it to the one next above, in order that he may correct it, holding the proper party responsible.

Art. 14. The appeals (*recursos*) against the decisions (*providencias*) of the Government or of the political Chiefs, in cases in which they can be taken, must be entered within the period of thirty days, in the form expressed in Article 6. This time having expired without the appeal (*recurso*) being proposed, the decision shall be final.

Art. 15. Whenever, according to the Law or to this Regulation, any corporation or person is to be heard, the original report (*dictamen*) shall be put into the *expediente*.

CHAPTER II.

Of the objects of Mining, and of the Mineral productions which do not pertain to it.

Art. 16. As established in Article 1 of the Law of the branch (*ramo*), all inorganic substances which are susceptible of being worked, whether metalic and combustible, saline, or precious stones, whatsoever the character of the deposits (*criaderos*) which contain them and the manner of their use, are the special object of Mining.

Art. 17. According to Article 3 of the Law, the mineral productions of an earthy nature are of common or private use, according to the ownership of the land in which they are found. To this class, which is not included in the branch of Mining, belong silecious or building stones, lime-stones and gypsum, ornamental stones, as serpentines, marbles, alabaster, porphyries and jaspers; lithographic stones; flint stones, common sands, marls; clays for porcelain, earthenware, pottery and fulling; epsom salts (*sal de Higuera*), and any other mineral substance not mentioned in Article 1 of the Law.

CHAPTER III.

Of the authorization to work mineral substances of an earthy nature.

Art. 18. Although Article 3 of the Law prohibits, as a general rule, the working, in the land of another and without the consent of the owner, the substances embraced in the first paragraph of that article; nevertheless, in the second paragraph, the power is reserved to the Government to supply this consent in two cases: 1st, when the said Government may find the said substances necessary for constructions of public interest; 2d, in case any one should desire to use any of said materials, applying them to pottery, the manufacture of earthenware, porcelain, fire-bricks, crystal or glass works, or any other branch of manufacturing industry. In both cases, if the owner refuse his permission, the Chief of the branch of public Administration, or the individual who requires the substances, may have recourse, the first officially, and the second in writing, to the political Chief by petition, for the authorization. They will allege, as the basis thereof, the construction of public interest, or the class of industry to which it is intended to apply the substances asked for, and the refusal of the owner. Finally, they will state the place where the said material is found, and the extent of land necessary for its working. The drawing up (*instruccion*) of the *Expediente* shall be in the following form: 1st, The political Chief will have indorsed upon the said petition the day and hour of its presentation, and the petition, together with its admission, shall also be entered in the book of registries, as is prescribed in Article 8 of this Regulation: 2d, There shall be issued to the claimant a certificate, in the terms prescribed in the said Article. 3d, The political Chief will transmit a copy of the communication or representation to the owner of the land, through the Alcalde of the Pueblo where he resides, and will give him a term of from 8 to 15 days within which he may, using the right reserved to him in Article 3 of the Law of Mines, show whether or not he wishes to make the working (*explotacion*) on his own account, and whether he has any cause of opposition to allege: 4th, As soon as the Alcalde receives said copy, he will have it delivered to the owner of the land, with administrative notification: 5th, Afterward he will return to the political Chief his official letter of transmission, indorsed (*diligenciado*), as stated in the preceding paragraph, in order that it may be annexed to the *expediente:* 6th, If the owner of the land shall desire to carry on the working on his own account, he will so inform the political Chief within the time fixed when he was notified of the petition, accompanied with a bond to commence the working within six months, or within the term which the political Chief may fix, in the name of the Government, if it be a matter of construction of public interest. In this case the *expediente* will be terminated, reserving to the person who petitioned for the authorization, the right of preference to obtain it, if the owner of the land shall not commence the working within the said period: 7th, If the owner of the land shall answer that he does not deem it expedient to carry on the working, on his own account, of the said materials, or if the time shall have expired without an answer, the political Chief will, within the period of six days, pass the *expediente* to an engineer of Mines, in order that he may report, after making the proper examination of the land; at this examination the parties interested shall be at liberty to be present, for which purpose they shall be summoned two days beforehand. If there be no engineer of Mines in the province, application shall be made to the nearest political Chief to furnish one:

8th, The report of the engineer of Mines being made, the political Chief will pass the *expediente* to the provincial Council, in order that it may make its report, which, being made, the said Chief shall remit the *expediente* with his own opinion to the Ministry of Commerce, Instruction and Public Works, in order that by it the authorization may be granted or refused. From this decision recourse may be had to the Royal Council.

Art. 19. When the Government grants the authorization, there shall be specified the extent and shape of the land which is to be included, and which shall not exceed 20,000 superficial *varas*. Moreover, there shall be imposed on the grantees, as requisite conditions, the following: 1st, That, before beginning the working, according to the provisions of Article 3 of the Law, the owner of the land is to be indemnified for its value, and, either one-fifth part more, or the injuries which may be caused to him, as he may elect, in consequence of the administrative notification which, for this purpose, shall be served on him, this proceeding being set forth in the *Expediente*. The appraisement of the value of the land, and of the damages which may be caused to its owner, when there is no agreement, belongs to the civil tribunals, in which case the political Chief will pass to them the whole proceedings (*actuaciones*), in order that they may proceed to determine it in the manner prescribed by the Law of July 17th, 1836: 2d, That the working shall be commenced within the period designated, which shall not exceed two months: 3d, That the substances shall be used for the purpose for which they were asked, and for no other: 4th, That the requisite works for forming the manufacturing establishment, in which these substances are to be employed, shall be begun and completed, if not previously existing, within the periods prescribed. These periods cannot be less than three months nor more than nine for beginning the works, nor more than two years for completing them.

Art. 20. When the conditions inserted in the authorization are not complied with, the extinction (*caducidad*) of this class of grants will be declared by the following proceedings: 1st, As soon as the political Chief receives notice, whether officially, or by written denouncement of the owner or of a third party, that the grantee has failed to perform the conditions imposed in his authorization, he will cause it to be entered in the book of denouncements, and the certificate (*resguardo*) to be delivered to the party interested, in the two last cases; and he will communicate it to the grantee, in order that, within the period of fifteen days, he may make such answer as he shall deem expedient. At the same time, he will direct such proceedings and examinations to be made, as he may judge proper, for the purpose of determining the truth of the alleged act. 2d, Having received the answer of the party interested, or the time allowed for making it having expired, and the formation of the *expediente* being completed, so that all the acts shall appear with exactness, the political Chief will declare whether or not there be cause for the extinction (*caducidad*). 3d, This declaration will be communicated to the interested parties; and the one who may consider himself aggrieved will be at liberty to make reclamation against it. 4th, In case the declaration shall be for the extinction (*caducidad*), the grantee may appeal therefrom to the provincial Council. The political Chief will defend, as a party, in the name of the Administration, his decision, pursuing the judicial proceedings and appeal, as prescribed in the first Article of chapter second of the Regulation on the mode of proceeding of the provincial Councils, in contentious matters of Administration. 5th, If the political Chief shall decide not to declare the extinction (*caducidad*,) an appeal may be taken to the Minister, and if he confirm the decision there can be no other recourse (*recurso*): but if the Minister shall declare the extinction (*caducidad*), then recourse may be had to the Royal Council. 6th, The termination (*caducidad*) being declared by the political Chief, or by the Minister in his case, without opposition, or even when opposition is made, if it has been confirmed by a final judgment (*sentencia ejecutoriada*), it shall be inserted in the *Boletin official* of the province, for notice to all, and particularly to the denouncer; whose denouncement shall be deemed a registry, and the term of one month from the publication of the extinction (*caducidad*) shall be granted to the party interested, so that within that time he may declare whether or not he insists on the registry, and may complete it.

Art. 21. The excavations (*labores*) for the working (*explotacion*) of the substances which are mentioned in Article 3 of the Law, shall not be subject to the dispositions of the present Regulation; but if they shall be carried on by means of shafts (*pozos*) or subterranean galleries, they shall, with respect to the rules of police, be subject to the vigilance of the engineers of the branch of Mines, under the authority, and by order of the political Chiefs, and, in their respective cases, of the civil Chiefs and of the Alcaldes. All the conditions imposed by this Chapter III upon those who shall obtain authorizations to work mineral substances of an earthy nature, must be complied with by the owners who work lands of their own, so far as they may be applicable.

CHAPTER IV.

Of the exploration of Mines.

SECTION FIRST.

Of trial pits.

Art. 22. He who may intend to open one or more trial pits (*calicatas*) in any land belonging to another person, even though it be not one of the cases in which, in accordance with the second paragraph of Article 7 of the Law, the explorer is required to have a permit, shall, nevertheless, have recourse to the Alcalde of the Pueblo in which the land is situate, petitioning him to notify administratively the owner, or his representative, in order that, if he deem it expedient, he may immediately adopt the proper measures to avoid damages. He who shall enter upon the land of another without having complied with this requisite, shall not be at liberty to use the right of making trial pits, and, moreover, he shall be liable to the penalties which the laws impose.

Art. 23. When the trial-pits have to be made at a less distance than fifty *varas* from a building, or in gardens, orchards, vineyards, enclosed or irrigable grounds, or in public servitudes, in which cases, according to Article 7 of the Law, it is necessary to obtain the permission of the owner or his representative, or, in case of his refusal, that of the political Chief, they will, in order to obtain this, pursue the following course of proceeding: 1st, He who purposes to make the trial-pit and has not obtained the consent of the owner, will move for a mutual agreement (*avenimiento*); and for this purpose he will make a written application to the Alcalde of the Pueblo where the land is situate, who will institute the corresponding proceeding of conciliation (*juicio de paz*). The Alcalde, in view of this petition, and, endorsing on it the day and hour of its presentation, will summon to his presence the petitioner and the owner of the land or his representative, each being accompanied by an *hombre bueno*. The statements which the parties may make being heard by the Alcalde, he will endeavor to bring them to an agreement; and if it be done, the act will be drawn up and authenticated by the Alcalde, he being charged with having the agreement made between the parties carried into execution. If, on the contrary, they do not come to an agreement, this will also be set forth in an act, an authenticated copy of which the said Alcalde will transmit to the political Chief, giving, in his official letter of transmission, his opinion, in full, as to whether or not there should be granted the permission to make trial-pits in the land of another. If the land, where it is proposed to make the trial-pit, be a public servitude, and consequently the Alcalde is the representative of that communal right, the proceedings for agreement will be carried on before the Alcalde of the nearest Pueblo. 2d, As soon as the political Chief shall have received the said copy of the act, he will command the party, who purposes to make the trial-pit, to designate the ground in which he wishes to make exploration, with the other circumstances necessary to show the expediency of making the exploration, and also to show the nature of the said land and its ownership, giving security to pay the damages and injuries. At the end of the writing there shall be added the petition for the permission of the political Chief, which shall also contain the dissent of the owner. 3d, The political Chief will also command the endorsements and annotations (*inscripciones*) to be made, and will give the certificate (*resguardo*) which is prescribed in Article 8 of this Regulation. 4th, This being done, he will pass a copy of the petition, within the period of the third day, to the owner of the land, appointing a time, which shall not exceed ten days, for him to make such representations as he may deem expedient, both upon the petition and upon the security. 5th, If the land, upon which it is proposed to make the trial-pit, be a public servitude, the proceedings will take place before the Alcalde of the jurisdictional district where it is situate. 6th, The answer being received, or the time for making it having expired, the political Chief will direct, that an engineer of Mines make, within a short period, an examination of the land, for which purpose the parties interested will be previously cited. 7th, Afterward the *expediente* will be passed for the report of the provincial Council; and its opinion being heard, the political Chief will refuse or grant the permission asked for, fixing the security in case it has not been approved by the owner. 8th, This resolution will be communicated to the parties interested: and in case the permission be granted, the security being given, there will be delivered to the petitioner a certificate of the Secretary of the political Government, indorsed by the Chief, inserting in it, besides the decision, an extract from the petition and the proceedings of the *expediente*. 9th, If any of the parties deem themselves injured by the decision (*providencia*) of the political Chief, they may appeal to the Ministry of Commerce, Instruction and Public Works, asking for its revocation or change.

Art. 24. Neither trial-pits nor other works of investigation will be permitted to be made: 1st, in high-ways and public roads; 2d, in rail-roads; 3d, within the precinct (*recinto*) of fortified places; 4th, in settlements (*poblaciones*) not rural; 5th, in edifices of private ownership, unless there precede the express and written consent of the owner, without which the permission cannot be given by any authority.

Art. 25. The permission shall expire, if use shall not be made of it within the period of two months. In this case, and in case it be not used for want of giving the requisite security, if there be other petitioners, the one next in order of the date of the presentation of the petitions, shall enter into the enjoyment of the same rights.

SECTION SECOND.

Of investigations by pits and galleries.

Art. 26. For the permission which, in accordance with Article 9 of the Law, must be solicited of the political Chief, whenever the explorer may wish to continue his investigations by means of shafts (*pozos*) or galleries, in any class of lands, there shall be drawn up an *expediente* in the form prescribed in Article 23. The same proceedings will be had whenever it is intended to carry on the investigation immediately, by means of shafts or galleries, in lands of private ownership.

Art. 27. If the land be of the character mentioned in the second paragraph of Article 7 of the Law, and its owner consents to the investigation by means of shafts or galleries, it must be shown by a document which accompanies the petition.

Art. 28. The owner, or, in his default, the political Chief, as prescribed in paragraphs 4th and 7th of Article 23 (of this Regulation), will approve the security given for indemnity of damages and injuries, and for compliance with the obligations which the grant may impose, which security is prescribed by Article 9 of the Law. This security being given, the permission cannot be refused, and it cannot be granted without the execution of such security, except in case of the consent of the owner of the land.

Art. 29. When shafts or galleries are to be opened within the radius of 1,500 *varas* of fortified places or points, for the previous permission which, as provided in the law, it is necessary to procure from the Minister of War, the political Chief will direct to him the petition, with his own report if there be anything to represent. The permission being obtained, it will be added to the *expediente.*

Art. 30. Notice will be given of the petition to the owner of the land, and to the owners of the adjoining (*colindantes*) Mines, if there be any, in order that they may make such representation as they shall deem expedient within the period which shall be assigned to them, and which shall not exceed fifteen days.

Art. 31. In the case where, according to Article 9 of the Law, the license of the Minister of the branch shall be requisite, for the reason that the land in which the wells or galleries are to be opened is included within the radius of 100 *varas* of settlements (*poblaciones*) not rural, the *expediente* being formed in the manner prescribed in the preceding paragraphs, it will be sent up to the Ministry of Commerce, Instruction and Public Works for its decision. Against this, recourse may be had to the Royal Council.

Art. 32. In cases where, according to the preceding Articles, it shall not be necessary to obtain the permission of the Government, the political Chief will grant or refuse it, according to what is expressed in Article 23. Against his decision reclamation may be made to the Government; and against its decision, to the Royal Council.

Art. 33. The permission of the owner, or of the political Chief in his case, being obtained, in order to procure the approval for the projected works, the *expediente* will be continued by the following proceedings: 1st, The party interested will, within the period of three months, designate the *pertenencia;* 2d, Then the engineer will mark it out, there being vacant land for it, without changing the designation made by the party interested. The owner of the land and the owners of the adjoining Mines will be cited three days before, in order that they may be present at the act; 3d, The formation of the expediente being completed in this way, the political Chief will grant or refuse his approval, communicating his decision to the parties interested; 4th, If he shall grant it, there will be delivered to the grantee a certificate of the Secretary of the political Government, with the endorsement (*visto bueno*) of the Chief, in which shall be stated the concession of the permission, and the designation and demarcation, stating with exactness the boundaries (*linderos*) of the *pertenencia.*

Art. 34. If, at the expiration of a year after the granting of the permission, the Miner shall petition to continue the works, the political Chief will direct that the engineer make an examination of those which have been executed, and hearing afterward the provincial Council, he will grant or refuse the extension of time, delivering to the party interested, in case he grants it, a certificate of the Secretary of the political Government with his indorsement (*visto bueno*), in which it shall be set forth. Against a refusal to extend the time, recourse may be had to the Government.

Art. 35. If the explorer shall not give the security which is prescribed in Article 9 of the Law, or shall permit the three months to expire as prescribed in Article 10, the termination (*caducidad*) of the permission or the concession, respectively, will be declared, by the proceedings prescribed in Article 20 of this Regulation.

Art. 36. The termination of this class of concessions, after the prolongation of the time, will also be declared by the proceedings prescribed in the said Article 20.

CHAPTER V.

Of the Concession of Mines.

SECTION FIRST.

Of the petition for registry, its proceedings, and preliminary examination.

Art. 37. In order to obtain the concession of a Mine, application by petition for registry will be made to the political Chief of the province. As this is an application for a concession of property or ownership (*propiedad*), it must be made with more minuteness (*mas circumstanciada*) than that for the registry of trial-pits, shafts and galleries. Therefore, it ought to express: 1st, the names, age, civil rank, place of birth, domicil (*vecindad*), residence, profession, employment or occupation of the parties interested, and of their representatives in the municipal district where the Mine is situate, in case they should wish to authorize any person to act for them in that capacity, which must always be done when the principal does not reside therein; 2d, The kind of mineral intended to be worked, accompanied with specimens of that which has been discovered; 3d, The place where the Mine is situate, the Pueblo or municipal district to which it belongs; all which shall be correctly and circumstantially stated; 4th, The adjoining Mines, if there be any, stating their names and owners, in a clear and precise manner; 5th, The name and residence of the owner of the land in which the Mine is situate, and the qualities thereof; 6th, The name it is wished to give to the Mine; 7th, The *pertenencias* which, in accordance with Article 11 of the Law, are claimed, and the reasons why the particular number of *pertenencias* is asked; 8th, Whether the *criadero* or mineral was discovered in mere trial-pits, or by means of shafts or galleries, with reference to the authorization, if there has been one for that purpose. With these particulars the petitions for registries of Mines will be made in the manner shown in Form No. 5.

Art. 38. Each petition shall not include more than one registry; and not more than two *pertenencias* can be asked for, in conformity with what is provided in Article 11 of the Law, except where three may be asked for, in conformity with the same Article, in the name of a company composed of four or more persons, in which case there must be presented the writing of formation (*escritura de fundacion*) of said company, or when a greater number of *pertenencias* may be asked for, which, in conformity with the said Article, can be granted in Mines of coal, lignite, or peat, or to the discoverer of an unknown vein, layer (*capa*) or pocket (*bolsada*). The extent which each *pertenencia* is to have will be that fixed in said Article 11 of the Law. The extent of the *pertenencias* of auriferous sands, the working of which has to be conducted in fixed establishments, and is not free according to Article 4 of the Law, shall be 30,000 square varas in a rectangular figure.

Art 39. At the time of the presentation of the writing, there shall be made the indorsements prescribed in Article 8 of this Regulation, the petition being disposed of by a decree for the preliminary examination by an engineer. In continuation, the certificate (*resguardo*) setting forth the whole, will be issued to the party interested, who will be cited for the examination. The form for this decree is given in form No 6.

Art. 40. In the registries which are presented by persons or companies (*empresas*) of well-known credit, and which are moreover indorsed by an engineer of Mines, the preliminary examination shall be dispensed with.

Art. 41. The engineer, on making the examinations for registry in a district (*comarca*), will do it with citation of the persons in charge of the bounding Mines, whether marked out or to be marked out. In order to do this, when the Mines are contiguous, he will rigorously follow the order of the priority of dates of the decrees, and at the foot of the same he will write out his reports, returning the petitions directly to the political Chief.

Art. 42. The engineer will state precisely in his report the conformity or differences of the specimens of the mineral presented with that of the deposit (*criadero*) examined, for which purpose he shall make, under responsibility, the necessary examination. If they should turn out to be different, the political Chief, attending to the circumstances of this fact, will proceed to what may be proper.

Art. 43. In case it be shown by the examination of the engineer that a deposit (*criadero*) or mineral has been discovered, and that this has been found in simple trial-pits, the land where it has been found being of private ownership, in order to comply with the provisions of Article 8, paragraph 3d, of the Law, this fact will be made known to the owner of the land by means of an administrative notification. Said owner may claim, within two months, the right of entering into partnership with the discoverers, for the tenth part of the profits and expenses; provided that in order to enjoy such participation he must re-imburse the Miner the tenth part of the expenses he has incurred in the discovery of the ore. In this case the following will be the mode of proceeding: 1st, The party interested will present his claim to the political Chief, who will indorse it, giving him the corresponding certificate (*resguardo*), in the terms prescribed in Article 8; 2d, A copy of this writing will be passed to the discoverer, in order that within the period of eight days he may make such answer as he may deem convenient; 3d, The reclamation of the owner of the land, and the answer of the discoverer of the ore, will be annexed to the *expediente* of registry. The proceedings provided in this Article are to form no obstacle to the continuation of the *expediente* of registry, the formation of which will not be suspended.

SECTION SECOND.

On the admission of the registry.

Art. 44. If the report of the engineer shall confirm the existence of the deposit (*criadero*) or ore, and show that there is vacant land for the *pertenencias* asked for, the political Chief will make the decree of admission according to the form given in *Form No.* 7, causing the edicts to be posted up in the Capital of the province on the board of notices of the political Government, in that of the Mining district, and in that of the municipal district where the Mine is situate, it being also published in the Official Gazette (*Boletin Oficial*). Of this proceeding a certificate (*resguardo*) will be given to the person making registry.

Art. 45. The edict posted up in the Capital of the province will remain exposed to public view for thirty days; that which is posted up in the municipal district of the Mine will be taken down at the end of nine days; and both, with their respective certificates, will be added to the *expediente*, or only the certificate, if the original notice should be lost. Moreover, there shall accompany every *expediente* of concession a copy of the Official Gazette (*Boletin Oficial*) of the province, in which the admission of the registry of discovery or denouncement has been published. To this end, the decree of admission shall be inserted in it for the period of six days, with the necessary punctuality and in preference to other matter; observing that, in addition, the registry or denouncement must be expressed with every particular.

Art. 46. If on the contrary it shall appear from the report of the engineer that the *criadero* or mineral does not exist, or that there is not land sufficient for a *pertenencia*, the political Chief will decree the refusal of the petition, making this known immediately to the party interested, or to his representative, according to *Form No.* 8.

SECTION THIRD.

Designation of the pertenencias.—Supply of the legal labor.

Art 47. The registry being admitted and published by the means indicated in Articles 44 and 45, the party interested will designate by formal writing, within the precise period of thirty days from the date of said admission, his *pertenencia* or *pertenencias*. The designation shall be made by expressing circumstantially, and with the greatest clearness, the point where is to be commenced the principal work or *labor legal*, starting from which there shall be determined, in Castillian *varas*, the length and width which are to be measured off, in order that there may result the rectangle of its *pertenencia* or concession, in compliance with Article 11 of the Law, without prejudice to others previously designated or marked out.

Art. 48. This designation being admitted by the political Chief, its essential part shall be copied into the certificate (*rèsguardo*) formerly given to the party interested, the Secretary of the political Government authenticating the copy with the indorsement of the political Chief. Afterward there shall be published a copy (*un tanto*) of the description on the board of notices of the political Government, where it shall remain exposed to public view, until the demarcation is made, or the registry abandoned.

Art. 49. The enterprises (*empresas*) which have an engineer, and which are entitled to the omission of the previous examination, in virtue of Article 40, will present with the designation an exact topographical plan, in duplicate, and on a scale of 1 to 3,600, of the space which they may designate, signed by the engineer and the owner or legal attorney of the enterprise. In this plan there must be marked down, not only the principal topographical objects of the space designated, but also with perfect exactness all the mouths (*bocas*), and the names of the Mines previously granted or designated which bound on it.

Art. 50. Within the period of four months, reckoned from the date of the admission of the registry, there shall be put in operation (*se habilitará*) a working-space, either in pit or gallery (*una labor de pozo ó galeria*), of at least ten Castillian *varas*, which shall be excavated on the discovered mineral. Such working space shall be known by the name of *labor legal.* In the registries for the working of auriferous sands, of which mention is made in the last paragraph of Article 37 of this Regulation, the legal working-space (*labor legal*) shall consist of a ditch (*zanja*) ten *varas* long and of sufficient depth to prove the discovery of the auriferous sands.

Art. 51. Said period having expired, the party interested will present new specimens of the mineral to the political Chief, setting forth in writing that he has made the prescribed working-space (*labor*), asking that it be examined by an engineer, and that, on due proof thereof, the *expediente* be sent up to the Ministry of Commerce, Instruction and Public Works.

Art. 52. The working-space (*labor*) of ten *varas*, prescribed in Article 50 ought to be made within the walls (*respaldos*) of the *criadero*, or within its cavity (*caja*) if it be a regular formation, and in others it will be established wherever most expedient according to their form.

SECTION FOURTH.

Opposition to the registry.

Art. 53. Whatsoever reclamation may be made in consequence of the edicts and publication in the Official Gazette (*Boletin Oficial*), shall be presented to the political Chief within the period (which cannot be prolonged) of sixty days, reckoned from the date of the said edicts of admission, and shall be annexed to the *expediente*. If those who present it allege any right previously acquired, the works shall cease as soon as the legal working-space (*labor legal*) is finished, the minerals extracted or their value being put in deposit, and the opposers being at liberty to place an inspector (*interventor*) in the works on account of whomsoever it may concern. Nevertheless, even in this case, and after the completion of the legal working-space (*labor legal*), the works may be continued, should the person making registry give security for the restoration of the minerals taken out, to him who may be declared the owner of the Mine. This security must be to the satisfaction of the claimants (*reclamantes*), or of the political Chief in his case, as provided in Article 23 of this Regulation.

SECTION FIFTH.

Examination of the Mine and the legal working-space (labor legal).—*The demarcation.*

Art. 54. The four months from the admission of the registry having expired, the political Chief shall order an engineer to examine the *labor* which has been made, and to mark off the *pertenencias*, provided the existence of the deposit (*criadero*) or mineral be ascertained, whether by the first examination, confirmed by the last, or by its appearing recently as a result of the legal working-space (*labor legal*), and that the land designated be vacant (*franco*), that is, not occupied in any part by Mines previously marked out, and which have not been declared subject to denouncement (*denunciables*).

Art. 55. The demarcation shall be made by notifying, six days beforehand, so that they may be present if they please, the parties interested and the owners of the adjacent Mines or their attorneys if any there be, these citations being shown in the *expediente*. Moreover, there shall also be cited on the land the persons in charge of said Mines.

Art. 56. If there be several registries in the same locality (*comarca*), and they be contiguous, the examinations and demarcations shall be made strictly according to the order of their dates, the oldest first.

Art. 57. On the day designated for the purpose, the examination and demarcation shall be proceeded with before a Notary.

Art. 58. If in making the examination the existence of the deposit (*criadero*) or mineral shall not be confirmed, or there be not land vacant (*franco*), or the working-space (*labor*) shall not have been made in due form, the engineer will suspend the demarcation, reporting to the political Chief, who will declare the *expediente* to be without effect; reserving however, to the party interested, in the first case i. e. when no deposit (*criadero*) or mineral is discovered, the right to continue the works, as works of investigation, provided that the requisites which for that purpose are prescribed in section second of chapter IV, have been or shall be complied with. From the decision of the political Chief, there may be an appeal to the Minister of Commerce, Instruction and Public Works, and from the decision of the latter, to the Royal Council.

Art. 59. If, on the contrary, the existence of the deposit (*criadero*) or mineral, and that of vacant land, and the construction of the legal working space (*labor legal*) be established, the demarcation shall be made in compliance with the provisions of the preceding Articles, and in the following manner: 1st, The *pertenencia* will be marked out by horizontal lines, whatsoever may be the configuration of the ground: 2d, The demarcations of the *pertenencias* shall be made, as a general rule, in the manner in which they have been designated, whether along the line of the deposit (*criadero*), or across it, or in any other way, provided they do not overlap each other in any part, and vacant spaces be not left unnecessarily between them. In the *pertenencias* of auriferous sands, mentioned in the last paragraph of Article 38 of this Regulation, it shall not be required that their sides shall have a constant relation to each other, but the width shall vary in proportion to the length, so that there will always result the *pertenencia* in the aforesaid rectangular figure. Care will also be taken that it shall bound on one of the contiguous *pertenencias*, if there be any, by at least one of its sides. This condition being complied with, and no vacant places or intermediate spaces being left, the *pertenencia* will be marked out in the form which shall best suit the parties interested. 3d, Very visible stakes will be fixed in the ground to designate the lines of the demarcation: 4th, There shall be drawn up an instrument (*acta*) signed by the engineer and those taking part therein, and authenticated by the Notary, in which shall be minutely stated what has been done, describing with exactness each one of the lines of the demarcation, and the places which the fixed stakes occupy to mark them out.

SECTION SIXTH.

Proceedings subsequent to the demarcation.

Art. 60. The *pertenencia* being marked out in the precise period of fifteen days, the original *expediente* will be transmitted to the Ministry of Commerce, Instruction and Public Works, together with, 1st, The *expedientes* relating to the oppositions made, if these shall not have been definitely settled, and to the reclamation, by the owner of the land, of the tenth part of the profits and expenses, as provided in Article 8 of the Law; 2d, Specimens of the ore of the Mine petitioned for; 3d, An exact plan of the demarcation of the Mines which bound on it respectively; which plan shall be drawn by the engineer; 4th, A succinct description, made by the same, of the working space (*labor*) and of the deposit (*criadero*), and of the different minerals which constitute it, its direction, inclination, and capacity, if it be a regular vein; the class of rocks in which it is found, and the other circumstances necessary to ascertain its importance; 5th and lastly, The accidental conditions which, in the opinion of the engineer, ought to be imposed in the concession, the political Chief giving his opinion with respect to these.

Art. 61. The original *expediente* being received in the Ministry of Commerce, Instruction and Public Works, and being added to, in the manner which may be thought convenient, the *Junta facultativa* of Mines, shall first be heard, and afterward the Section of Commerce, Instruction, and Public Works of the Royal Council, as provided in Article 5 of the Law. The Section as well as the *Junta* will give their opinions (*Consultas*) with all possible dispatch.

Art. 62. The formation of the *Expediente* being completed, the Minister of Commerce, Instruction and Public Works will decide upon it. Against his decision the party considering himself aggrieved, may have recourse to the Royal Council.

SECTION SEVENTH.

Of the concession and its conditions.—Issue of the title of property.

Art. 63. By Article 2 of the Law of the 11th of April last, the ownership (*propiedad*) of all the substances which are the special object of Mining (*Mineria*), pertains to the State; and there is no private ownership (*dominio particular*) in this branch, which is not derived by concession from the State, made, in its name, by the Government. Wherefore, no one shall be at liberty to explore or work Mines, even in his own lands, unless such concession has been previously made, with the formalities (*tramites*) prescribed for obtaining it; and every Mine which, without this requisite, may be discovered or opened by the proprietor of the land, may be registered by any other person.

Art. 64. If the decision shall have been to grant the Mine, the conditions of the concession shall be communicated to the party interested, and he having expressed his acceptance according to Article 5 of the Law, there shall be given to him the corresponding title of property (*titulo de propiedad*). This shall be issued in the name of Her Majesty the Queen, and countersigned by the Minister of

Commerce, Instruction and Public Works, being drawn up according to Form No. 9. The fees for issuing the title shall be 60 *reales vellon* for each *pertenencia*, besides the cost of the stamped paper (*papel de ilustres*) upon which it is to be written out.

Art. 65. The conditions under which the concession is made shall be stated in the title. These conditons are general or particular (*accidental*). The concession cannot be made without containing all the general conditions, and besides, according to the tenor of what is provided in the Law, it shall contain the particular conditions which, of those expressed in this Regulation, correspond to each particular case.

Art. 66. The general conditions are either derived from the Law or from the said Regulation. The first are the following. 1st, Obligation to work the Mine conformably to the rules of art, its owners and workers submitting themselves to the rules of police, which the regulations establish, as is provided in Article 21 of the Law. 2d, To respond to all damages and injuries which may result to a third party on account of the working (*explotacion*), in accordance with what is provided in Article 14 of said Law. 3d, That the Miner, in case of using the waters found within his Mine, will make compensation for the damages and injuries which shall be caused to a third party by its exit, conveyance or incorporation into rivers, creeks or drains, conformably to said Article. 4th, That compensation be also made to the neighbors for damages done to them by the waters accumulated in his workings, if, on being required, he shall not drain them off within the time which may be prescribed, as provided in Article 15 of the Law. 5th, That he will contribute, in proportion to the benefit he shall receive by the draining of the adjacent Mines and for the general adits of drainage or of transportation, when, with the authorization of the Government, they shall be opened for the group of *pertenencias*, or for those of the whole Mining district (*comarca*) in which the Mine granted is situate, in accordance with the same Article. 6th, That he will begin the works within the period of six months from the date of the concession, or within eight months, if it be for old rubbish and slag-heaps,-not being prevented by superior force, as is provided, with respect to Mines in the second number and last paragraph of Article 24 of the Law, and, with respect to rubbish and slag-heaps in the second number of Article 31 of said Law. 7th, To keep the Mine or slag-heap occupied or worked by at least four regular operatives for each *pertenencia*, conformably to Articles 22 and 30 of said Law. 8th, To not leave the Mine unoccupied (*despoblada*) for four consecutive months, nor eight interrupted months in the course of one year, not being prevented by superior force, as provided in the 3d number and last paragraph of Article 21 of said Law. 9th, That, if the concession be of rubbish or slag-heaps, he will not interrupt the operations of reduction for more than two months, no superior force intervening, in accordance with what is provided in the 3d number of Article 31 of said Law. 10th, To secure the Mine within the time which may be prescribed, when, on account of mal-direction of the works, it threatens to fall in, superior force not preventing, as provided in the 4th number and last paragraph of Article 24 of the Law. 11th, Not to render difficult or impossible the ulterior enjoyment of the mineral, by a greedy working of it, as mentioned in the 5th number of said Article 24. 12th, Not to suspend the working of the Mine with the intention of abandoning it, without first giving notice to the political Chief, and to leave it well secured, in accordance with what is provided in Article 23 of the Law. 13th, And lastly, to pay for the Mine and its products the taxes which the laws now, or may hereafter, impose, as provided in the 6th of the Transitory Dispositions of the said Law.

The general conditions of the Regulation are: 1st, Obligation to establish the works necessary for the security and health of the settlements (*poblaciones*) and of the operatives. These works shall be those which the political Chief may direct, hearing the engineer; and in case the proprietors (*empresarios*) do not agree thereto, those which the Government shall direct, hearing the *Junta facultativa* of the branch. From the decision of the Government in these matters, there can be, from its very nature, no appeal (*recurso*). 2d, To execute the works which, in the terms set forth in the preceding condition, shall be prescribed as necessary in order to avoid the loss of waters and of irrigations.

Art. 67. Moreover, according to the particular circumstances of the Mine, there may be imposed one or more of the following particular conditions: 1st, Obligation to occupy the Mine with a greater number of operatives than is provided in Article 22 of the Law. This will be required only when the Mine contains a substance which the State requires for its security and defense, and in case the workings permit it, hearing the report of the *Junta facultativa.* 2d, To permit the intervention of the military authority in Mines which are situate within 1,500 *varas* of fortified places, and in the works (*labores*) made for the purpose of investigation, which, by shafts or galleries, are opened with the permission of the Minister of war, within the said distance from fortified places and points. 3d, To observe the dispositions which the political Chief may make, hearing the engineers of roads, when the works of the Mines are executed within the belt of 30 *varas* on each side of highways and canals. With respect to these works, in case of the non-conformity of the Miner, there will be observed what is prescribed in the first of the general conditions of the Regulation, Article 66. 4th, To deliver into the storehouses of the State, the quicksilver and the salt which, in use of the right conferred by Article 6 of the Law, they may work purposely, or the salt which they may accidentally find, which delivery must be made in conformity with said Article, so long as said articles continue to be a monopoly in favor of the public Treasury, and at the prices and with the formalities which may be established. 5th, To permit the intervention which the Treasury may deem expedient to establish in these Mines of monopolized goods (*efectos estancados*), in order to reconcile the exercise of industry with the interests of the State.

Art. 68. If the person making registry oppose the concession by not admitting one or more of the general or particular conditions, it will be immediately published in the *Gaceta*, the *Boletin Oficial* of the Ministry, and in that of the province where the Mine is situate, stating the condition which is resisted. If, in view of this publication, any other company (*empresa*) or individual shall desire to take the Mine with the said resisted condition, the matter will be proceeded with in the following manner: 1st, Petition shall be made in writing to the political Chief, making out in full the indorsements, registry, and certificate (*resguardo*) for the party interested, provided in Article 8. 2d, A copy of the writing shall be communicated to the grantee who has resisted the condition, in order that, within the precise term of fifteen days, he may make answer whether he withdraws his opposition to the resisted condition or conditions, or whether he yields the right of preference which the law gives him. If he do not answer within that period, his silence shall be construed to be a surrender of his right. 3d, The answer of the grantee being received, or the said period for making it having expired, the political Chief will transmit, with his report, the *expediente* to the Ministry of Commerce, Instruction, and Public Works, for a decision with respect to the concession to the new petitioner.

SECTION EIGHTH.

Of the taking of possession.

Art. 69. The title of property being issued, the party interested will have recourse to the political Chief, representing that he has received the title, and asking that in virtue thereof possession of the Mine be given to him. This act (*acto*) shall be executed in the form following: 1st, The owners or representatives of the adjacent Mines, if there be any, shall be cited three days beforehand, in order that they may be present thereat. This citation shall set forth the demarcation of the boundaries of the Mine of which the possession is to be given, for which purpose *Form No.* 10 will be followed. 2d, On the day and hour appointed, the landmarks (*mojones*) of the *pertenencia*, which the party interested will have ready for that purpose, will be definitely fixed, placing them exactly in the points where the stakes stuck in making the demarcation are found. 3d, Then, the grantee shall be placed in the possession of the Mine, with all the legal formalities. 4th, A document (*diligencia*) shall be made out, in which the act is stated, and it shall be signed by the party interested and the others present, and be authenticated by the Notary.

Art. 70. The landmarks (*mojones*) being once fixed, with the solemnity prescribed in the preceding Article, they cannot be moved without a previous public *expediente*, approved by the Ministry of Commerce, Instruction and

Public Works; and the grantees shall be obliged to preserve them always erect and very visible, under the penalty of a fine of from 400 to 1,000 *reales*.

SECTION NINTH.

Of the division of pertenencias and adjudication of the surplus.

Art. 71. When the concession of a Mine includes two or more *pertenencias*, and the party interested, availing himself of the right which is conferred by Article 12 of the Law, wishes to divide them, he will petition the political Chief, who will ask a report of an engineer, afterward transmitting the *expediente*, with his own report, to the Ministry of Commerce, Instruction and Public Works. The latter, in view thereof, and completing the formation of the *expediente* if deemed necessary, will grant or refuse the authorization for the division asked for.

Art. 72. For the fulfillment of Article 13 of the Law, whenever there is, between two or more *pertenencias*, a space which has at least a rectangular superficies equal to, or greater than, two-thirds of the extent of an ordinary *pertenencia*, there will be formed and granted a new *pertenencia*, if any person should petition for it. If no one shall ask for it, or if the space be less than above prescribed, it will be adjudicated as surplusage to the owners of the adjacent Mines, in proportion to the lines of contact.

Art. 73. Therefore the whole vacant space cannot be adjudicated to one *colindante*, even though he alone asks for it, without administrative notification of the petition to the others, and their express renouncement, or the renouncement implied by allowing the ten days to expire without answering. Every renouncement of any part of the surplus shall be considered as made in favor of the applicant for the surplus, provided, the land which corresponds thereto has lines of contact with his *pertenencia*.

Art. 74. In the application for and adjudication of these vacant spaces, the proceedings which are to be observed, are as follows: 1st, Petition in writing to the political Chief, registry and certificate (*resguardo*) in accordance with Article 8. 2d, Administrative notification, with the period of ten days to the owners of the adjacent Mines, there also being inserted in the *Boletin Oficial* of the province an edict announcing the application, and giving notice of the said term of ten days within which any one interested may reply. 3d, The ten days after the notification having expired, on a day which shall be appointed, and after citation to all the applicants for the surplus, an engineer will, by order of the political Chief, make the examination. If it be thus ascertained that, in conformity with the law, it ought to be granted, it will be divided between the adjacent Mines in proportion to the lines of contact, having regard to what is provided in Article 72 of this Regulation, the boundaries of each *pertenencia* being designated by very visible stakes. 4th, This being done, a document will be written out in which it shall be so stated, which document shall be signed by the engineer and those taking part in the proceeding, and be authenticated by a Notary. 5th, The engineer will then transmit the *expediente*, with his report, to the political Chief, and the latter will send it up to the Minister of Commerce, Instruction and Public Works for his decision; from which decision appeal (*recurso*) may be taken to the Royal Council. 6th, The surplus being granted, if the Mines which have a right to it shall not all, or any of them, have been yet marked out, the part which is to be added to them as surplus shall be included in their respective demarcations, making mention of this circumstance in the proceedings (*diligencias*) of the same, and stating it in the title of property, when that is issued. 7th, To the owners of Mines previously granted there shall be issued new titles of property of the *pertenencias*, and possession shall be given of the surplus, in the terms provided in Articles 64 and 69.

Art. 75. The grants of *pertenencias* of Mines shall be announced in the *Gaceta*, in the *Boletin oficial* of the Ministry, and in that of the province where the Mine is situate.

CHAPTER VI.

Of the working and enjoyment of Mines.

SECTION FIRST.

Of the waters found in Mines.

Art. 76. The ownership (*propiedad*) of waters found in a Mine belongs to the owner of the land, according to the common law; but the owner of the Mine has a servitude in them for their enjoyment, so far as they may be necessary for all the uses of the working, while he is in the possession of the Mine. All this is in fulfillment of what is provided in Article 14 of the Law, with the obligations which it imposes. If the owner of the land shall propose to use the surplus waters which are not applied to the working of the Mine, the necessary works for that purpose shall be on his account.

Art. 77. When the breaking out (*aparicion*) of the waters, their conveyance or incorporation into the rivers or creeks, or their accumulation in the workings (*labores*) of the Mine, cause damages which, according to Articles 14 and 15 of the Law, the Miner has to pay, the political Chief, hearing an engineer, will, by virtue of his office, or on petition of a party, require him either to drain them off or prevent the injury, constructing the works requisite for this purpose, within the period which shall be assigned to him. If the Miner shall not do this, in addition to indemnity for the damages, the political Chief, using the power conferred on him by Article 21 of the Law, will impose on him a fine, according to the gravity of the case, of from 400 to 2,000 *reales*, and double that sum in case of a repetition of the offense.

Art. 78. The cognizance of questions relating to the appraisment or indemnification of injuries in the cases which are designated in Article 15 of the Law, there being no agreement between the parties, belongs to the civil tribunals, with the proceedings prescribed in the first paragraph of Article 19.

SECTION SECOND.

Of general galleries of drainage or of transportation, and those of investigation.

Art. 79. When an individual or a company (*empresa*) desires to open general galleries of drainage or of transportation, for a group of *pertenencias* or for those of an entire Mining district (*comarca*), for the observance of the said Article 15 of the Law, the following proceedings will be observed: 1st, The authorization of the political Chief will be asked for, in writing, to open said galleries, accompanying the plan with a project and detailed estimate of the works and a report in which these works are minutely described, forming also a calculation of their benefits. This project and report must be drawn up and signed by an engineer. 2d, The political Chief, having admitted the application, will cause to be inserted three times in the *Boletin Official* an edict, announcing the project, stating that the report, plans and estimates are in the office of the Secretary of the political Government, where any one who wishes may examine them, within a period which he shall designate, and which shall not exceed thirty days, during which time there shall be admitted all the protests or objections (*oposiciones*) which the owners, or parties interested in the Mining district who are affected by the work, or their representatives, may present. Within the said period there shall be admitted all the proposals for changing or improving the project, which any one may present. 3d, Notice of the project shall be given, administratively, to all the owners of Mines of the group or district, for which the general gallery of transportation or drainage is proposed to be opened, in order that, within the said period, they may examine it and represent what may be deemed expedient for their right. 4th, The period having expired, with examination of the answers or objections, if any have been made, and of the propositions presented, the engineer, having previously made an examination of the ground, will make his report. In this report the objections will be examined, the manner of obviating them as much as possible explained, and finally, an opinion shall be given on the project presented, if there be only one, or which is to be preferred, if there be several, establishing the conditions on which the proposal which is deemed preferable ought to be acceeded to. 5th, The political Chief, hearing the provincial Council will then send up, with his opinion, the *expediente* to the Ministry of Commerce, Instruction and Public Works, which, hearing the *Junta facultativa* of the branch, and completing the information on the matter in any other manner, if deemed necessary, will decide upon the authorization asked for. 6th, In this authorization the conditions on which it is granted shall be stated, which conditions shall be established according to the provisions of the following Articles. 7th, From the decision of the Minister appeal may be taken to the Royal Council.

Art. 80. The concession being made in accordance with Article 15 of the Law, the owners of the Mines interested in the general gallery of drainage or transportation, are not only obliged to consent to the works, but also to bear the expenses thereof, in proportion to the benefit they may have received or shall in future receive, while their workings (*labores*) continue.

Art. 81. The dimensions of a principal shaft (*pozo*) of drainage, in which the machines or apparatus for draining are to be established, shall not exceed the maximum of 24 by 10 feet, without reckoning the thickness of the masonry, in said maximum. These dimensions shall be determined in each particular case. The shaft shall, as a general rule, be sunk perpendicularly or vertically, unless economy and the consent of the owner of the *pertenencia* in which it is constructed require that it be inclined.

Art. 82. The dimensions of a shaft (*lumbrera*) for giving ventilation to the galleries of drainage shall not exceed the maximum of 10 by 6 feet, without reckoning the masonry or lining (*entibacion*), within which maximum they shall be fixed in each particular case. With respect to its direction the provision of the preceding Article shall govern.

Art. 83. In the galleries of drainage, whether starting from a suitable place on the surface, or from the interior of the principal shaft for drainage, the maximum hight for single galleries shall be 11 feet, by 5 feet wide. Double galleries, when it is expedient to establish them, shall have as a maximum only 8 feet hight by 12 feet width, constructing, in such a case, a division-wall two feet thick.

These dimensions shall be understood as in the clear, and not reckoning the thickness of the masonry or supports, but including the spaces for the circulation of the water and air.

Art. 84. The inclination of the galleries shall be such as to prevent the waters from standing. When in a district (*comarca*) of general drainage there may be productive workings lower than the level of the draining galleries, these shall have water-tight wooden troughs, such as the engineer shall deem necessary, in order to prevent the infiltration.

Art. 85. The times for cleaning, and the measures requisite to prevent fragments and dirt from getting into the galleries and machines for draining, shall be prescribed in each particular case.

Art. 86. If particular Mining enterprises (*empresas*) should desire to make use of the draining works to take out, with greater economy, their ore and rubbish, they may agree with the draining enterprise (*empresa*) upon the conditions. The same disposition shall govern, with respect to communications for the purpose of ventilation.

Art. 87. If the most economical prosecution of the draining works shall require the said communications for ventilation or for the extraction of rubbish, they shall be made of the smallest dimensions which, in the opinion of the engineer, may be expedient.

Art. 88. If a draining undertaking (*empresa*) fail to fulfill its object, or to comply with one of the clauses expressed in its act of authorization, it shall be liable to denouncement, the same as any other particular Mine, for non-compliance with the Law or any of the conditions of its concession.

Art 89. If an individual or an enterprise (*empresa*) shall desire to open adits or general galleries of investigation, they will petition therefor to the political Chief, accompanying their application with a topographical and geological map of the ground which it is proposed to cross; and, in case it is to cross *pertenencias* already granted, the consent in writing of the owners of these, which, according to Article 18 of the Law, is indispensable; consequently, when this requisite does not accompany the petitions, the matter will not be proceeded with. In other respects the *expediente* will pass through the same forms as are prescribed in Article 79 for concessions of authorization for the establishment of drainage and transportation galleries.

SECTION THIRD.

Of the workings (labores) of Mines.

Art. 90. Mines requiring to be worked according to the rules of the art, as is prescribed in Article 21 of the Law, their owners are obliged to keep them clean, drained, ventilated and well secured, under the penalty of a fine of from 400 to 2,000 *reales*, and double that amount in case of repetition, and the payment, in all cases, of damages and injuries.

Art. 91. In order that the preceding Article may be duly complied with, and for the observance of all the dispositions of Article 21 of the Law, and the Regulations of the branch, the engineers of Mines will exercise a direct vigilance over them, under the authority of the political Chief, and also under his authority and that of the civil Chiefs and Alcaldes in their respective cases, in all matters belonging to the branches of police, health, and security of the same.

Art. 92. The local authority, in order to dictate any disposition on this subject, must hear the engineer, if there be one. But he may, on his own responsibility, act against the engineer's opinion, rendering immediately an account to the political Chief. The latter may do the same, in his case respectively, communicating it to the Minister of Commerce, Instruction and Public Works.

Art. 93. An engineer will visit each Mine, at least once a year, in order to examine its condition, and the disposition and security of its works. The engineer who makes the visit will give to the owners of the Mines, or to those in charge, the instructions which he may consider expedient for the better direction of the workings (*labores*). He will state the defects which he notices, and the means of correcting them, bringing all to the knowledge of the political Chief, in order that he may oblige the owners of the Mines to execute these precautions, under the corresponding fines, in accordance with Article 21 of the Law, and within its limits.

Art. 94. Each engineer will keep a book of visits, in which he will note all the visits which he shall make. Moreover, of each visit he shall write out and enter in the book which, for that purpose, the owner of the Mine or the person in charge shall keep, a record of his proceedings, (*acta*) stating therein the condition of the *labores*, the observations which he may have made on the same, and the instructions which he may give. The owner, or the person in charge of the Mine, will also sign the act (*acta*) in said books, in proof that the said instructions have been communicated to him.

Art 95. The object of these duplicate books is to furnish evidence that the engineer (*facultativo*) and the Miner respectively perform their duties, for which purpose the political Chief may examine them whenever he shall deem it expedient.

Art. 96. The engineer will commence his visit by examining whether the directions which he gave on his former visit have been complied with. The result of this examination will be stated in the *acta* and in the books of visit. If negligence or omission shall appear, he will immediately report it to the political Chief. He will then continue his visit in the manner prescribed in the preceding Articles.

Art. 97. In the reducing offices he cannot inspect the secret of the process employed; but if the owners or those in charge shall ask the intervention of the engineer, he will give the instructions which he shall deem expedient. The public authority and its agents shall not intervene in these establishments, except for approved causes of health or public order, under their responsibility.

Art. 98. Besides the annual visits, those which may be necessary will be made at any time, with the same formalities, whenever the engineer shall deem it expedient, or whenever the political chief, in virtue of his office or on petition of a party, shall direct.

CHAPTER VII.

Of cases of forfeiture of the property of Mines.

SECTION FIRST.

Of the abandonment of a Mine.

Art. 99. It being the duty of the grantee of a Mine to return it to the State without deterioration, when he does not find it expedient to continue working it, the following dispositions for the abandonment will be observed: 1st, The person interested will give notice thereof to the political Chief, 15 days beforehand, by means of a petition, stating his motives for the abandonment. He will state therein whether he has taken care to close up all the mouths of the Mine, except the entrance, and will accompany it with a map of the interior of the Mine. 2d, The political Chief will, without delay, acknowledge the receipt of this notice, for the security of the party interested. 3d, He will then direct an engineer to immediately examine the Mine, and report on the correctness of the map, and on the acts which are stated in the first paragraph. 4th, If it result that these have not been done, the necessary works will be executed at the expense of the Mine-worker, if, by his fault, the abandonment has become necessary. At the expense of the same, the entrance to the Mine will also be closed up. 5th, The political Chief will then direct that the abandonment be announced in the *Boletin Oficial*, in order that any other company (*empresa*) or individual may petition for the property.

Art. 100. The owner of a Mine who suspends the works with the intention of abandoning it, without complying with the requisites of previous notice, incurs, according to Article 23 of the law, a fine of from 400 to 2,000 *reales*, and he shall be responsible for all damages and injuries which the suspension of the works may cause to the said Mining *pertenencias*, or to a third party, and for the taxes which may become due until the abandonment is legally declared.

Art. 101. As soon as the abandonment of a Mine or reducing office, or of a *pertenencia* of a slag-heap, without the requisite of previous notice having been complied with, comes, through the report of an engineer, or any authority or functionary, or by denouncement of a party, or by any other means, to the knowledge of the political Chief, he will direct to be made the examination prescribed in the third paragraph of Article 99, and on the report which the engineer shall give, he will make the official declaration of the abandonment, requiring of him who abandoned the Mine the responsibility prescribed in the fourth paragraph of said Article 99. In case the party interested shall deny the abandonment, the *expediente* will be continued by the proceedings (*tramites*) which are prescribed for those of termination (*caducidad*) of Mining *pertenencias.*

SECTION SECOND.

Of Denouncements.

Art. 102. When a grantee of Mines shall come within any one of the five cases, mentioned in Article 24 of the Law, in which the right to a Mine is forfeited (*se pierde*), the political Chief, either by virtue of his office or by denouncement of a party, will make declaration of the termination (*caducidad*) of the grant, by the proceedings prescribed in Article 20 of the Regulation.

Art. 103. When a denouncement of a *pertenencia* is presented, in addition to the aforesaid proceedings, the following shall be observed: 1st, In the writing of denouncement there shall be stated the name and situation of the Mine, the name and residence of its owners, and whichever of the cases, mentioned in Article 24 of the Law this one may happen to be, all in accordance with *Form No.* 11. 2d, Note shall be taken of the registry, and the certificate (*resguardo*), prescribed in Article 8, shall be given. 3d, A copy of the writing of denouncement shall be communicated, by administrative notification, to the grantee of the Mine, in order that he may allege what he shall deem proper. 4th, If he deny the facts alleged, the political Chief, commissioning an engineer, will take cognizance of them. And if the political Chief shall think he ought to insist, the matter will be contentious-administrative (*contencioso-administrativo*); it being conducted in the provincial Council between the Administration and the grantee, in the form prescribed in the 4th paragraph of Article 20. In this trial the denouncer cannot be a party, no right of his being involved, until, the termination (*caducidad*) having been declared, the registry of the Mine which he has denounced may be admitted. 5th, Nevertheless, when the political Chief shall decide against the denouncement, the denouncer may have recourse to the Minister. 6th, The termination (*caducidad*) being declared by the political Chief without opposition, or being established by final decision, the denouncer will be notified, in order that he may solicit, within the precise term of thirty days, the grant of the forfeited Mine (*la Mina caducada*). 7th, If he shall not wish to apply for it, the forfeiture (*caducidad*) will be announced in the *Boletin Oficial* of the province, in order that any other enterprise (*empresa*) or individual may ask for it. 8th, In either case, the proceedings of the *expediente* of concession will be those prescribed in Chapter V for registries, in the book of which, the new petition for the Mine shall be noted, omitting, as unnecessary in this case, the proceedings prescribed in order to ascertain whether the mineral has been found.

Art. 104. The forfeiture (*caducidad*) of the concession of a Mine cannot be declared, nor shall the Mine be considered liable to denouncement, for neglect to pay imposts, or for any other cause than those mentioned in Article 24 of the Law.

CHAPTER VIII.

On the concession and enjoyment of old heaps of slag and rubbish.

Art. 105. He who shall propose to acquire a slag or rubbish-heap formed from old abandoned Mines, whose slag and refuse-heaps may, in conformity with Article 27 of the Law, be liable to denouncement, shall request in writing the concession of the political Chief, the *expediente* being made with the formalities, established for registries of Mines, with the abbreviations and variations which shall be stated.

Art. 106. A professional examination of the ground shall be made by the engineer, observing the following dispositions: 1st, Citation shall be made, three days beforehand, by administrative notification, of the party interested and the owners of the adjacent *pertenencias*, if there be any, in order that they may be present at the examination. 2d, There shall be marked on the ground three or more spots where the parties interested must open, within the period of thirty days reckoned from the date of the examination, an equal number of pits or trenches, of the depth necessary to form an idea of the nature of the rubbish-heap or the slag. 3d, Specimens taken from different points of the slag or rubbish shall be collected. 4th, There shall be drawn by the engineer an exact map, in duplicate, of the whole extent and figure of the slag or refuse-heap. These maps shall contain the following circumstances: *First*, they shall be drawn on a scale of 1 to 3,600; *Second*, they shall have marked on them the natural circumference of the heap by a continuous curved line, and the boundaries of the grant asked for; *Third*, They shall have the name of the slag-heap or place, and the provisional number of the petition; *Fourth*, They shall contain a particular explanation of the locality, and of its lines and adjacent places, and the demarcation of the three or more points designated in order to ascertain the depth of the heap; *Fifth*, They shall be signed by the engineer, and the parties interested, and the others present.

Art. 107. The examination having been made, the engineer will send up to the political Chief the maps and the specimens of slag or rubbish, reporting circumstantially the result of the act.

Art. 108. The thirty days designated for the opening of the pits or trenches, mentioned in the 2d paragraph of Article 106, having expired, an examination of these works will be made, and they will proceed to make the demarcation of the *pertenencia*. This demarcation shall be made in accordance with what is provided in Article 29 of the Law, in the rectilinear polygonal figure which the petitioner shall select, provided the extent shall not exceed 80,000 superficial *varas*, and there be vacant land for it. The formalities for this class of demarcations shall be those prescribed in the 5th Section of Chapter V of this Regulation, for those of Mines.

Art. 109. If, on the second examination, the works ordered to be made in the first shall not be found completed, and any one shall allege this nullity, the engineer shall suspend the demarcation, reporting the matter to the political Chief, who, in view thereof, will declare the *expediente* of concession to be without effect. For the new demarcation, he who protested shall have the priority, if he make a formal denouncement. There being no protest, the political Chief may grant a period for completing said works, not to exceed fifteen days, and when this has been done, the new examination and demarcation shall be made as provided in the preceding Article.

Art. 110. The *pertenencia* being marked out, the political Chief will transmit the original *expediente* to the Ministry of Commerce, Instruction and Public Works, within the period of twelve days.

Art. 111. The abandonment or forfeiture (*caducidad*) of grants of old slag or refuse-heaps, shall be declared in the cases provided in Article 31 of the Law, and in the manner prescribed in the First Section of Chapter VII, and in Article 20 of this Regulation.

SPECIAL AND TRANSITORY DISPOSITIONS.

1st. The Law of Mining of April 11th, 1849, will, in accordance with the 5th of its Transitory Dispositions, begin to be in force from the publication of the present Regulation in the *Gaceta*, and after the expiration of the several periods required by the existing law, to make it obligatory in each locality.

2d, If the owners of Mines granted prior to the time when the Law and present Regulation take effect, wish to increase the dimensions of their *pertenencias* to 300 varas long by 200 varas wide, measured horizontally, according to Article 11 of the Law, provided that there be vacant land, they will petition the political Chief therefor, and the *expediente* of augmentation will pass through the same proceedings as those prescribed in this Regulation for the registries, omitting those which are had for the purpose of proving the existence of the deposit (*criadero*) or mineral.

3d. The grantees will continue in the enjoyment of the rights which they have acquired in conformity with the laws and dispositions in force heretofore; but in matters of police and direction of the works of the Mines, in applications for augmentations by the vacant spaces, and with respect to jurisdiction, proceeding (*tramitacion*) of the *expedientes* on matters relating to their *pertenencias*, and in all else which may not be civil rights, they will be subject to the provisions of the present Law and of the Regulations for its execution.

4th. Any person who may wish to establish reducing works with high furnaces (*altos hornos*) or catalan forges (*forjas catalanas*), in which vegetable fuel is to be used, will, in conformity with the Law, apply for the proper authorization through the political Chief, who will hear the councils (*Ayuntamientos*) of the Pueblos where the charcoal is to be made, and the Commissioner of forests of the district, and then transmit the *expediente*, with his report, to the Government for the proper decision.

5th. The Superior Tribunal, and the Direction General of Mines, are abolished. The tribunal and the inspections of districts will cease to exercise jurisdiction of the branch, immediately transferring, for continuation, the pending matters, according to their condition and nature, to the competent tribunals, in conformity with the Law.

6th. The Direction General of Mines will transmit to the Ministry of Commerce, Instruction and Public Works, the *expedientes* of registries and denouncements, and the adminstrative *expedientes* which are before it. The *expedientes* of inchoate registries and denouncements, according to the prior laws, will be continued according to the provisions of the same laws, the political Chief doing what was incumbent on the inspectors of district, and the Ministry of Commerce exercising the functions of the abolished Direction General.

7th. Until the passage of a special law regulating imposts on Mines and their products, the following will be collected:

Mines granted prior to the present law will pay the same contribution of superficies (*de superficie*) as they paid before its publication.

Every Mine which shall hereafter be granted, whose dimensions may he 300 *varas* by 200, will pay, as the tax of superficies, 600 *reales* per annum.

Mines of coal, lignite or peat, which have a length of 600 *varas* by width of 300 *varas*, will pay the same as has been heretofore assessed upon *pertenencias* of the same class.

When Mines have less dimensions than those designated in Article 11 of the Law, they will pay the tax of superficies in proportion to their extent.

Besides the tax of superficies, there will be paid as heretofore five per cent. of the total products, at the price which they bear at the places of production.

8th. The corps of Engineers of Mines, the schools of the branch, and the Mining establishments belonging to the State, shall continue to be governed by special Regulations, which will be drawn up in conformity with the Law and this Regulation, and in the meantime they will observe the old Regulations so far as they do not conflict with these.

Approved by Her Majesty, July 31st, 1849.

BRAVO MURILLO.

Form No. 1.

Receipt or Resguardo of Petitions for grants.

Political Government of the Province of ——.

I, A— B—, secretary of the same, certify that C— D—, citizen of ——, residing in ——, on the — day of ——, 18—, [the date to be in writing] presented to this political government a petition in writing dated ———— [here state particularly and clearly what has been asked for, the name of the Mine, pertenencia, or slag-heap, the locality where it is found, and its boundaries in case a registry is asked for; if the petition is for denouncement on account of forfeiture or abandonment, there will also be stated the name of the owner of the Mine, and the act or cause on which the denouncement is based; if the petition be for grant of permission to work substances not included in Mining, there shall be stated the class of the substances asked for, in what land they are situate, who is the owner thereof, and for what establishment or manufacturing industry they are intended, referring strictly to the terms in which the petition is expressed.]

And for the security of the party interested, and in order that he may use it as evidence where and when it may please him, I give him this present, with the *visto bueno* of the Political Chief of the Province, in conformity with the provisions of Article 8 of the Regulation for the execution of the Mining Law of April 11th, 1849.

Dated ————.

A— B—, *Secretary.*

(*Visto Bueno* of the Political Chief.)

FORM No. 2.

Page of the Day Book (*Libro Diario*).

1849

July 7th.

Registry. Book of Registry, page	*Quebradilla* (Mine of), by C— D— resident of ——. It is situate in ——. He asks —— pertenencias.

July 8th.

Demarcation.	*Esperanza y Concepcion* (Mine of). Registered by E. F. in Book —— page ——. Examined on the —— by the engineer D— his communication being received this day.
Inspection of District.	Engineer D— sent to the same, and to the Province of —— by Royal Order of ——.

July 16th.

Registry.	*San Teodoro* (Mine of). For giving possession. M— and N— owners of the adjacent Mines *Recreo* and *Narciso*, cited by administrative notification, directed to the Alcalde of—.

Surplus (*Demasia*).	*Hernan Cortes* Mine; its owner Don M— solicits the vacant space. Inserted to-day in the *Boletin Oficial* of the Province, No—, notifying administratively N— an owner bounding on it.

Form No. 3.

Page of Book of Registry.

Book — of Registry No. — page — Mine of (class of mineral), A— B— (the person registering), Diario of Mines, Book — page — (Here insert name of Mine, and name of person registering).

A—B— citizen of —— residing in —— on the (here insert the date with the hour, in writing) presented a petition in writing for the Mine of (class of mineral) called —— situate in the Pueblo of —— in the Municipality of ——. He asked —— pertenencias in virtue of (here state the case of Article 11 of the Law; and, if the registry is asked by a company, that it is accompanied by the writing of partnership). The agent A— B— Don M— residing in —— his domicil in ——. There was issued to the party interested a *resguardo* below.

(In continuation there will be stated the other proceedings of the *expediente* in the order of their dates, placing the headings in the margin.)

[The alphabetical index which the books of registries should have, is to be at the end of the book. In it there shall be noted, beginning with the first letter of the name of the Mine and of the name of the person making the registry, the number of the registry and the page, as follows:

Amistad, (Mine called *de la*). Vide No. 1, page 60.

Arias (Don Juan). Vide No. 1, page 60.

The Books of registries shall also have, after the alphabetical index, a numerical index in the following form:

No. 1. Mine *Amistad*, registered by Don Juan Arias, page 60.

No. 2. Mine *Consuelo*, registered by Don Pedro Fernandez, page 61.

(RESGUARDO.)

Mine *Amistad*, Registry No. 1, Book A, page 60.

POLITICAL GOVERNMENT OF THE PROVINCE OF ——.

I, A. B., Secretary of the same, certify that Don *Juan Arias*, citizen of ———, residing in ———, on the ——— day of ———, A. D. 18—, at 2 o'clock, P. M., presented to this political Government a petition in writing, dated ——, (here state clearly and particularly what is asked for, the name of the Mine, *pertenencia* or slag-heap, the place where it is found and its boundaries, and the other circumstances of registry, as stated in the petition and in the folio from which this *resguardo* is taken.)

And for the security of the party interested and in order that he may use it when and where he may please, I give him this present, with the countersign (*visto bueno*) of the Political Chief of the Province, in conformity with what is prescribed in Article 8 of the Regulation for the execution of the Mining Law of April 11th, 1849.

Dated —————.

A. B., *Secretary*.

(*Visto Bueno* of the Political Chief.)

FORM No. 4.

Page of the Book of Denouncements.

Book of Denouncements No. 4, page 20, Mine *Amistad* of (metal), situate in ———, registered on the (date), by

Don Juan Arias, in Book A, of Registry No. 1, page 60.

Don Juan ——, citizen of ———, residing in ——, on the (date) presented in writing a petition of denouncement of the Mine of (metal) called *Amistad*, the property of *Don Juan Arias*, situate in ———, of the Municipality of ——. He founds the denouncement on the —— case of those included in Article 24 of the Law, saying, &c.

[There will be given below the *resguardo* of this denouncement, and in case the defeasance (*caducidad*) be declared, the denouncement shall be deemed a registry, and thirty days shall be allowed for making it in due form.]

(RESGUARDO.)

Mine *Amistad*, No. 1 of denouncements, book B, p. 20.

POLITICAL GOVERNMENT OF THE PROVINCE OF——.

I, A— B—, Secretary of the same, certify that Don Juan Blanco, citizen of——, residing in—— on the (date) at— o'clock, P. M., presented in writing, dated the——denouncing the Mine called *Amistad*, situate in the——, of which Don Juan Arias is in possession, based on its having come within the case (or cases) of Article 24 of the Law; which Mine is recorded in page——of Book——of the Diario, and in page——of Book——of registries, being therein numbered No.——

And for the security of the party interested and in order that it may serve him when and where he may please to use it, I give him this present, with the *visto bueno* of the

Political Chief of the Province, in conformity with the provisions of Article 8 of the Regulation for the execution of the Mining Law.

Dated ——————

A— B—, *Secretary.*

(*Visto Bueno* of the Political Chief.)

FORM NO. 5.

Petition for Registry.

I, Don Juan Arias, —— years of age, (of civil rank), native of —— citizen of —— resident in ——, of (profession or employment). (These same circumstances will also be stated of the representative of the party interested, in the municipal district, when there is such a representation, observing that the person who registers, when he does not reside in the municipal district where the Mine is, must have such an agent, &c.) Before Your Honor represents: That I desire to acquire in accordance with the Mining law the ownership (*propiedad*) of—— pertenencias of the Mine of (mineral) situate in the (place) of the Pueblo of —— in the municipal district of ——. The Mine petitioned for will be called by the name of *Amistad.*

The land in which it is found is the property of (state name, residence and circumstances of the owner), bounded by, (state names and owners of adjacent Mines, clearly and precisely, or say, bounded by no other Mining pertenencia, but by ——). The said *criadero* or mineral, of which specimens are herewith presented, was discovered in (simple trial-pits or in investigations by pits or galleries, in virtue of the corresponding concession dated ——). (In case more than two *pertenencias* are asked, the reason why they are so asked, will be stated, in accordance with Article 11 of the law, sending therewith the writing of partnership, when, on the ground of the company being composed of four or more persons, three *pertenencias* are asked for).

Wherefore I pray Your Honor (V. S.) to be so good as to admit the present petition, causing it to be inserted in the registry of Mines of the Province, and note to be taken of it in the Diario of Mines, and giving me the proper *resguardo*. And the proceedings prescribed in the Law and Regulation of the branch being previously had, to send up the *expediente* to the Ministry of Commerce, Instruction, and Public Works, in order that the grant may be made to me, and that there may be issued to me the corresponding title of property in conformity with the Law and Regulation of the branch.

Dated ————————

JUAN ARIOS.

To the Political Chief of the Province of ——.

FORM No. 6.

Admission of the Petition for Registry.

The petition of registry having been presented; let it be noted in the *Diario de Minas*, and in the *Registro de Minas* of the Province, giving to the party interested the corresponding *resguardo* of the same leaf in which it is registered, in which, state the day and hour of its presentation; and pass it to the engineer of the branch in order that he may make the preliminary examination of the Mine (with citation of those in charge of the adjacent Mines, marked out or to be marked out, if any there be); reporting at the foot of this document whether the *criadero* or mineral really exists, and if there be sufficient vacant land for the concession; whether it was found by simple trial-pits, and if the specimens presented are of the same or a different class from those which he finds in the Mine, the engineer then returning the *expediente* to this political Government for the resolution which may be required.

Dated ——————

C— D—, *Political Chief.*

FORM No. 7.

Admission of the Registry.

In view of the preceding report, by which it is shown that the *criadero* or mineral exists in the point registered, and vacant land for the concession asked for, the petition of registry is admitted; let note be taken of it in the books of *Diario* and of Registry; give to the party interested the proper document for his security; and let the edicts be posted, and the notice be given in the *Boletin Official*, in the manner prescribed in Articles 44 and 45 of the Regulation for the execution of the Law of the branch. (The form of the *resguardo* may be seen in Form No. 1, modifying it, to suit the particular circumstances of the particular case.)

Dated ——————

C— D—, *Political Chief.*

Form No. 8.

Refusal of the Registry.

In view of the preceding report, and that (there is no *criadero* or mineral in the place registered, or there is no vacant land for the designation of the pertenencia) the present registry cannot be admitted. Let note be taken in the books of the *Diario* and of Registry of Mines of this Province, and immediately give thereof administrative notification to the party interested, or to his representative.

Dated ———.

C— D—, *Political Chief.*

Form No. 9.

Titles of property of Mines and slag-heaps.

I.—TITLE OF PROPERTY OF MINES.

Doña Isabel II, by the grace of God and the Constitution of the Spanish Monarchy, Queen of Spain: Whereas I have deemed it proper to grant to Don A— B— by Royal order of the ———, the ownership (*propiedad*) of the Mine of ——— called ——— situate in the place ——— of the Pueblo ——— Municipal district of ——— in the Province of ———, with the conditions which were stated in said Royal Order, and were accepted by the party interested: I have resolved, (with the date ———) that there be issued to him the present title of property, in conformity with what is prescribed in Article 5 of the Law of Mining, with the insertion of the following conditions:

First. The Mine is composed of —— *pertenencias*, each one of which is composed of a solid having a rectangular base, —— *varas* long by —— *varas* wide, of an indefinite depth in a vertical direction, without including the superficies. Its demarcation is as follows: ————————

————————————————————

Second. The grantee accepts and promises to comply with the following general conditions, which the Law imposes on him:

1st. That of working the Mine conformably to the rules of the art, and subjecting himself and his workmen to the rules of police which the regulations may prescribe, as is provided in Article 21 of the Law.

2d. That of responding to all damages and injuries which, from working the Mine, may be caused to a third party, according to the provisions of Article 14 of the said Law.

3d. That of making indemnity, in case that he uses the waters found within his Mine, for the damages and injuries which their breaking out (*aparicion*), conveyance, and incorporation into rivers, creeks or drains, may cause to a third party, conformably to said Article.

4th. That of making indemnity also to his neighbors for the injuries which he may cause them by the waters accumulated in his workings, if, on being so required, he does not drain them within the time which may be assigned him, as is provided in Article 15 of the Law.

5th. That of contributing in proportion to the benefit he may receive by the drainage of the adjacent Mines, and by the general galleries of drainage or of transportation, when, by the authorization of the government, they shall be opened for a group of pertenencias or for the whole Mining district (*comarca*) where the Mine is situate, in accordance with the said Article.

6th. That of beginning the works within the period of six months from this concession, not being prevented by superior force, as is provided in the second number and the last paragraph of Article 24 of the Law.

7th. That of keeping the Mine occupied (*poplada*) and in working with at least four regular operatives in proportion to each *pertenencia*, conformably to Article 22 of the said Law.

8th. That of not leaving the Mine unoccupied (*despoblada*) for four consecutive months, nor eight interrupted months in the course of one year, not being impeded by superior force, as is provided in the third number and last paragraph of Art. 24 of the said Law.

9th. That of securing the Mine in the time which may be assigned to him, when by the improper direction of the works it threatens to fall in, not being impeded by superior force, as is provided in the fourth number and last paragraph of Article 24 of the Law.

10th. That of not rendering difficult or impossible the ulterior enjoyment of the mineral by a greedy working of it, as is determined in the fifth number of said Article.

11th. That of not suspending the works of the Mine with the intention of abandoning it, without giving previous notice to the political Chief, and that of leaving its supports in good condition, according to what is provided in Article 23 of the Law.

12th. That of paying the duties which the laws impose, or

shall impose, on the Mine and its produce, comformably to the sixth of the Dispositions of the said Law, called Transitory.

Third. The grantee also accepts and obligates himself to comply with the general conditions of the Regulation for the execution of the Mining Law, approved July 31st 1849, to wit:

1st. That of establishing the necessary works for the security and health of the settlements, or of the operatives, in the terms prescribed in Article 66 of the Regution.

2d. That of executing the works necessary for preventing the waste of waters and of irrigations, according to the 2d of the General Conditions of the Regulation, which are contained in said Article 66.

Fourth. The grantee also accepts, and promises to comply with the following particular conditions, included in Article 67 of said Regulation.

(Here insert the one or more of these conditions which shall have been imposed in the concession.)

Wherefore, in virtue of this Royal Title, I grant to Don A—— B—— the ownership (*propiedad*) of the said Mine of ———, called ————, for an unlimited time (*por tiempo ilimitado*), while he complies with the preceding conditions, in order that he may work it, enjoy its products, and dispose of them freely, alienating it in such manner as he may please; all in subjection to the laws. And in order that the contents of the said conditions may be complied with and observed punctually, both by the said grantee and by the authorities, tribunals, corporations and individuals whom it may concern, I have ordered to be issued this present title of ownership, signed by my Royal hand, sealed with the corresponding seal, and countersigned by the undersigned Minister of Commerce, Instruction and Public Works.

Given at ——————

[L. S.] I, THE QUEEN.

The Minister of Commerce, Instruction and Public Works.

(Signature of the Minister.)

Her Majesty issues in favor of Don A— B—, the title of property of the Mine of ———, called ———, situate in the place ——— of the Pueblo, in the Municipal district of ———, in the Province of ———.

Registered at folio — of the corresponding book, at number ————.

II.—TITLE OF PROPERTY OF A SLAG-HEAP.

Doña Isabel II, by the grace of God and the Constitution of the Spanish Monarchy, Queen of Spain: Whereas I have been pleased to grant to Don A— B—, by Royal Order of ——— the ownership (*propiedad*) of the Slag-heap called ————, situate in the point —— of the Pueblo of ———, Municipal district of ———, in the Province of ———, with the conditions which were stated in said Royal Order, and were accepted by the party interested, I have resolved with the date —— that there be issued to him the present title of property, in conformity with what is prescribed in Article 5 of the Mining Law, with the insertion of the following conditions:

First. The slag-heap is composed of ——— *partenencias*, each one of which is composed of a solid having a rectangular base of —— superficial varas. Its demarcation is as follows: ————————————————

Second. The grantee accepts and promises to comply with the following general conditions which the Law imposes on him :

1st. That of working the slag or rubbish-heap conformably to the rules of the art, subjecting himself and his operatives to the rules of police which the regulations may prescribe, as is provided in Article 21 of the Law.

2d. That of responding to all damages and injuries which from working it may be caused to a third party, according to the provisions of Article 14 of the said Law.

3d. That of beginning the working of the slag or rubbish-heap within the period of eight months, reckoned from the day of its concession, as is provided in the second number of Article 31 of said Law.

4th. That of keeping the slag-heap occupied (*poblada*) with at least four operatives, as prescribed in Article 30 of said Law.

5th. That of not interrupting the operations of working for more than two months, no superior force intervening, as is determined in the third number of Article 31 of said Law.

6th. That of not suspending the working of the slag-heap without giving previous notice to the political Chief, according to the provision of Article 23 of the Law.

7th and last, That of paying the taxes which the laws have imposed or shall impose on the slag or rubbish-heap and its produce, conformably to the sixth of the Dispositions of the said Law, called Transitory.

Third. The grantee also accepts, and obligates himself to comply with the general conditions of the Regulation for the execution of the Mining Law, approved July 31st, 1849, to wit:

1st. That of establishing the necessary works for the security and health of the settlements or of the operatives, in the terms prescribed in Article 66 of the Regulation.

2d. To execute the works necessary to prevent the waste of the waters and irrigations, according to the second of the general conditions of the Regulation, which are contained in said Article 66.

Fourth. The grantee accepts and promises to comply with the following particular conditions, included in Article 67 of the said Regulation.

(Here insert the one or more of the conditions which shall have been imposed in the concession.)

Wherefore, in virtue of this Royal Title, I grant to Don A—— B—— the ownership (*propiedad*) of the said slag-heap, called ——, for an unlimited time (*por tiempo ilimitado*), while he complies with the preceding conditions, in order that he may work it, enjoy its products, and dispose of them freely, alienating it in such manner as he may please; all in subjection to the laws. And in order that the contents of the said conditions may be complied with and observed punctually, both by the said grantee and by the authorities, tribunals, corporations and individuals, whom it may concern, I have ordered to be issued this present title (*titulo de propiedad*), signed by my Royal hand, sealed with the corresponding seal, and countersigned by the undersigned Minister of Commerce, Instruction and Public Works.

Given at ————

[L. S.] I, THE QUEEN.

The Minister of Commerce, Instruction and Public Works.
(Signature of the Minister.)

Her Majesty issues in favor of Don A—— B—— the title of property of the slag-heap called —— situate in the place —— of the Pueblo of —— in the municipal district of —— in the Province of —— Registered at folio —— of the corresponding book at number ——.

FORM No. 10.

Official Order to the Alcaldes to cite the owners of adjacent Mines to be present at the giving of possession.

POLITICAL GOVERNMENT OF ——

The title of property of the Mine of —— called —— having been issued to Don A—— B—— and it being proper to proceed to give him formal possession of it, I have appointed for the performance of that act the —— day of —— at —— o'clock in ——.

Of which, as is provided in Article 69 of the Regulation for the execution of the Law of Mines, you will give notice administratively to Don C—— D—— owner of —— adjacent called —— in order that, if he please, he may concur in the act, giving him a copy of the demarcation of the said Mine, which is as follows: (Here insert it.)

Dated ———

M— N— *Political Chief.*

To the Civil Chief or Alcalde of ——

Form No. 11.

Petition of Denouncement.

Don A—— B—— of the age of — years, of (state civil rank), native of —— citizen of —— resident of —— of (state profession or employment). (Also there shall be stated these circumstances of the representative of the party interested, in the municipal district, when he has one). To Your Honor (V. S.) represents that the Mine of —— which Don C—— D—— resident in —— situate in the place —— of the Pueblo of —— in the municipal district of —— (here shall be clearly stated the act or acts which give cause for the denouncement).

Being therefore included in —— paragraph —— of Article 24 of the Law of Mining.

I pray Your Honor (V. S.), the proper proceedings being had, to declare the forfeiture (*caducidad*) of the said Mine

---

admitting for me then the present denouncement, and issuing to me the proper *resguardo*, in order to secure the use of my right when the time comes.

Dated ——————.

C—— D——.

To the Political Chief of ——.

APPENDIX No. II.

MINING ORDINANCES OF PERU,

1785—1786.

[From Appendix to Mexican "Ordenanzas de Mineria."]

Royal Order, December 8*th*, 1785, *directing that the Mining Ordinances of New Spain, dated May* 22*d*, 1783, *be modified and adapted to the Viceroyalty of Peru.*

The King, convinced of the deteriorated condition into which the important branch of Mining of that Kingdom has fallen, from a want of method in governing the *Reales* of Mines, and also on account of the frequent and troublesome litigations in which the individuals of this useful profession are involved, causing them enormous expenses and distractions from their business, by requiring them to reside in the Capital and in other places where they go in the prosecution of their law-suits; and His Majesty desiring to give to this important branch all the encouragement which it requires and is susceptible of; having tried the good effects produced in the Kingdom of New Spain by the practice and execution of the General Ordinance of

Mining enacted for it on the 22d of May, 1783, of which several copies have been transmitted to Your Honor (V. S.); the necessity of like provisions and rules being in fact most urgent in that Kingdom; the King wills that Your Honor (V. S.) immediately proceed to put in practice and to adapt the said Ordinance to its local circumstances, establishing a Royal Tribunal General in that Capital, and the Courts of Appeals and territorial Deputations which you may deem necessary, in the localities and places most proper for the purpose and object of said Ordinance, proceeding in concert with the President of Chili with respect to what relates to that Kingdom, forasmuch as its limited extent of territory cannot sustain a separate Tribunal General, and as it may be sufficient to have therein one or more Courts of Appeal.

Your Honor (V. S.) will also take care to regulate the territory, in which the Royal Tribunal General is to exercise the contentious jurisdiction which Article 2, Title III, of the said Ordinance gives to that of Mexico, without necessarily limiting it to the twenty-five leagues in extent which is there prescribed, for this matter is to be regulated by Your Honor (V. S.) with prudence, so that said Tribunal General shall not be embarrassed so much with contentious matters as not to be able to attend to its economical and directive duties, nor, on the other hand, be without a suitable territory within which to exercise its contentious authority.

With respect to appeals from the decisions which the Royal Tribunal General may give, of which Article 13 of Title III treats,. the King wills, that, in place of the Judge (*Oidor*) who, as there provided is to be the Judge of Appeals, Your Honor (V. S.), as Superintendent of Royal Treasury, and also your successors in that office, shall constitute the head of said Court of Appeals, and that, in the form of hearing appeals and in the other rules which the said Article and the succeeding ones of that Title provide, punctual observance and attention be given to what is there prescribed.

His Majesty has also resolved, that, in place of the two-thirds of a *real* per mark, which, in Article 1, Title XVI, of the said Ordinance, he grants to the Royal Tribunal of Mexico for its endowment and expenses, on all the bullion which shall be delivered to that Mint and to any others which may in future be established in the said Kingdom, or which shall be sent to Spain on account of private owners, remitting the ancient double duty of one *real* on each mark, which was paid to the Royal Treasury under the title of seigniorage (*señoreage*), the Miners of that Kingdom contribute, for the expenses of the Body, one *real* per mark of that coin, in the same form and under the rules as prescribed for New Spain with respect to said two-thirds of a *real*.

Although Article 4, Title I, of the said Ordinance provides that the offices of Administrator-General and Director-General of the Tribunal shall be for life, as in that determination he was principally influenced by the personal circumstances of the two individuals who in Mexico filled these offices at the time the Tribunal was established, Your Honer (V. S.) will provide that, in that Kingdom, that proceedings be immediately had to appoint for both offices, in the regular form which the Ordinance provides for such appointments, after the death of the Administrator-General and Director-General who now hold office in Mexico.

Finally, Your Honer (V. S.) is advised that the King hopes favorable results from this establishment, which results Your Honor (V. S.) will report with all possible accuracy and clearness, mentioning the doubts or difficulties which may occur to You, and the means which You may deem best calculated to remove them, since it is the will of His Majesty that there be established in Peru, as soon as possible, the same practice as in New Spain, and he confides this matter entirely to the care and activity of Your Honor (V. S.)

God preserve Your Honor (V. S.) many years.

Madrid, December 8th, 1785.

MARQUES DE SONORA.

To the Superintendent Subdelegate of the Royal Treasury of Peru.

August 1st, 1786.

Decree, directing that the foregoing Royal Order be obeyed.

Let what His Majesty commands be observed and complied with, and, inasmuch as, in order not to retard His Majesty's worthy intentions, the proper directions (*providencias*) have already been communicated to the Intendants, it is necessary to send to them the Ordinance of Mining of New Spain, with suitable declarations and directions, so as to remove doubts and to facilitate its adaptation to, and observance in Peru, said Ordinance will be reprinted, placing, at the end, the foregoing Royal Order and this Decree, and the Declarations which are to be made in the order of its several Titles, to the end that being promulgated, and all persons being able to acquire a knowledge of these rules and notices, they may be made effective, and that, without substantially varying that wise plan, its effects may be as rapid and advantageous as has been experienced in Mexico; and inasmuch as this first step is the foundation of the good which is to result to Mining (*Mineria*), the printing will be paid for out of the *real* per mark which is commanded to be paid, for this expenditure is very conformable to the objects to which it is applied, and is made to forward them; it being required that the necessary number of copies be sent to all the Deputations and Mining places (*Minerales*), and that the remainder be preserved in the Archives of the said Tribunal which is to be erected.

Lima, August 1st, 1786.

JORJE ESCOBEDO.

October 7th, 1786.

Dispositions, explaining and changing the Ordinances of New Spain, so as to adapt them to the Mining of Peru.

Art. 1. (*Title* I, *Article* 1, *of the Ordinance of New Spain*). The Royal Tribunal General of the Important Body of Mining of Peru (which it is thus to be called) shall be held and esteemed by all, with the same respect and consideration as is declared by His Majesty for that of Mexico.

Art. 2. (*Title* I, *Article* 2.) Two Deputies-General being sufficient for the present condition of the Mining of Peru, and for the funds with which the Establishment now commences, those who are to compose the Tribunal are limited to that number, but, besides these, the Tribunal shall always have, and be composed of, an Administrator-General, who shall be its President, and a Director-General; and if the *Reales* or *Asientos* of Mines should be increased, then the Tribunal may add another Deputy, making three, which number can never be increased.

Art. 3. In the said form, the Tribunal must begin its duties with the coming year, for which purpose, on the 31st day of December of this year, the elections will be held, to which end, by the circular order of June 12th, there were then given to all the *Intendencias* the instructions (*providencias*) which will be continued now, passing, with the printed copies of this Ordinance, the proper official letters and notices to His Excellency the Viceroy, to the Royal *Audiencia*, and to the other tribunals and offices which are entitled to receive them.

Art. 4. (*Title* I, *Articles* 4, 8 *and* 13.) All these terms of office, beginning with the persons who are now elected, shall only continue for the time specified in Articles 8 and 13 of Title I of the Ordinance of Mexico; and, although His Majesty has declared, that, inasmuch as the same circumstances do not occur here as in Mexico, they shall be hereafter elective, as there could not have come to his Royal notice the appointment of Directors, made beforehand, in the persons of Don José Coquett and Don Santiago Urquizu, and as, moreover, these persons are now considered the most suitable, on account of their intelligence and zeal, and the knowledge which they have acquired in the Commission of Mining with which they were charged, they will continue for the present as Directors, being governed in every respect, as to their powers and duties, by what is prescribed in the Ordinance of Mexico, and without, on that account, separating themselves from the establishment and instruction of reduction by smelting, which has been the most important of their duties.

Art. 5. (*Title* I, *Art.* 5.) In the said circular order of June 12th it is prescribed, as an important matter, that the Deputies take part in the elections, in order to be personally instructed and informed of what is most expedient for the encouragement of the Mines, and reference is also made to the power which they ought to transmit with the necessary instruction, in case that, on account of the distance or other just reasons, it should not be convenient for them to come, and, in order to prevent the disorder and abuses which may result therefrom, it is added that the individuals empowered to act in place of the owners and suppliers of Mines, may be private persons or employes of the Royal Treasury, provided none of them have been voted for in the elections; and the same will be observed in all the subsequent elections, in order that, being literally regulated by Article 3 of Title I of the Ordinances of Mexico, these offices may not be filled by persons who reside in Lima, and who either never have worked Mines, or have abandoned the business, and are not practical, intelligent and expert Miners.

Art. 6. (*Title* I, *Art.* 6 *and* 7.) In Article 15 of these Declarations, the Deputations which have been ordered to be now established are mentioned, because it is not possible to have them in all the Mining establishments (*Minerales*), on account of their poverty and on account of the geographical character of the Kingdom and the severity of its climate, especially when they are at a great distance and in unpopulated places; and under these circumstances said Deputations shall have a vote in the elections, and also the others which may be established, there concurring in them the circumstances prescribed in Article 6, Title I, of the Ordinance of Mexico; for this must be the general rule in future, as also that no one shall have more than one vote, unless some Mining place (*Mineral*) shall be so noted and distinguished for its good order, riches, works, and a punctual observance of these Ordinances, as to merit in future the preference which, in Mexico, has been given to a few; which matter will be determined by the Tribunal, subject to the approval of the Superintendency of the Royal Treasury.

Art. 7. (*Title* I, *Art.* 8.) The scrutinies mentioned in Article 8 of Title I of the Ordinances of Mexico, will be held in the house of the Superintendency, at present and until the Tribunal shall have a house of its own, and there also will be held the other *Juntas* for the conference and hearing of the Deputies, who will meet as often as may be deemed conducive to the welfare and encouragement of Mining and the punctual execution of this new plan.

Art. 8. (*Title* I, *Art.* 15.) In attention to the notable difference between this Viceroyalty and that of Mexico, as well in the extent of land as in the number of Mines, Miners and Mine-suppliers (*aviadores*), the number of Advisers (*consultores*) will be reduced to six, at least for the present, and two or three of these must be Suppliers (*Aviadores*), or Miners resident in Lima, or in Places (*Lugares*) near by, in order that use may be made of those persons who, although they have given up the business, have the instruction and other necessary requisites, as provided in Article 15 of the Ordinance, which, in other respects, will be literally observed.

Art. 9. (*Title* I, *Arts.* 17 *and* 19.) The just respect and attention due to His Excellency the Viceroy, as first Chief of the Kingdom, require that the scrutinies and elections shall not be proceeded in without his permission, and that he be notified of them, the persons elected immediately presenting themselves to him, if they are present in this Capital, and if not, as soon as they arrive in it; and the same obligation and formalities are to be observed toward the Superintendency ot Royal Treasury, to whom His Majesty has especially confided the subject of Mining as being so immediately connected with the government and regulation of his Royal exchequer, and by the hands of the said Superintendency the annual report will also be transmitted, and whatever else may occur respecting the working of the Mines and the condition of things pertaining to the Body of Miners.

Art. 10. (*Title* I, *Art.* 20.) The Tribunal will appoint, now and in future, the agent (*apoderado*) who is to represent its affairs at the Court of Madrid, and if it be necessary to send a person of greater confidence for any matter or occurrence, it must represent the cause before the Superintendency of Royal Treasuary, and having its approbation, will ask, through the Superintendency, the permission of His Excellency the Viceroy.

Art. 11. The Superintendency of the Royal Treasury will be at liberty, whenever deemed expedient, to examine the archives and books of the Tribunal, to see if everything be kept and made out with the proper method, clearness and forms, and will watch over the faithful observance and execution of these ordinances, as well with respect to the Tribunal and its members, as to the territorial Deputations, in order to prevent abuses which, especially at first, may be introduced, notwithstanding the just confidence with which it is expected all will devote themselves to the punctual execution of the new plan, and the greatest success and progress in Mining.

Art. 12. (*Title* II, *Art.* 2.) It is already stated in Article 6, that it is not possible to place Deputies in each Mining place (*Mineral*), but not for this reason shall all the Miners neglect to enroll themselves, and in order that this may be done, it is commanded in the circular order of June 12th, and the command is now repeated as a general rule, that those who live and work in the small Mining places (*Minerales*) which, from their poverty and want of population, have no Deputations, shall be annexed to the *Real* or *Asiento* of Mines which may be nearest and in which there may be Deputations, so that, the book of enrollment being drawn up and kept in said *Asiento*, it will include the other Miners scattered about in that district, and all may form the body which has to be recognized as its territorial head, the Deputation, whose jurisdiction and care shall extend to such Mining places (*Minerales*) and Miners, with the same zeal and attention as to those of the principal *Asiento* in which it resides, extending to them the same privileges, voice and vote as to the others.

Art. 13. (*Title* II, *Art.* 3.) Although the election of Deputies ought to be held in the beginning of January of each year, it will be anticipated, for this time only, in order that they may, by this means, concur in voting for the officers of the Tribunal, which voting must take place in December of this year, and those elected must have, as far as possible and present circumstances will permit, the qualifications prescribed by the Ordinances of New Spain, and they shall hold their offices all the following year and until, in January, 1788, the new Deputy, who is to come in, is elected, and the same alternation shall be continued in the subsequent years; as, for this end, has already been directed to all the *Intendencias* in the circular order of June 12th.

Art. 14. (*Title* II, *Art.* 7.) In order that the elections may be held with the proper effect, legality and tranquility, it will be very consonant to the zeal and care of the Intendants to issue to the Sub-delegates the proper instructions, directing them to be personally present at the elections and to conduct themselves with the greatest impartiality, not restraining the liberty of the Miners nor exerting their influence directly or indirectly in favor of any one, nor receiving any pay or fees which they are not entitled to receive, not even under the name of presents which those elected may desire to make to them; and if it be proved that any one has obtained the office by means of agreements, or other disgraceful measures and presents made to the Judge or to other electors, in addition to the punishment which shall be inflicted according to the nature of the offense, they shall all be incompetent for six years, either to vote or to be elected; the same being understood with respect to the elections of the Tribunal, in order to avoid in all the elections, the abuses with which they may be vitiated, and also the pursuit of trade and the promotion of private interest, in view of which, forgetting the welfare and encouragement of Mining, persons may aspire to these offices.

Art. 15. (*Title* II, *Art.* 8.) The multitude and quality of Mining places spread over the entire Kingdom and mostly in positions unpopulated and of the most severe climate, do not permit that there be established Deputies in each one; and as this is in future to be regulated by the said Tribunal, which, according to its practical knowledge, and the greater or less opulence and number of Mining places, will propose to the Superintendency those in which Deputies ought to be established, or whether those already established ought to be removed, in order that there may be some immediately established who can concur in the formation of the Tribunal and in exercising the proper jurisdiction, taking care to have the enrollment and other matters and necessary notices attended to. For the regulation and progress of Mining in each Intendancy there are designated for the present, in that of Lima, the *partido* of Huarochiri; in that of

Tarma, the *Mineral* of Pasco; in that of Guamanga, the *Mineral* of Lucanas; in that of Guancavelica, Castro-Vireina; in that of Cuzco, the *Mineral* of Curaguasi, of the *Partido* of Abancai; in that of Arequipa, Calleoma and Huantajaya; and in that of Trujillo, El Cerro de Chota or Gualgayoc, in order that in these eight *Asientos* there be established immediately their respective Deputations to which the other *Minerales* of the district will attach themselves as directed in Declaration 12, and without thereby preventing the establishment of any other additional Deputation, if it shall be deemed necessary by the Intendants, to whose zeal and local information it is left to be determined, taking care to give notice of it, and that in such case the enrollment be divided in such a manner that each book and Deputation shall include the *Minerales* nearest to it, and may have the four Substitutes which are necessary for the important ends provided by Article 9 of Title III of the Ordinances of Mexico.

Art. 16. (*Title* II, *Articles* 11 *and* 14.) Inasmuch as, for the present want of a Tribunal, it has been now directed to all the *Intendencias* that the election of Deputies being held, they will report it to this Superintendency for its approval, it will be understood that in future this must be obtained by means of the said Tribunal to whom the report will directly be made with a letter signed by the Judge who may preside at the election and by the Deputy or Deputies who go out and by those who are newly elected, and in the same manner these will have recourse to the Tribunal if there shall happen the case spoken of in Article 11.

Art. 17. (*Title* II, *Article* 15; *Title* III, *Article* 36.) In order that the Deputations may be administered faithfully and with more application, comfort and pleasure, a moderate salary will be assigned the members, provided that this do not come out of the Royal Treasury nor from the fund of a *real* per mark which the Miners are to contribute, and which is to be applied to more interesting uses; and with this view the Deputies, after the enrollment has been made, will have an understanding with all the Miners embraced in it, in order to examine into the most suitable and proper means which they may deem most secure and opportune for these salaries, and will propose them to the respective *Intendencia* in order that, through it and with its report, they may come to this Superintendency, where the matter will be determined as may seem best, after hearing the Royal Tribunal General and substantiating the *expediente* according to its nature.

Art. 18. (*Title* II, *Article* 16.) The Deputies who shall now meet in Lima at this first election, will take care to come informed of all the matters prescribed in Article 16, in order to move in the *Juntas* which shall be held what may be useful and expedient for the regulation which is desired, and the encouragement which is expected to be given to all the *Minerales;* and the same will be done by their successors; without on that account omitting on any pretext the detailed report which they must make in February of each year to the Royal Tribunal, in order that it may report to the Superintendency, which will take care to inform His Majesty; and to this end they will be furnished by the courts, Royal exchequers (*cajas*) and other offices with the documents which may be asked, without the necessity of a decree or order of the *Intendencia* or other court, or of directing to it the said report, since this would be a useless work, when its Chief, by his near locality and the superiority which belongs to him in the offices, will readily obtain these notices which are suited to their zeal, and which to the Deputations it would be very laborious to duplicate.

68

Art. 19. It follows from what has been said, and is conformable also to Royal Orders, that in no Mining place (*Mineral*) ought there now to be a Chief Alcalde of Mines, even when he is not paid out of the Royal Treasury, but by voluntary agreement and election of the Miners, as besides the inutility and other inconveniences of these offices, their functions are embodied in those which belong respectively to the territorial Judge and to the Deputations, as provided in this Article and the Declarations by which they are to be regulated.

Art. 20. (*Title* III, *Arts.* 1, 2 *and* 3.) It being understood that both the Royal Tribunal General and the Deputations are to have in Peru the same jurisdiction and powers as in Mexico in matters of government, direction and economy, as also that in contentious matters, they shall all be regulated in these matters by the letter of that Ordinance; it being understood that in the *Intendencia* where there may be only one Deputation, there shall devolve upon it all the powers and cares of that district; but where there are two Deputations, each one shall exercise its functions with entire independence of the other, both limiting themselves to those Mining places (*Minerales*) which may have been joined to them respectively in the enrollment, and with the strict and inseparable subordination which they must all observe to the Tribunal General, which, with respect to contentious jurisdiction will exercise it only in the territory of the *Intendencia* of Lima, which has been marked out to it conformably to the Royal Order and will of His Majesty.

Art. 21. (*Title* III, *Arts.* 4 *and* 5.) As in the *Partidos* of the Intendencia of Lima there are hardly one or two in which any Mines are worked, and as in fact only in that of Huarochiri can it be said that their working is carried on with any formality and encouragement, if the contentious jurisdiction declared to the Tribunal should not be extended to this *Partido*, it would be useless, for it would rarely be exercised; but as in that *Partido* there is also a Deputation, in order that there may be no embarrassments or disputes, it is directed that, notwithstanding what is provided in Declarations 12 and 15, the enrollment and Deputation of Huarochiri must be limited to only the Mining places (*Minerales*) and Miners of its own *Partido*, since those of Canta, Yauyos, Chancai, Santa, Cañete, Ica, and Cercado, will have direct understanding in everything with the Tribunal, which will take charge of the enrollments, taking the requisite notices, and giving to the Subdelegates the proper advices, and under this supposition the contentious jurisdiction of the Tribunal in these seven *Partidos* although of little use, will be provided for; and in order that it shall be in that of Huarochiri and that its Deputies may also exercise it in the same manner, it is declared that the Miners shall resort to these in verbal demands even though they shall exceed two hundred dollars, and if the parties acquiesce in what said Deputies shall determine, the suit (*recurso*) shall here end; but if they do not agree the contentious jurisdiction of the Deputies shall end, and that of the Tribunal shall begin, in order that if the matter do not exceed two hundred dollars it may hear and oblige the execution of what may be verbally decided, and if it exceed that sum, the interested parties not agreeing, the petitions in writing will be admitted before the said Tribunal, following in all respects the Ordinance of Mexico.

Art. 22. (*Title* III, *Art.* 12.) It being impossible in all the Mining Places (*Minerales*) to pay a Notary, constables, (*alguaciles*) and other officers, the Deputations will avail themselves of the services of those who, according to the practice of the Kingdom, the territorial Judges employ for the execution of their orders and decrees, or will give a special commission to a certain person according to the importance of the business, it being understood that the office of Notary of Mines and Registries being sold (*rematado*), such will have to reside strictly in the *Asiento* where the Deputies are, who for this purpose shall apply to the respective Intendant, as also to propose to him in case there be no such office vendable and renunciable, that it may be created anew, in order that the *expediente* being drawn up it may be determined in the Superior *Junta;* but, in all events, it shall be the duty of the Deputies to deliver to their successors by inventory the papers and proceedings which to this end shall be kept with due fidelity and legality, taking care that the Notaries, where any there be, perform the duties of their offices in the same way, and to report to the Intendants whatever may require to be remedied by their provisions (*providencias*).

Art. 23. (*Title* III, *Art.* 13). In the causes which, as directed in the Ordinance of New Spain, admit of appeal, this shall be taken to the Tribunal or respective Court of Appeals (*Jusgado de Alzadas*) which in conformity with the Royal Order of December 8th, 1785, will be composed, in Lima, of the Superintendent of Royal Treasury, as its President, of the Director and of a Miner who shall be elected for this purpose in the general *Junta* every three years, and as he cannot now be had with the qualifications which are prescribed in Article 13 of the Ordinance of Mexico, it will be sufficient that he have those which are prescribed for the Advisers (*consultores*) for the same reason in Declaration 8, and, observing that there are now two Directors, and that neither can the court exceed the three votes which ought to constitute it, nor can the Miners be deprived of what is conceded to them, it will be understood that when both Directors are in Lima they must alternate by months in sitting in the court, with the proviso that when a cause is heard by one it must be decided by the same.

Art. 24. (*Title* III, *Art.* 13.) What is said in the preceding Declaration only refers to appeals which are interposed from the decrees and sentences of the Royal Tribunal General, since for the other territorial Deputations, the Court of Appeals will be composed of the Intendant of the Province and the two Miners, Substitutes, nearest to the capital, observing in their election Article 9 of Title II of the Ordinance of Mexico, and the preference which for the same, in Article 13 of Title III, is granted to the *Consultores* of the Tribunal for these cases.

Art. 25. (*Title* III, *Art.* 27.) As the functions of the Tribunal and Deputations, and the contentious jurisdiction which is assigned to them respectively, ought to begin with the coming year, there will be passed to them the causes begun which, according to their nature and the provisions of the Ordinance of Mexico, belong to them, and in those which are pending in other courts, for the reasons which are expressed in Article 27 of Title III, there will be observed what is in it prescribed, as for this end there will be directed, as is already stated in Declaration 3, the copies of this Ordinance and official orders to all the Tribunals.

Art. 26. (*Title* III, *Arts.* 29 *and* 30.) In Declaration 21 is explained the contentious jurisdiction which in the Partido of Huarochiri its Deputies are to exercise without prejudice to that which is granted to the Royal Tribunal; and that there may arise no doubt in the criminal causes which are treated of in Article 29 of Title III, and that in them the Miners may have prompt proceedings without the necessity of going to Lima in search of them, it is now added, that the territorial Deputation only shall take cognizance of the said causes without any interference by the Royal Tribunal, which may be wanting in that practical knowledge necessary to apply without delay the remedy which is so important for the regulation of the work and for the good order of the Mining establishment; but if an appeal be interposed, it shall be admitted to carry it before the Court of Appeals of this Capital, composed as already stated in Declaration 23.

Art. 27. (*Title* III, *Art.* 31.) The Superior *Junta* of the Royal Treasury will decide the questions of jurisdiction (*Competencias*) which may arise between the Tribunal General of Mining and its territorial courts and other courts, not being of the said *Junta* or Royal *Audiencia*, since in these two cases recourse will be had to His Excellency the Viceroy, conformably to Article 75 of the Royal Ordinance of Intendants.

Art. 28. (*Title* III, *Article* 35; *Title* XIII, *Articles* 5, 7, 8, 9 *and* 11). In matters of supply (*Abastos*), works, and roads, the Tribunal General and the territorial Deputations will have recourse to the respective Intendants, as this is one of the duties with which, by their particular ordinance, they are specially charged.

Art. 29. (*Title* III, *Article* 36). What is said in Article 17 must be equally understood of any other duties, charges or taxes (*arbitrios, cargas ó gavelas*) which are attempted to be imposed for the benefit and encouragement of Mining and for the other objects indicated in Article 36 (of the Ordinance of Mexico), but if the Tribunal General should attempt to establish any tax or charge on the whole body of Miners of the Viceroyalty, it must propose it to the territorial Deputations, in order that, the subject being examined and discussed there, they may report upon it, and, the *expediente* being formed, the Tribunal will pass it, with these documents and such others as it may deem proper, to the Superintendent of Royal Treasury in order that he may add the proceedings which its nature may require, and that, with the resolution which is taken, he may solicit what more may be required, of the Sovereign pleasure of His Majesty.

Art. 30. (*Title* III, *Article* 37). Considering the scarcity of funds with which the Tribunal of Peru commences, without being able to count on more than the *real* per mark which His Majesty has commanded that all the Miners shall contribute, the salaries and employes will for the present be regulated by the plan which is given at the end of these Declarations, notwithstanding that in Mexico it is ordered that the proper Tribunal shall form it, because that Tribunal is to be composed of the authors who shall have organized it, and who with their distinguished zeal have already made the calculations and ascertained the funds on which they can rely; but if those of Peru shall hereafter permit many additions, and more employes shall be necessary for the said objects of the establishment, these and their salaries may be increased by the proper Tribunal, reporting it to the Superintendency in order to obtain the Royal approval, which it will also solicit for those which are immediately to be designated.

Art. 31. His Majesty commanding that the Ordinance of Mexico be immediately put in force and adapted to this kingdom, it follows that in future that must be the rule by which Mining and all its matters are to be governed, nevertheless those of the dispositions contained in the ancient Ordinance of Peru will be observed wherein they are not opposed to the Ordinance of New Spain or wherein that does not provide, and this declaration is here the most opportune that can be made for the understanding of what in various of its Titles is wisely provided in favor of the Miners, with respect to the mode of bringing suits, the ownership (*dominio*) of Mines, their aquisition, registries, measurements, surplus-spaces (*demasias*), method of working them, draining them, etc., in which it will be sufficient to particularize some points where, although easy to reconcile both Ordinances, doubts may arise in practice.

Art. 32. (*Title* VI, *Article* 1.) That there may be no doubt concerning the Mine (*estaca*) of the King, of which Article 19 of Title I of the Ordinances of Peru makes mention, designating it precisely between those which are called the first discovered and the alternate (*descubridora y salteada*), it is directed that in future it ought to be located after the *pertenencias* which are granted to the discoverer, so that if he take them contiguously, the Mine (*estaca*) of the King follows immediately after them ; and if he select them not continuous, this will be located where the discoverer may designate, provided that it be measured between the *pertenencia* of the discoverer and that of any other interested party, and all being governed by the dimensions which the goodness of the King has enlarged in favor of the Miners; this rule will be followed until in the *Reales* of Mines there shall be the professional Experts (*peritos facultativos*) spoken of in Title XVII of the Ordinance of Mexico, for when there are such with the qualifications which are there prescribed, it will belong to them to designate the Mine (*estaca*) of the King, without prejudice to the rights of the discoverer, with whom they will concur for that purpose after he has selected his *pertenencias*.

Art. 33 (*Title* VI, *Article* 15). In the case provided in Article 15, Title VI, recourse must be had to the respective Intendancy, as it is specially charged with the police of settlements and their buildings, and through that magistracy recourse will be had to the Superintendency of Royal

Treasury, in order that, hearing the Royal Tribunal General, it will report to the Superior Government, and with its assent will determine what is most proper.

Art. 34. (*Title* VI, *Article* 17 ; *Title* XI, *Article* 1.) Those who please may form companies with the certainty of being protected and assisted as much as possible; but if such companies shall be formed for extraordinary undertakings, as for the supply of many abandoned Mines, and in order that these may be adjudicated to them, or in order that there be granted to them other uncommon favors, aids and exemptions, they must apply for it to the Intendant of the Province where the parties interested reside, in order that without being under the necessity of going in person to Lima, the application may be conducted before that magistrate, who, it being prepared, shall remit it with his report to the Superintendency in order that it may be by that officer passed to the Royal Tribunal which will report upon the merit and circumstances of the undertaking (*empresa*) and the privileges which they ask for, in order that, with this information, the Superintendent may decide, and report be made to His Majesty, if exemptions and favors are asked which are not within the ordinary authority of the Superintendency itself, even with the aid of the high powers of His Excellency the Viceroy.

Art. 35. (*Title* VI, *Article* 22.) Mines of quicksilver deserve particular attention, and as His Majesty has repeatedly charged this Superintendency General of Royal Treasury to promote their discovery and working, and with this object there have been made various grants, resort will be had to the same Superintendency as heretofore, in order that, examining the matter according to its importance and to the circumstances of the Treasury, there may be decreed and determined what shall be expedient; with the understanding that, for four years reckoned from the day on which permission shall be given to any individual to work Mines of quicksilver, there shall be paid to him, for what he may deliver and enter in the Royal Stores, the price of seventy-nine dollars and three reales, without prejudice to what His Majesty may determine, nor to what, after the expiration of that time, may seem just.

Art. 36. (*Title* VII, *Article* 2.) The period within which, agreeably to what is declared in this Article, secular ecclesiastics must sell and place in the hands of lay subjects Mines and reducing *Haciendas*, is to be fixed by the respective Intendant of the Province in whose district the Mine may be, and these magistrates, in the cases of this class which may occur, may fix said term without delaying their proceedings for the purpose of waiting for or asking report of the Tribunal General of Mining, for it will be sufficient to take that of the territorial Deputation, and that they report what they may direct and do to the Superintendency general, in order that through it the Tribunal of Mining may have the notice proper for the object of its institution.

Art. 37. (*Title* XII, *Article* 1.) Nothing is more important than that there be an abundance of operatives in the working of the Mines, but as they will, if their labors are not properly remunerated, forcibly withdraw from the work, this will be attended to by the territorial Judges and Deputations with the most efficacious care and impartial justice; and inasmuch as, whether from the disordered condition of Mining in Peru, or from its poverty, or other causes it is much to be feared that in this matter there may exist no legitimate custom, and that abuses, both on the part of the owners and of the operatives, have introduced a vicious system, in regard to the daily wages, the mode of paying them, and the hours of work, it will be understood that where such custom exists, justly and legitimately established, it ought to be observed, as well in the Mines as in the mills and *Haciendas* for grinding and reducing the ores; but where there is no such custom, the pay will be regulated by Ordinance 18, Title I, Book 3, of those of Peru, which will also be observed with respect to the hours of work, both day and night, but voluntary agreements between the owners and the operatives, to pay them more, are not by this provision prohibited, and the same shall be done when the situation and depth of the Mines render it just, for the said Ordinance is only to be the rule in order to prevent a reduction of the pay below that which it prescribes, and in order that the most active and proper care be taken that neither the greediness of the owners may underrate the value of labor, nor that of the operatives overrate it and render it unprofitable, by curtailing the hours.

Art. 38. What is said in the preceding Declaration must be understood for the present, for when the Tribunal General is established one of its first cares will be to require all the Deputations to give it an exact and detailed report of the day-wages and hours of work which, whether from legitimate custom or from practice well or ill introduced, may be observed in the Mining places (*minerales*) of their respective enrollments (*matricula*), in order that, with this information and that which the Deputations themselves may add to it of what they may deem just on both points, under the circumstances of the country, the Tribunal may receive such information as will enable them, with reason and justice to propose to the Superintendency the ways and means which appear to them best for the regulation of those points, on which in great part depends the deterioration of Mining by the want of operatives.

Art. 39. (*Title* XII, *Articles* 3, 6 *and* 9.) Both the Ordinances of Peru and of Mexico have always prohibited with the greatest rigor that payment be made in goods, fruits, provisions and other effects; but as the non-observance of these dispositions and of those which also prohibit the binding of the Indians by loans advanced, has been proved by long and sad experience, the territorial Deputations and Judges of the *Partidos* are charged to constantly watch over, without the least dissimulation, the observance of Articles 3, 6 and 9 of Title XII of the Ordinance of New Spain, with the understanding that they will be held responsible and will be punished accordingly, as soon as any neglect shall be observed or proved, which the Intendants will take care to enforce with all rigor, procuring information on the matter secretly, and omitting no means to prevent concealed and unpunished violations of these Ordinances, which, by Article 133 of the Ordinance of *Intendencias*, are already specially charged upon the said Magistrates, to whose distinguished zeal and judgment it is left to prevent the abuses which, may be introduced under pretext of the permission given in Article 6, to pay partly in provisions, for that permission ought only to be used so far as may be sufficient and requisite for supplying the necessities of the workmen, and to provide for their natural subsistence, without there being required, with loss to themselves and to their owners, to go and seek them in distant places; nor on the other extreme ought this to be a mere matter of bargain by which they may become bound for all their lives and even for the lives of their children and descendants, a thing not to be tolerated and which may be prevented, by taking care that the owners settle with punctuality their accounts with the workmen, and that, besides using correct weights and measures, they be limited to equitable prices, according to the quality of the supplies, and to the distance, risks, expenses and other circumstances of their transportation.

Art. 40. (*Title* XII, *Article* 13.) The said Ordinance of Intendants, in Articles 56 and 127, also provides that idle persons, vagrants, and even delinquents, without excepting the tributary class, may be set at useful works and works in Mines; and the means which are now permitted to the owners of Mines, of keeping gatherers (*recojedores*) of such persons, being so well adapted to effect that important object, they (the owners) having a license from the territorial Deputation to use such means, shall solicit a similar license from the respective Intendant, who will take care that they do not exceed the limits of the license.

Art. 41. (*Title* XIII, *Article* 13.) The scarcity of forests in the Mining places of this kingdom makes more important the observance of the directions contained in this Article for the preservation of the few that exist, and for the increase, as much as possible, of this auxiliary, which is required for the smeltings by the Miners and the Royal Mints; and for the same reason, and because all the smelters find the charcoal to be of bad quality, the Tribunal of Mining will form the Regulation which is provided for in this Article, with attention to all these objects, and when formed, will pass it to this Superintendency in order that it may be approved in due form, observing that new discoveries of mineral coal ought to be attended to by the said Tribunal, on account of the good effects which this material, being worked, will produce in the smelting of the ores.

Art. 42. (*Title XVI*, *Article* 1.) The said Royal Order which determines the establishment of the Tribunal, and the adaptation of the Ordinance of New Spain, justly provides that for the expenses and other purposes for which, in Mexico, the two-thirds of a *real* are used, there be collected here a *real* on each Mark of bullion which the Mines produce, and in compliance with this sovereign disposition, the first day of August of the present year was fixed, in the Circular Order of June 12th, for beginning the exaction with equality, and provision made to avoid the concealments or frauds which might be committed, and, with the said object and that of increasing the fund, said collection is ordered to be made in the Royal Treasuries (*cajas Reales*), which shall continue to be done in future, a separate book being kept in them for this branch, which has nothing to do with the Royal Revenue, and for this work neither the officers of the Treasuries, nor the subalterns, nor the assayer nor smelter, shall ask any salary, emolument or reward.

Art. 43. It being well known, that, although all the silver does not go to be coined in the Royal Mint, it cannot be used for any other purpose without being assayed and the duty paid (*quintada*) as this is done only in the Royal Treasuries (*cajas Reales*), that office has been designated for such collection, and for the same reason it ought to watch that they do not smuggle or carry it away without these requisites, and besides the obligation of the Judges and officers of the Royal Revenue to prevent said frauds, the territorial Deputations may also examine into them and report to the Intendants, because the common good and encouragement of Mining are interested that the fund be not deprived of its legitimate incomes, which, the more they increase, will afford proportionally greater and more advantageous aids.

Art. 44. The collection of said *real* must be understood as per mark of silver after smelting, in order to avoid the injury to the Miner of paying it on what the bullion (*piña*) loses in that operation, and it is remarked, in order to remove doubts, that the *real* is not to be diminished by the Royal dues, nor the latter by the *real*, since both are to be collected according to their respective rules, and according to the entire weight of the bar or piece immediately after the smelting.

Art. 45. (*Title* XV, *Article* 10; *Title* XVI, *Article* 10.) For the present, the said collection shall be made only in silver, without any Miner, merchant, or supplier being exempt from it, for even when, on account of the distance of the respective Royal Treasury (*caja Real*), or other lawful reason, it is permitted, in favor of some Mining places, to carry their silver to be smelted out of the district of the *Partido*, Province or Viceroyalty, they must secure the corresponding payment to the satisfaction of the Deputies, if there be such, and of the Royal Judge, and in default of both, of the Receiver of *alcabalas*, for which purpose they shall present the certificate of the Treasury (*caja*) where they carried it to be smelted, (which certificate shall be given without any charge,) and shall state that the *real* per mark has been paid there; but if it be given in the Treasuries (*cajas*) of any other Viceroyalty, it shall contain the number of marks which the piece or pieces had after being smelted, in order that with this notice the collection may be made without delay, which shall be for the entire weight which the bullion (*plata en piña*) had at the time it was taken away, if the said document be not presented within the proper time which, according to the distance, may be prescribed for bringing the return-permit (*tornaguia*) and cancelling the charge of the permit (*guia*) of the Receiver of *Alcabalas*, without which, even though the bond be given, it shall not be transported, and all the bullion (*pastas*) of gold and silver which shall be so found shall be irremissibly confiscated, and the permits (*guias*) of the Indian Alcaldes shall not be sufficient, on account of the frauds which in this matter have been committed, and which are to be prevented by the appointment, in all the Mining places, of Receivers and Commissaries who will issue the permits (*guias*), without a greater expense than four *reales* which will be paid by the party interested in all the Mining districts; and by this rule those of Guantajaya and Lucanas will be governed in the permits which they have already given to carry bullion to Potosi and Lima to be smelted, it being left to the charge of said Receivers to take care, by virtue of their office, of the mutual correspondence and notices of permits which are issued.

Art. 46. (*Title XVI, Art.* 2.) The payment of this *real* will be received from the Miners in coin, or in bullion at its proper fineness and value, and whatever is paid in either way will be kept in deposit and remitted in the same form to the Treasury General of Lima, where, deducting the expense of its transportation at the same rates as those paid by the King, the net remainder will be immediately placed at the order of the Tribunal, to which body official notice thereof will be sent by the officers of the Royal Treasury, in order that it may come and receive it, and the said officers will pass, in Lima, to the Superintendency, a statement or account of the amount of each one of these deliveries, distinguishing between what is coin and what is bullion (*pasta*); and the officers of Royal Revenue of the other Treasuries (*cajas*) and *Intendencias* will remit monthly a like statement of what is deposited, but for no reason or urgency however commendable it may be, can this fund be touched, or its transmission be deferred when there are secure and fit opportunities, except what the Tribunal of Mining may draw upon in said Treasury or Treasuries (*cajas*), as that must be paid, and made to appear by the drafts which must be remitted alone, or with the remaining funds, so that their amount may be included by the Treasury of Lima in the total of those funds which it has to deliver to the Tribunal, the same being so set forth in the account which has to be presented to the Superintendency.

Art. 47. (*Title XVI, Articles* 3 *and* 11.) Even supposing there be made an exact collection in the said form, nevertheless the produce of the one *real* per mark will be very small for the supplies (*habilitaciones*) and other important objects mentioned in Title XVI of the Ordinance of New Spain, if, agreeably to what is inferred from its Article 11, capital is not received by said fund on interest, for which purpose the Tribunal of Mining, shall take the measures which it deems proper, restricting itself to the surplus which may be in the fund after deducting expenses. And on this principle, and because notwithstanding the security and profit which such capital will evidently enjoy, it will be very difficult to find it on account of the notorious poverty and want of funds in the whole Kingdom, this Superintendency has addressed in anticipation, the proper official communications, exciting the pastoral zeal of the Illustrious Bishops, and Regular Prelates, to the end that, with their distinguished love to the King and Country, they may cause the funds belonging to pious works, and to foundations, and the surplus moneys of the Communities, to be immediately put out at interest in the Tribunal of Mining, from which it is expected that some relief may be obtained.

Art. 48. From the statement which has been submitted of the number of marks of silver melted in three years, it is calculated that, even in the present decayed condition of the Mining districts, the contribution of a *real* may amount to $45,000 a year, and as many of the expenses which are designated in the plan do not involve immediate and positive outlays, because some time is required to set up the establishment, and employ the persons, which occasion the expenses, it follows that even when forty thousand dollars are paid for interest at the rate of five per cent., (which is the highest that ought to be paid), the interest would be secured without the least contingency, and also the capital would be free from it, as, should misfortune, which is not to be expected, frustrate the anxious desires of the Royal intentions, the return of the capital, and the payment of the interest which may have accrued, shall alone be attended to, dedicating to this single object the *real* per mark, which is more than sufficient, and is certain, to cover the debt; all which is added for the greater clearness and satisfaction of the foregoing measure.

Art. 49. (*Title* XVI, *Article* 3 *et seq.*) If the fund should prove to be as prosperous as is calculated, (even should it not amount to the sum mentioned), as soon as there is a sufficient amount, it shall be invested with the other sums which from time to time may be received, in the establishment of the bank mentioned in Title XVI of the Ordinance of New Spain, the regulations which are attended with no difficulty in Peru; unless it be considered more profitable, safe and effective, to commence with the establishment of a general bank for the purchase of bullion (*un banco general de rescates*), which, by means of an agent in the mineral districts where there are Deputations, who has the necessary funds, and has given security for his good conduct, shall purchase silver bullion, after it has been burned a second time, in order that, having purchased it with the necessary precautions, at the rate of seven dollars, less one *real*, which shall be retained for the contribution, observing the formalities prescribed in Article 45, he shall remit it to the bank, which shall run it into bars on its own account; and having taken out the *real* per mark, for greater clearness in the accounts, and also the profits which may remain, after paying expenses and the Royal dues, the bank will increase its funds by the double operation of supplying Mines, which investment (the Mining interest having been benefitted by the aid of money placed within its reach, and the higher price paid for bullion), will be more safe, and free from the risks which at present may be apprehended.

Art. 50. What has been said in the preceding declaration is solely for the satisfaction of the public and of the Miners, in regard to the security on which the first may rely for the money lent on interest, and that the latter may be persuaded that the investment will be made for their relief and encouragement; but as to the manner of putting this in practice nothing is said, until the Tribunal of Mining shall examine the matter, and with its better information resolve what destination shall be given to the funds collected; and according to the plan which shall be adopted, and having consulted the Superintendency, the orders and regulations which shall appear most just and necessary shall be given, extending these to provide an abundance of quicksilver, to regulate its retail sale, and other matters relating to its distribution, which by reason of their connection with the purchase of bullion and the supplying of Mines, cannot be treated of at present, until the most preferable course, and that most favorable to the whole Body of Mining, shall be adopted; and as the private banks, measures, and means, by which some Mining districts have contrived to procure funds, produce only beneficial results, they cannot injure the general bank and fund and the contribution of one *real* per mark, and under such circumstances their establishment shall be promoted, the Tribunal of Mining itself encouraging them, and relying for the necessary support on the respective Superintendency and *Intendencia*, in which much has already been done by the activity and zeal of their Chiefs, especially in Chota and Pasco.

Art. 51. (*Title* XVII, *Arts.* 7 *and* 10, *Tit.* IX, *Art.* 10.) The Professional Experts of Mines and of Reduction, mentioned in Tit. XVII of the Ordinance of New Spain, have been, until now, unknown in Peru, where both professions have been followed by persons who, on account of their practical knowledge, have been considered competent, and it being difficult at present to find Experts with the qualifications prescribed in said Title, this practice must be tolerated some time longer, and until, by a collegiate education, and the arrival of the persons whom His Majesty, desirous of the progress of Mining, has sent to Saxony for, there may be persons who may practice these professions with the requisite knowledge; but in the mean time the Tribunal may use with prudence the measure proposed in Art. 10, and if any persons deserving a diploma, should voluntarily present themselves for examination in the district where these are established, they shall be preferred, and there shall be enforced against those not thus qualified the penalties designated in the same Article; and in like manner the Tribunal, without delay, shall form the Tariff spoken of in Article 9, and refer it to the Superintendency, to the end that, the matter having been examined according to its nature, the Miners may escape the expenses and vexations which they have suffered by the visits, the Deputations being governed as regards them by the said Tariff, and by what is provided in Article 10 of Title IX for the due regulation and usefulness of these acts.

Art. 52. (*Tit.* XVIII, *Arts.* 1, 4 *and* 5.) The Tribunal being established, it shall be one of its first objects to regulate the College, according to the plan which is proposed provisionally for officers and salaries, and to what His Majesty may resolve on the arrival of the Professors, whom His Royal benevolence has sent for, as has been noticed; but should any be found here, capable of teaching, the commencement of it, on such terms as may be agreed upon, shall not be deferred.

Art. 53. (*Tit.* XVIII, *Arts.* 2 *and* 3.) That such education may be more universal and certain, and that those who acquire it may not refuse to go and live in the Mineral districts, on account of their distance or climate, it will be very proper for the territorial Deputations to use their best influence to the end that all the enrolled Miners in their district elect two youths from their provinces or *partidos* and pay their traveling expenses and board, who, having the necessary qualifications may be educated in the College; for which purpose they shall contribute annually three hundred dollars for each, and in this manner, there being eight Deputations, there will be obtained from the whole Viceroyalty sixteen students, and the expenses of the Seminary will bear less heavily on the fund; the Seminary at its own cost will maintain eight more students, and will give gratuitous instruction to all who attend its public lectures.

Art. 54. (*Tit.* XVIII, *Art.* 7, *et seq.*) The method of teaching, the sciences to be taught, and the text books shall not be determined, till after the regulations provided for in Article 7 are formed, when that which is most proper will be decided, in view of what His Majesty may resolve after having seen the documents which have already been drawn up and transmitted to him relative to this subject.

Art. 55. (*Title* XIX, *Arts.* 1 to 9.) As the Miners of Peru are to enjoy the same privileges as the Miners of New Spain, they shall be protected in those which are expressed in Article 19 of that Ordinance; and all persons being persuaded of the protection which His Excellency the Viceroy extends to that commendable pursuit, they may rely upon it, and on that of the respective *Intendencias*, and of this Superintendency which will not fail to recommend to His Majesty the Miners who may most distinguish themselves, and who for their industry, poverty and other circumstances may deserve to experience, in their own persons and families, the Royal benevolence.

Art. 56. Finally, in Article 31, it has already been stated, that His Majesty commands that the Ordinance of New Spain shall be observed in this Kingdom, and he desires that it be put immediately in operation, adapting it to local circumstances; and this being the object for which, to facilitate the first steps, the preceding Declarations have been formed, they shall all be observed and understood in the same manner as the Articles to which they refer, and if in other matters any doubts should arise, they shall be referred to the Superintendency, and for this reason the mention of other less obvious points, has been studiously omitted, leaving them to what time and experience may dictate, so as not to cause confusion by providing for them beforehand; and these Declarations being observed for the present, they shall be annexed to the copies of the Ordinance of New Spain which have been ordered to be printed, and transmitted, with the respective official communications to His Excellency the Viceroy, and the Royal *Audiencia*, as has been stated in Declaration third, and they shall be also transmitted to the Royal Tribunal of accounts, and to the offices of the Royal Treasury in this Capital, where they shall be placed in the archives to be delivered to their successors, and the like shall be done in the offices of the other *Intendencias*, and by the Judges of the *Partidos* and Territorial Deputations, for which purpose there shall be sent to the Intendants the necessary copies, the remainder being kept to be sold to the public at a fair price, for the benefit and reimbursement of the said Mining fund, and the whole shall be reported to His Majesty, who shall be advised of other matters deserving his notice.

Lima, October 7th, 1786.

JORJE ESCOBEDO.

GLOSSARY.

GLOSSARY

OF WORDS AND TERMS USED IN THE SPANISH AND MEXICAN MINING LAWS, AND IN COMMENTARIES ON THESE LAWS.

A.

Abadengo. Abbatial. Things belonging to abbots. This term is applied to ecclesiastical property, or to lands belonging to the church.

Abastos. Supplies. Provisions, &c., supplied to a town or Mining establishment.

Abras. Clefts or fissures in the earth. These fissures, and particularly the spar found in them, are said to indicate veins of ore.

Achicar. To diminish or lower. Used to express the diminishing or draining of the water in any part of a Mine.

Achicadores. Workmen employed to reduce the water in a Mine.

Achichinques. Workmen who with hide buckets collect the water in the lower parts of a Mine and empty it into cisterns, or into the sump of the pit.

Acta. A written account or record of the proceedings or acts of any body or public authority.

Acto. An act or proceeding. In judicial matters this term is applied to any of the proceedings, orders, decrees, or sentences of the court.

Actuaciones. Proceedings. Sometimes applied to the drawing up or formation of any part of the proceedings in a case.

Ademar. To line or secure with timber the interior workings of a Mine, the shafts, pits, adits, chambers, pillars, &c.

Ademador. The liner, or workman who is employed in the business of securing and strengthening the walls and pillars of Mines.

Adema. Timber used for the purpose of securing or strengthening the different parts of Mines.

Ademes. The supports and coverings of timber for the security of Mines.

Afinar. To refine. To separate the dross which is combined with the metal in the process of smelting.

Afinacion. Refining.

Afinador. Refiner.

Agravio. Grievance. An application to a higher court to revise the decision of the Tribunal below, on account of injury done to the party. Usually applied to matters of account, but anciently to every kind of appeal.

Agrimensor. A surveyor of land.

Aquadrillados. Organized into bands. Applied to Indians who are enrolled in the working band (*quadrilla*) of a Mine.

Aguage. Current. Also applied to a watering place for cattle.

Alcribis. A funnel, into which the nozzle of the bellows of a smelting furnace is fitted, to conduct the blast.

Alcabala. Excise duty. This is a duty on the transportation and consumption of goods and effects, the sales of property, &c. The term is also applied to certain import duties.

Alzada. Appeal. Name given to the Mining court of appeal.

Aparejo. Apparatus, harness, gear. This term when used in the singular is sometimes applied to a machine for raising the timber-linings of a Mine. In the plural it is applied to the tools and implements used in Mining, and also to saddles aud gear of pack-animals.

Apartar. To separate. Applied to the separation of metals, by smelting or otherwise.

Apartador. One who separates metals.

Apartado. Office for separating metals; a smelting-house.

Aperador. The workman who has charge of Mining implements.

Aperos. Implements. It is applied to all kinds of implements, tools and utensils used in the various operations of Mining.

A Pique. Downward. To work *à pique* is to work perpendicularly downwards.

Apoderado. Agent, proxy, attorney-in-fact.

Apurar. To clean or drain; also, to glean.

Apuradores. Gleaners. Persons who pick up particles of metal in the refuse ores and rubbish.

Arbitrios. Duties or taxes imposed on provisions exposed for sale; means, expedients, discretion. *Arbitrio de Juez*, discretionary power of the Judge. *Propios y Arbitrios*, ways and means.

Arancel. Tariff; list of rates, fees, taxes or duties to be charged.

Arrastre. Act of dragging. Applied to a machine for grinding and mixing ores.

Asesor. Counsellor, law-adviser. A lawyer appointed to assist the ordinary judge in his law proceedings.

Asiento. Seat; place where a Court is held. This term is applied to a small Mining place or village where a seat of justice is established.

Atacador. Rammer, used in blasting.

Atajador. Interceptor. The person who brings the relief-animals for working the mills and machinery.

Atecas. The workmen who bale the water from the lower levels of a Mine into the tubs or shins, in order to be raised through the shaft.

Atierres. Rubbish which impedes the working of a Mine.

Auto. Decision, decree, sentence. In the plural applied to the pleadings, and sometimes to any of the proceedings in a suit.

Aviador. A supplier. One who supplies a Mine, or contracts to supply a Miner with the funds or means for carrying on his Mining operations.

Aviado, Aviada. A person or Mine supplied by an *Aviador*.

Aviar. To supply a Mine with funds or means for its working. Also to contract for such supplies, or for the working of a Mine.

Avios. Supplies, in money or otherwise, furnished for working a Mine. The term *Avio* is sometimes applied to the contract or agreement for supplies.

Azar. Hazard, chance. A game of chance.

Azogueria. The place where quicksilver is incorporated with metals. This word is also used to express the process of reducing ores by means of quicksilver.

Azoguero. Amalgamater. The person who works ores by means of quicksilver.

B.

Badana. A dressed sheep-skin. Also directions, instructions.

Bando. Proclamation. An Edict or law solemnly published.

Bancos. Banks. This term is applied to the rock which contracts a vein, changes its direction, or causes faults.

Barra. A crow, or iron bar used in Mining. This term is also applied to the shares into which a Mine is divided. Every Mine is divided into 24 shares or *barras*.

Barrear. To fortify or strengthen.

Barrena. Borer. A bar or tool used for boring the rock, preparatory to blasting.

Barreno. A hole bored for blasting. Also a communication between two Mines.

Barrenar. To pierce or bore through, so as to form an underground communication.

Barretero. Bar-man. A working Miner who blasts or digs out the ore, as distinguished from those who are engaged in other operations of Mining.

Beneficiar. To benefit. To work or improve a Mine; to reduce its ores; to derive profit or advantage from working it.

Beneficio. Benefit, favor, profit, advantage, labor, work, reduction. This term is applied to the working of a Mine, the reduction of its ores, and the profit derived. *Hacienda de beneficio* is an establishment for reducing ores, or extracting the metal from them.

Boca. Mouth. The first opening made on a vein.

Bocamina. Mouth, or principal entrance to a Mine.

Boca mejora. Mouth improved. Pit made to communicate with the *estaca fixa.*

Bocado. Mouthful, morsel. A charge or fee levied in the Mint. This charge was abolished by the Decree of February 7th, 1822.

Bochorno. Heat. Applied to the high temperature in a Mine, resulting from defective ventilation.

Bombero. Pumpman. A workman who pumps out the water.

Bonanza. Fair weather. A Mine is said to be *en bonanza* when it is in a prosperous condition, that is, yielding a profit.

Borrasca. Foul weather. A Mine is said to be *en borrasca* when it yields little or no metal, or is unprofitable, not paying the expenses of working it.

Botas. Buckets made of hide, for drawing off the water from a Mine.

Bruto. Rough. *En bruto*, in a rough state, unworked.

Buitron. A furnace of particular construction in which silver ore is reduced.

Buscar. To search. To go about searching for indications of ore. In California, *to prospect* for ores or Mines.

Buscon. One who goes about searching for ore; in California, *a prospector.*

Bula. Bull. An apostolic letter; ecclesiastical tax.

C.

Caballo. Horse. A mass of firm hard rock.

Cacique. An Indian prince or nobleman; chief of village or tribe.

Caducar. To grow old, to be worn out, to terminate, to become extinct, to be forfeited. This term is applied to a legacy or fiduciary estate which terminates, and to a Mine which is forfeited for non-compliance with the conditions imposed by the law or grant which carry that penalty.

Caducidad. Decay, forfeiture, termination, extinction.

Caja, Caxa. Chest, box, or vault where money is kept. This term is applied to the Treasuries or offices where the public or Royal moneys are kept.

Cala. A small pit or experimental hole.

Calentar. To heat. Also, to smelt.

Calentadura. The first ingot reduced in a smelting furnace.

Calicata. A digging, or trial pit.

Camino. Road, way. This term is also applied to bags or sacks of ore.

Camara. Chamber, exchequer.

Cañon. Passage. Applied to small underground passages or galleries.

Capa. Cover, stratum, layer, lamina.

Capataz. Overseer, foreman, head-Miner.

Capellina. A vessel employed in separating the quicksilver from the amalgum.

Capitulares. Members of a chapter or body, as a *Regidor* in an *Ayuntamiento.*

Capitula. Chapter, head. Also, a charge or accusation; *capitulo de residencia*, cause of accusation.

Catar. To taste, to try, to make experiment, to examine.

Cata. Taste, trial, experiment. Applied to a Mine of small depth.

Catalan. A particular kind of forge.

Catear. To search or try for a Mine; to discover.

Cateador. One who tries for, or seeks to discover a vein or Mine.

Catedra. Chair, office, professorship.

Cebar. To feed; to supply a furnace with ore or ingots, as the metal goes off. Also to keep up the fire.

Cedula. Edict, law, order. Also a vote.

Cendrada. Ashes, or smeltings found at the bottom of a furnace. These are valuable for use in other smeltings.

Cendradilla. A small refining furnace for testing rich ores; also called *galeme.*

Cesante. A public officer who receives pay, but does not exercise the functions of his office.

Chacra. A small farm, or habitation.

Charco. Pool of standing water.

Charqueo interior. Draining off the water from pools into the pit.

Chancilleria. Chancery. A certain superior tribunal of justice.

Chiflon. To work *à chiflon* is to extend the work both in length and width at the same time.

Ciego. Blind. Applied to a passage which is choked or shut up.

Cielo. Sky. Working *de cielo,* is working at the vault or roof of a passage or chamber.

Clerigo, Clergyman; an officer of particular rank in the church.

Cofradia. Society, association, fraternity.

Cohetazo. Blast. Mode of blasting rocks.

Cohete. Rocket; charge of powder, wrapped in paper, which is put in the *barreno.*

Color. Color. The shade or tint of the earth or rock which indicates ore.

Colindante. Coterminous, bounding with, adjacent; a near neighbor.

Comarca. A particular territory or district. A political or judicial district of country.

Comer. To eat. This term is applied to digging away the pillars or supports of a Mine.

Competencia. The right of a Court or Judge to take cognizance of a cause; right of jurisdiction. This term is also applied to conflicts of jurisdiction, and to the cognizance of cases.

Comun. Common. That which belongs to no one individual, but to several or to a community. The common lands of a town or corporation are for the common use of the whole community.

Conclusion. End, termination, close. This term is sometimes, in the old laws, applied to stopping the proceedings in a case without a final decision; putting off the final determination of a case, by an interlocutory decree.

Constancia. The final result of judicial proceedings. Used sometimes in the plural for the proceedings themselves.

Consulado. A tribunal having cognizance of commercial affairs.

Consultores. Advisers. Certain Miners selected to advise in matters of business relating to Mining.

Contador. Auditor, accountant. This term is also applied to the cashier, or person who counts out money.

Contaduria. Office of the Auditor or *Contador.*

Contramina. Countermine, tunnel, adit.

Cornadillos. Small piece of money of little value; a contribution. *Cornadillos de cofradias*, society contributions.

Corregidor. Corrector; a mayor of a town; a magistrate having jurisdiction of certain special cases, as prescribed in the laws.

Corregimiento. The office of *corregidor*, or place of holding his court.

Corriente. Current. A Mine which is worked is said to be *en corriente.*

Costal. Sack, bag for holding ores.

Coto. Inclosure. *Cotos*, lands or pasture grounds inclosed.

Criadero A deposit of ore, a Mine.

Cuarton. A large beam; *Cuartones*, frame-work.

Cuaderno, *Cuadernillo.* Sheets of paper stitched together; a book, code, treatise.

Cuota. Part, proportion, rate, duty.

D.

Dados. Dice; a game prohibited by the Mining Ordinances.

Dedo. Finger. In measurement, the forty-eighth part of a *vara.*

Dehesas. Pastures. *Dehesa concejil*, a common.

Demasia. Surplus. Vacant spaces between the *pertenencias* of Mines.

Denunciar. To denounce. To give information that a Mine is forfeited for being insufficiently worked, or for a violation of some condition which imposes that penalty. This term is sometimes applied to the giving notice of a discovery, for the purpose of registry.

Denuncio. A denouncement.

Denunciable. Subject to denouncement.

Derrumbar. To fall, to cave in.

Derrumbe, *Derrumbamiento.* A falling down, or caving in of a Mine.

Desague. Drain, channel, outlet; drainage.

Descargue. Discharge. Also, the last ingot reduced in a smelting furnace.

Desechaderos. Refuse ores, rubbish.

Desmonte. Cut-away. Refuse matter.

Despensa. A room for storing rich ores.

Despilaramiento. Want of support.

Despoblar. To desert, to abandon, to leave a Mine unworked, or without the number of operatives required by law.

Despoblada. Unpeopled. A Mine is said to be *despoblada* when not occupied with the number of workmen specified in the Ordinances.

Destajo. A job or task.

Dilacion. Delay, postponement, procrastination.

Diligencias. Proceedings, operations.

Dinero. An old Spanish copper coin; money.

Dominio. Dominion, property, ownership. This right includes the enjoyment, free disposition, and exclusion of all other persons.

Dotacion. Endowment, salary, pay.

E.

Echadero. Resting place. A place for cleaning and weighing the ores, and for loading the mules.

Echado. Inclination, slope, bearing.

Ejidos or Egidos. Uncultivated land on the outskirts of a town, for the common use of its inhabitants.

Embarazado. Embarrassed, obstructed, encumbered. This term is sometimes applied, in the old laws, to simple occupation or possession.

Embargar. To seize, attach, levy on.

Emborrascarse. To be unproductive, unprofitable. Applied to a Mine when the vein loses itself.

Empresario. A person who undertakes to construct some work, or to perform some business of great importance, on his own account. Also a contractor.

Empresa. An important enterprise, work, or business. Also applied to a company which undertakes works or operations of great importance.

Encomendero. An agent. A person placed in charge of Indians, for their management and protection.

Ensayes. Assays. Trial of ore to determine the richness, or of bullion to ascertain its fineness.

Entibacion. The proping or lining of a Mine.

Escaleras. Ladders. Round timbers or poles with notches, for the Miners to ascend and descend.

Escombros. Rubbish, fragments, refuse matter.

Escorial. Scoria, slag, a slag-heap.

Escribano. Notary; clerk of a Court.

Escribania. Office of a Notary, or clerk of Court.

Espejuelo. A kind of spar.

Estaca. Stake. *Estaca fixa.* Fixed stake. The principal pit by which a Mine is registered and its location fixed. This term is sometimes used for the Mine itself.

Estacar. To stake out, locate, establish boundaries.

Estado. State, condition. A measure used in Mining.

Estanco. Monopoly, prohibition.

Expediente. Proceedings. An *Expediente* is usually composed of the original papers of proceedings, whether administrative or judicial. In Mexico they often constitute the only records which are kept of such proceedings.

Exploracion. Exploration, examination, investigation; search for ore.

Explotacion. The working of a Mine.

F.

Fabrica. Manufactory; manufacture; work-shop. This term is usually applied in the plural to reducing works.

Faena. Work, fatigue. This term is more usually applied in Mining to works not in ore or on the vein. *Faenas muertas*, dead works.

Facultativo. Professional; belonging to some faculty or art. Applied to persons who are masters or experts in some business, and hold commissions or diplomas.

Feble. Deficiency. Used in the mint to express deficiency in the weight or quality of coin.

Fiel. Trusty. Applied to an officer appointed to weigh gold or silver.

Fierros. Irons. Tools. This term is also applied to the dross removed from the ore, or ingots, when smelted.

Fincas. Improved landed property; buildings, houses, tenements.

Fiscal. A ministerial or law officer, who represents the Government.

Fisco. Fisc, exchequer.

Fomentar. To protect, encourage, patronize. In Mining to encourage by a supply of money, or a grant of privileges or favors.

Foraneas. Foreign. Sometimes applied to frontier or provincial places where foreign articles are introduced.

Forera. Lawful. *Moneda forera,* a tribute or tax, to be paid in money.

Franco. Free, vacant, unoccupied.

Franquicia. Immunity; exemption from taxes, imposts or duties.

Fronton. Wall. A work which the Miners carry straight forward.

Fuero. A statute-law, charter, privilege.

Fuslina. Office or place where ore is reduced.

Fusline. A particular kind of smelting furnace.

G.

Galeme. A small furnace for testing ores.

Galera. A shed used for the ores.

Gallo. Cock. This term is applied to rich specimens of gold or silver ore, with threads or grains of the metal.

Glosa. Note, comment, statement.

Golpeador. Striker.

Grangerias. Agricultural property.

Gravamen. Charge, encumbrance, mortgage.

Grassas. Scum or scoria.

Grassero. Slag-heap.

Greta. A flux used in smelting; principally an oxide of lead. It is sometimes called chalk, from its whitish color.

Guarda. Guard. Rock at the sides and top of a vein.

Guarda-raya. Limit or boundary mark.

Guadrilla. Band. Applied to the working parties of Indians assigned to work in Mines.

Guia. Guide, permit, mark, bound, limit.

Guija. A crystalline substance which is said to indicate ore.

Guijo. An iron spike which holds the mortar in its place.

H.

Habilitar. To provide, to supply, to put in order, to enable. In Mining it usually refers to the supply of funds for carrying on the work.

Hacienda. Exchequer, treasury, public revenue, capital, funds, wealth, landed estate, establishment. In Mining, it is usually applied to the offices, principal buildings, and works, for reducing the ores.

Hueco. Hollow, vacant space; interval between Mining *pertenencias.*

Hundido. Sinking or falling in of a Mine.

I.

Incorporadero. Place where the ore is mixed with quicksilver in reducing it by amalgamation.

Ingenio. Mill, machinery; a manufactory.

Intendencia. Intendency. The office or district of an *Intendente.*

Intendente. Intendant. The first chief of the Treasury or Royal Revenues of a province.

Interventor. An inspector placed in a Mine to watch over and report upon the works carried, and upon the use made of the supplies.

J.

Jaboncillos. A whitish unctuous substance, said to be an indication of rich ore.

Jalsontles. Portions of ore not properly ground. Also the slime or dust from the washing vats of amalgamating works.

Jornales. Day-wages. Rates of wages paid to operatives in Mines.

Junta. A meeting, assembly, congress. Also applied to a commission or board, as *Junta de Comercio*, Board of Trade; *Junta de Mineria*, Board of Mining.

Justicia. Ordinary Courts, as distinguished from special tribunals.

L.

Labor. Labor, work, working. This term is applied in Mining to the work which is actually going on and to the spaces which have been dug out. It includes *galleries, cavities* and *shafts*. *Labor legal*, legal space; principal work.

Lamas. Slime. The earthy matter from the vats in the amalgamating works.

Lamero. Place where the *lamas* are deposited.

Lampazo. A kind of broom or mop made of green boughs and used to moderate the heat of the smelting furnaces.

Lanza. Lance, spear. A duty or tax paid in lieu of military service.

Lavadero. Washing place. A large vat in which the ore is washed to separate the earthy particles.

Llave. Key. Timber with notches to support the lining or covering in a Mine.

Laxas. Loosening, opening. Applied to a fee charged for opening flasks of quicksilver.

Lazadores. Persons employed to collect workmen for a Mine.

Leñador. Woodman. One who supplies fuel for the smelting furnaces.

Lesion. A term used in civil law to signify the injury suffered by one who does not receive an equivalent for what he gives in a commutative contract. There are several kinds of lesion mentioned in the Spanish laws.

Letrado. Learned. This term is applied to a lawyer who has received his diploma, and to a judge who is also a professional lawyer.

Ley. Proportion of metal in the ore; fineness of bullion; also an alloy or base metal.

Liga. Alloy.

Limadura. Filings. A film with which the metal becomes coated in the assays.

Lis. The silver is said to form *lis*, when the quicksilver is resolved into almost imperceptible particles, in the course of the reduction by amalgamation.

Lumbrera. Sky-light. A shaft sunk for the purpose of ventilation.

Lugar. Place. This term is applied to a Mining village or town. It differs from Asiento in that the latter must be a seat of justice.

M.

Magistral. A powder of roasted copper pyrites, used in the reduction of ores.

Malacate. Whim. A machine used to raise ores or water from a Mine, moved by men or animals.

Mandon. An overseer, foreman.

Manta. Blanket; sack of ore.

Manto. Cloak. A vein which spreads out horizontally.

Maquilla. A mill where the ore is ground on shares.

Maravedi. The smallest Spanish coin.

Matricula. Enrollment, register, list.

Matriculados. Enrolled Miners; members of the body of Mining.

Mecha. Match used for firing the blast.

Media-Anata. A particular kind of tax.

Medida. Measurement, dimension, extent, boundary. Applied to the pertenencia of a Mine.

Memorias. Memoirs, accounts, rolls. Applied to the weekly or monthly rolls and accounts of a Mine.

Merced. A gift. This term is applied to a grant which is made without any valuable consideration.

Mestizo. Mixed race; the offspring of a white and an Indian.

Metal. This term is applied both to the ore and to the metal extracted from it. It is sometimes used for vein, and even for a Mine itself. *Metal en piedra*, ore in the rough state. *Metal ordinario*, common ore. *Metal pepena*, selected ore. *Metal de ayuda*, ore used to assist the smelting of other ores.

Mina. Mine. This term, in its more technical sense, implies subterranean workings, as distinguished from a *Criadero*, *Placer*, *Rebosa*, &c.

Minero. Miner. This term is not limited to those who work Mines, but includes their owners, and all who have the qualifications prescribed in the Ordinances, and are enrolled as members of the Body or Craft. Many of the laborers who work in Mines are not technically speaking Miners. This term is sometimes used in the old laws for Mine.

Mineria. Mining. This term embraces the whole subject, including both Mines and Miners, and also the operations of working Mines and of reducing their ores. It however is often used in a more restricted sense.

Mineta. A little Mine; a chamber, or cavity.

Ministro. Minister. *Ministro de Hacienda,* Minister of Treasury; *Ministro de Relationes Exteriores,* Minister of Foreign Affairs. This term is also applied to subordinate officers and employes in any department.

Ministerio. Ministry, office of Minister. An administrative department of the Government. This term is also frequently applied to the Minister or Chief of the department.

Mita. The number of Indians subject to compulsory labor by turns, in conformity with the law of that name.

Mogrollo. A very rich specimen of ore.

Mojon. Heap, pile. A land-mark or boundary.

Molonque. A piece of ore of uniform richness, being at least one-half silver.

Moneda. Money, coin, coinage; *casa de Moneda,* Mint.

Monton. Heap, pile. It usually contains about 20 *quintals.*

Morteros. Stamps, for crushing the ore.

N.

Nata. Scum, Scoria. The dross thrown off in the smelting furnace.

Negociacion. Business, trade, management. This term is also applied to a general enterprise in Mining, and to the company and organization for carrying it on.

Negrillo. A silver ore. Black sulphuret of silver.

O.

Ochava. A small Spanish copper coin, valued at two *maravedies,* or half a quarter.

Ordenamiento. A law. The Spanish laws all emanated from, or were issued by the authority of the King, but took different names, according to their form, and the mode of promulgation. They were called *Autos Acordados, Cedulas, Cartas, Decretos, Ordenamientos, Pragmaticas, Reglamentos, Ordenanzas,* and *Resoluciones,* according to their form, object, and mode of promulgation, or the authorities who issued them or to whom they were directed.

Ordinarios. Ordinary, as distinguished from special. *Justicias ordinarias* are courts of general jurisdiction, as distinguised from the special tribunals which have cognizance only in the particular cases assigned to them by law.

Otorgar. To declare, to consent, to grant.

P.

Palmo. A hand. A measure used in Mines.

Palero. A ditcher, drainer.

Panino. The appearance of the ground indicating ore, or a vein.

Parcionero. Partner, part-owner of a Mine.

Partido. A district; a territorial division of a country for political or judicial purposes. A share. Division of ore among partners, or between the owners and the operatives. To work *á partido*, is to work on shares.

Papel. Paper. The pass-paper kept by the operatives has marked on it the number of their days work in strokes or lines and half-lines, and their accounts in circles and half-circles, so that they can readily understand them, although unable to read.

Patio. The yard where the ores are cleaned and assorted. Also the amalgamation floor.

Peonadas. The wages or pay of laborers or *Peones.*

Perder. To lose, to forfeit.

Perdida. Lost, forfeited.

Perito. An expert; a person skilled in any profession or art; as *Perito de Minas*, a Mining Expert; *Perito de Beneficio*, a reducing expert.

Pellas. An amalgam.

Pepena. A select piece of ore.

Pepeñadores. Workmen who assort the ore into classes.

Piedra. Stone, ore. *Piedra de mano.* One of good quality set aside for a special purpose.

Pilar. Pillar. A part of the natural earth or rock left as a support.

Pileta. In a Mine, a cistern where the water is collected. In a smelting furnace, the breast-pan or vessel into which the melted metal flows.

Piña. Pine-apple. Silver in the shape of a pine-apple. *Plata-Piña.* An amalgam of silver before all the quicksilver is driven off.

Pinole. Ground parched corn, used as food by mixing it with water and sugar.

Pinta. Spot or mark, indicating the presence of metal.

Placer. Deposit of gold or silver, found in particles in the earth or beds of streams.

Planes. The lowest workings of a Mine.

Planchera. A place or mold made of white earth, in which the ingots are formed.

Plata. Silver. *Platas.* Bullion.

Plomo. Lead. *Plomo-Plata,* lead-silver.

Plomillos. Particles charged with lead, which are carried off by the scum or scoria.

Poblacion. A settlement, or small town. The population or inhabitants of a town or place.

Poblar. To people, to occupy, to work.

Poblada. Peopled, occupied, worked. *Mina poblada,* is a Mine occupied or worked with the number of operatives required by the Ordinances.

Pozo. A shaft, pit or well.

Postura. An offer or bid.

Pragmatica. Edict, vide *Ordinamiento.*

Pre-eminencia. Pre-eminence; a right; also a grant.

Privativo. Privative, exclusive. *Jurisdiccion privativa,* exclusive jurisdiction.

Privilegiado. A privileged person; one entitled to certain rights and privileges specified in the laws, as a minor, an absentee, &c.

Propiedad. Dominion, ownership, right of property. Also, landed estate. This term is used to express an exclusive right of property, in contradistinction to the right of use or usufruct.

Propietario. Proprietor, owner, landlord.

Providencia. An order, disposition, instruction.

Puertas. Gates. Rocks which conceal the vein.

Puja. A better offer, or a higher bid.

Q.

Quadra, cuadra. Properly a square; but used in the Mining laws to designate a rectangular space.

Quadrilla. A working band of Indians assigned to work in a Mine.

Quatequil. A compulsory assignment of Indians to work. Vide *Mita.*

Quemazon. Burning. Light metallic dross, vesicular and scorched, which is said to be one of the indications of a vein.

Quintar. To take a fifth. To deduct the duty on gold and silver.

Quinto. One fifth. The Royal due on silver and gold.

Quita-pepena. He who attends at the entrance of a Mine and to the getting out the ore, to guard against theft.

R.

Raya. Tally, time-roll of the workmen in a Mine.

Rayador. He who keeps the tallies, or time-rolls, and the amount of ore produced by tak-work.

Real. Royal. A town in which there are Mines of silver.

Reboltura, Rebolturon. A mixture of ground ore with other substances preparatory to smelting.

Rebosadero. A place or pocket where the ore or metal is found in the greatest abundance.

Rebotallero. A person who searches for particles of metal amongst the refuse-ores and rubbish.

Recogedor. A person authorized to go out and collect workmen for a Mine, and to bring in idlers. Same as *Lazador*.

Recurso. A resort to another Judge or court in case of notorious injustice, or from second supplication. *Recursos* are divided into ordinary and extraordinary. This term is also used in the general sense of appeal.

Recusacion. An exception taken to the competency of a Judge or officer to act in a case or suit.

Recursado. The Judge or officer excepted to.

Recursar. To except to a Judge or officer, or to his jurisdiction.

Refaccion. The restoration or repair of a Mine.

Refaccionario. A restorer, or one who supplies the means for putting a Mine in working order.

Regalia. This term is applied both to things and to rights to things, *corpora* and *jura*. It is sometimes confounded with sovereignty, or *jura majestatis*, but it is not inherent to or inseparable from the sovereign power. "Upon the breaking up of the Roman Empire," says Gamboa, "the Princes and cities, which declared themselves independent, appropriated to themselves those parts in which nature, most rich and liberal, yields extraordinary products. These portions, or reserved rights, were called *Regalias*."

Registro. Registry, registration, register. This term is applied to the act of registration, to the registry made, and to the book in which it is made.

Regular. A person belonging to a religious order.

Religioso. Religious. A person who dedicates his life to religion; used in the old laws in the sense of *clerigo*.

Repartimiento. A distribution or allotment of Indians to work in a Mine.

Repasar. To tread out the mixture of ground ore and other ingredients.

Repaso. The treading out, or mixing up of the *montons* of ore with the magistral and quicksilver, in the process of reduction.

Reposadero. A stone placed at the bottom of the smelting furnace.

Rescatador. A person who purchases ore from the Mine-owner or from the workmen, or exchanges bullion for such ore.

Rescate. The purchase of ore.

Resguardo. A certificate given to the party who presents a petition, or files a paper, in which the substance of the act, or a summary of the document is set forth.

Resocàs. Residium.

Restitucion. Restitution. Restoration of a minor or other privileged person to his rights and actions. There are various kinds of restitution in the Spanish law

Rielle. Linget; a small ingot of unrefined gold or silver.

Riscos. A substance partly crystalline and partly granular, like a cauliflower.

S.

Saca. Sack; also a sack of ore. This term is also applied to the produce. A Mine is said to give a good *saca,* when it yields plenty of good ore.

Salera. Rock salt; a salt pit.

Salina. Saline; a salt spring or well; salt works.

Saltierra. A mixture of salt and earthy matters, used in reducing ores.

Señoreaje. Seignorage. A duty belonging to the King on the coining of money.

Señorio. Seignory, lordship, dominion.

Sindico procurador. A law officer, whose duties are similar to those of an attorney-general or a district-attorney.

Sitio. Place, locality, tract of land, a farm. This term is applied to the site, or legal extent of land, for a *Hacienda* or reducing works.

Socabon. An adit, or tunnel. A subterranean passage driven, from the side of a hill, into a Mine for the purpose of drainage, or for getting out the ores.

Sumaria. A judicial proceeding; preliminary examination.

Suplicacion. A petition to the same, or to a higher court, to review a decision or judgment which has been rendered. *Suplicacion segunda,* is a second petition of review which was formerly allowed in certain cases. It seems that the order of these several proceedings, after decision in Second Instance, was, at least in some cases, as follows: appeal, supplication, second supplication, ordinary recourse, and extraordinary recourse. These supplications have now no place in judicial proceedings.

T.

Tabas. A game played by the Indians; sheep's shanks.

Talas. The felling of trees.

Tanda. Turn; an assignment of time for working and relief in Mines.

Tanate, or Tenate. A sack of hide, used for carrying the ore or rubbish out of the Mine. Those who use those sacks are called *Tanateros*, or, more commonly, *Tenateros.*

Tanteo. Valuation; also price bid.

Tapextle. A wooden platform for the workmen to stand on.

Targea. Aqueduct; conveyance of water.

Tarea. Task; task-work.

Temescuitate. The earthy part of ground ore.

Tentadura. An assay of the mixture of ore and quicksilver, made in a cup, to ascertain if the ingredients are in proper proportions.

Tepetate. Earthy matter, which is taken from a Mine, and is destitute of ore.

Tequesquite. A saline substance, used in the reduction of ores.

Tequio. A duty, tax, task. A certain portion of ore which the workman is obliged to deliver during working hours. The remainder which he takes out, is called *partido.*

Terrero. A rubbish-heap. Also the place where refuse ore is thrown.

Territorio. A district of country; the space included within the boundaries of a city, town or place (*lugar*).

Testera. A beam, used to support the roofing of a Mine.

Testimonio. An authenticated copy; an instrument legalized by a Notary.

Texear. A furnace is said to *texear bién*, when it throws off fine, and brittle dross or scoriæ.

Tiro. A pit, or perpendicular shaft, for raising the ore, rubbish, or water. A *Tiro* is sometimes slightly inclined.

Titulo. A title, diploma, commission.

Titulado. One having a diploma or commission; duly authorized to act in any particular basiness or employment.

Tornaguia. Return-permit. A certificate showing that the goods mentioned in the permit (*guia*) had been entered or delivered.

Tovera. Funnel for the nozzle of a bellows.

Tramites. Proceedings, forms, course of judicial process.

Torta. Cake. A large flat circular collection of pulverized ore, prepared for the mixing *repaso.* A *torta* sometimes contains 60 *montons*, and even more.

Trapiche. A rude machine for grinding ores, composed of two stones, the upper one being fastened to a long pole.

Tributo. Tribute; personal tax. *Tributos*, public stores of tributes or contributions collected.

Trompa. Trumpet. This term is applied when the blast from the bellows makes no noise, and the ore collects at the *tovera* or tweer, in the form of a trumpet.

V.

Valdio. Vacant. *Terreno valdio,* is land which has no private owner, and is therefore susceptible of being granted.

Vapor. This term is applied to the foul air and noxious vapors in a Mine.

Velador. A watchman; night-guard. One who is charged with the care and delivery of the candles.

Vellon. A particular kind of money. This term is used like the English word *sterling.*

Vena. Vein. This term is applied to the branches of the main vein (*veta*), or to mere threads of ore.

Vergas. A bow or yard. Bars of metal.

Veta. A vein of metalic ore. Various names are applied to it, according to its size, shape, and direction.

Vocal. Voter. A member of a board, society or assembly, where votes are taken.

Vuelta. Turn. The silver in the refining furnace is said to *dar vuelta* when the dross is driven off and it is of a red color.

X.

Xacal. A hut. The roofing made over a pit or ore-shed is so called.

www.ingramcontent.com/pod-product-compliance
Lightning Source LLC
LaVergne TN
LVHW052206200926
848481LV00001B/36

* 9 7 8 1 4 2 5 5 6 8 7 1 9 *